Maximum Likelihood Estimation with Stata

Fourth Edition

D1585589

Maximum Likelihood Estimation with Stata

Fourth Edition

WILLIAM GOULD
StataCorp

JEFFREY PITBLADO
StataCorp

BRIAN POI
StataCorp

A Stata Press Publication
StataCorp LP
College Station, Texas

Published by Stata Press, 4905 Lakeway Drive, College Station, Texas 77845
Typeset in LaTeX 2_ε
Printed in the United States of America

10 9 8 7 6 5 4 3 2 1

ISBN-10: 1-59718-078-5
ISBN-13: 978-1-59718-078-8

Library of Congress Control Number: 2010935284

Contents

Tables

Figures

Preface to the fourth edition

Maximum Likelihood Estimation with Stata, Fourth Edition is written for researchers in all disciplines who need to compute maximum likelihood estimators that are not available as prepackaged routines. To get the most from this book, you should be familiar with Stata, but you will not need any special programming skills, except in chapters 13 and 14, which detail how to take an estimation technique you have written and add it as a new *command* to Stata. No special theoretical knowledge is needed either, other than an understanding of the likelihood function that will be maximized.

Stata's ml command was greatly enhanced in Stata 11, prescribing the need for a new edition of this book. The optimization engine underlying ml was reimplemented in Mata, Stata's matrix programming language. That allowed us to provide a suite of commands (not discussed in this book) that Mata programmers can use to implement maximum likelihood estimators in a matrix programming language environment; see [M-5] **moptimize( )**. More important to users of ml, the transition to Mata provided us the opportunity to simplify and refine the syntax of various ml commands and likelihood evaluators; and it allowed us to provide a framework whereby users could write their likelihood-evaluator functions using Mata while still capitalizing on the features of ml.

Previous versions of ml had just two types of likelihood evaluators. Method-lf evaluators were used for simple models that satisfied the linear-form restrictions and for which you did not want to supply analytic derivatives. d-family evaluators were for everything else. Now ml has more evaluator types with both long and short names:

Short name	Long name
lf	linearform
lf0	linearform0
lf1	linearform1
lf1debug	linearform1debug
lf2	linearform2
lf2debug	linearform2debug
d0	derivative0
d1	derivative1
d1debug	derivative1debug
d2	derivative2
d2debug	derivative2debug
gf0	generalform0

You can specify either name when setting up your model using `ml model`; however, out of habit, we use the short name in this book and in our own software development work. Method `lf`, as in previous versions, does not require derivatives and is particularly easier to use.

Chapter 1 provides a general overview of maximum likelihood estimation theory and numerical optimization methods, with an emphasis on the practical implications of each for applied work. Chapter 2 provides an introduction to getting Stata to fit your model by maximum likelihood. Chapter 3 is an overview of the `ml` command and the notation used throughout the rest of the book. Chapters 4–10 detail, step by step, how to use Stata to maximize user-written likelihood functions. Chapter 11 shows how to write your likelihood evaluators in Mata. Chapter 12 describes how to package all the user-written code in a do-file so that it can be conveniently reapplied to different datasets and model specifications. Chapter 13 details how to structure the code in an ado-file to create a new Stata estimation command. Chapter 14 shows how to add survey estimation features to existing `ml`-based estimation commands.

Chapter 15, the final chapter, provides examples. For a set of estimation problems, we derive the log-likelihood function, show the derivatives that make up the gradient and Hessian, write one or more likelihood-evaluation programs, and so provide a fully functional estimation command. We use the estimation command to fit the model to a dataset. An estimation command is developed for each of the following:

- Logit and probit models
- Linear regression
- Weibull regression
- Cox proportional hazards model
- Random-effects linear regression for panel data
- Seemingly unrelated regression

Appendices contain full syntax diagrams for all the `ml` subroutines, useful checklists for implementing each maximization method, and program listings of each estimation command covered in chapter 15.

We acknowledge William Sribney as one of the original developers of `ml` and the principal author of the first edition of this book.

College Station, TX   William Gould
September 2010   Jeffrey Pitblado
                                                                    Brian Poi

Versions of Stata

This book was written for Stata 11. Regardless of what version of Stata you are using, verify that your copy of Stata is up to date and obtain any free updates; to do this, enter Stata, type

```
. update query
```

and follow the instructions.

Having done that, if you are still using a version older than 11—such as Stata 10.0—you will need to purchase an upgrade to use the methods described in this book.

So, now we will assume that you are running Stata 11 or perhaps an even newer version.

All the programs in this book follow the outline

```
program myprog
        version 11
        ...
    end
```

Because Stata 11 is the current release of Stata at the time this book was written, we write `version 11` at the top of our programs. You could omit the line, but we recommend that you include it because Stata is continually being developed and sometimes details of syntax change. Placing `version 11` at the top of your program tells Stata that, if anything has changed, you want the version 11 interpretation.

Coding `version 11` at the top of your programs ensures they will continue to work in the future.

But what about programs you write in the future? Perhaps the here and now for you is Stata 11.5, or Stata 12, or even Stata 14. Using this book, should you put `version 11` at the top of your programs, or should you put `version 11.5`, `version 12`, or `version 14`?

Probably, you should substitute the more modern version number. The only reason you would not want to make the substitution is because the syntax of `ml` itself has changed, and in that case, you will want to obtain the updated version of this book.

Anyway, if you are using a version more recent than 11, type `help whatsnew` to see a complete listing of what has changed. That will help you decide what to code at the top of your programs: unless the listing clearly states that `ml`'s syntax has changed, substitute the more recent version number.

Notation and typography

In this book, we assume that you are somewhat familiar with Stata. You should know how to input data and to use previously created datasets, create new variables, run regressions, and the like.

We designed this book for you to learn by doing, so we expect you to read this book while sitting at a computer and trying to use the sequences of commands contained in the book to replicate our results. In this way, you will be able to generalize these sequences to suit your own needs.

Generally, we use the `typewriter font` to refer to Stata commands, syntax, and variables. A "dot" prompt (.) followed by a command indicates that you can type verbatim what is displayed after the dot (in context) to replicate the results in the book.

Except for some very small expository datasets, all the data we use in this book are freely available for you to download, using a net-aware Stata, from the Stata Press web site, http://www.stata-press.com. In fact, when we introduce new datasets, we load them into Stata the same way that you would. For example,

```
. use http://www.stata-press.com/data/ml4/tablef7-1
```

Try it. Also, the ado-files (not the do-files) used may be obtained by typing

```
. net from http://www.stata-press.com/data/ml4
(output omitted )
. net install ml4_ado
(output omitted )
. ml4_ado
(output omitted )
```

The `ml4_ado` command will load the ado-files in your current directory, where you can look at them and use them.

This text complements but does not replace the material in the Stata manuals, so we often refer to the Stata manuals using [R], [P], etc. For example, [R] **logit** refers to the *Stata Base Reference Manual* entry for `logit`, and [P] **syntax** refers to the entry for `syntax` in the *Stata Programming Reference Manual*.

The following mathematical notation is used throughout this book:

- $F()$ is a cumulative probability distribution function.
- $f()$ is a probability density function.

- $L()$ is the likelihood function.

- ℓ_j is the likelihood function for the jth observation or group.

- g_{ij} is the gradient for the ith parameter and jth observation or group (i is suppressed in single-parameter models).

- H_{ikj} is the Hessian with respect to the ith and kth parameters and the jth observation or group (i and k are suppressed in single-parameter models).

- μ, σ, η, γ, and π denote parameters for specific probability models (we generically refer to the ith parameter as θ_i).

- $\boldsymbol{\beta}_i$ is the coefficient vector for the ith ml equation.

When we show the derivatives of the log-likelihood function for a model, we will use one of two forms. For models that meet the linear-form restrictions (see section 4.1), we will take derivatives with respect to (possibly functions of) the parameters of the probability model

$$H_{ikj} = \frac{\partial^2 \ln \ell_j}{\partial \theta_{ij} \partial \theta_{kj}} = \frac{\partial^2 \ln \ell_j}{\partial \mathbf{x}_j \boldsymbol{\beta}_i \partial \mathbf{x}_j \boldsymbol{\beta}_k} = \dots$$

We will then use some programming utilities to perform the chain-rule calculations that result in the gradient vector and Hessian matrix. For the other models, we may have to take derivatives with respect to the coefficient vector of an ml equation

$$\mathbf{H}_{ikj} = \frac{\partial^2 \ln \ell_j}{\partial \boldsymbol{\beta}_i' \partial \boldsymbol{\beta}_k} = \dots$$

In either case, we will always identify what it is we are taking derivatives with respect to.

1 Theory and practice

Stata can fit user-defined models using the method of maximum likelihood (ML) through Stata's `ml` command. `ml` has a formidable syntax diagram (see appendix A) but is surprisingly easy to use. Here we use it to implement probit regression and to fit a particular model:

```
────────────────────────────────────────────── begin myprobit_lf.ado ──────────
program myprobit_lf
        version 11
        args lnfj xb
        quietly replace `lnfj' = ln(normal( `xb')) if $ML_y1 == 1
        quietly replace `lnfj' = ln(normal(-`xb')) if $ML_y1 == 0
end
──────────────────────────────────────────────── end myprobit_lf.ado ──────────

. sysuse cancer
(Patient Survival in Drug Trial)

. ml model lf myprobit_lf (died = i.drug age)

. ml maximize
initial:       log likelihood = -33.271065
alternative:   log likelihood = -31.427839
rescale:       log likelihood = -31.424556
Iteration 0:   log likelihood = -31.424556
Iteration 1:   log likelihood = -21.883453
Iteration 2:   log likelihood = -21.710899
Iteration 3:   log likelihood = -21.710799
Iteration 4:   log likelihood = -21.710799
```

			Number of obs	=	48	
			Wald chi2(3)	=	13.58	
Log likelihood = -21.710799			Prob > chi2	=	0.0035	

died	Coef.	Std. Err.	z	P>|z|	[95% Conf. Interval]	
drug						
2	-1.941275	.597057	-3.25	0.001	-3.111485	-.7710644
3	-1.792924	.5845829	-3.07	0.002	-2.938686	-.6471626
age	.0666733	.0410666	1.62	0.104	-.0138159	.1471624
_cons	-2.044876	2.284283	-0.90	0.371	-6.521988	2.432235

For our example, we entered a four-line program into an ado-file (`myprobit_lf.ado`) and then typed two more lines of code after reading some data into memory. That was all that was required to program and fit a probit model. Moreover, our four-line program is good enough to fit any probit model. This program reports conventional

1

(inverse, negative second-derivative) variance estimates, but by specifying an option to `ml model`, we could obtain the outer product of the gradients or Huber/White/sandwich robust variance estimates, all without changing our simple four-line program.

We will discuss `ml` and how to use it in chapter 3, but first we discuss the theory and practice of maximizing likelihood functions.

Also, we will discuss theory so that we can use terms such as *conventional (inverse, negative second-derivative) variance estimates*, *outer product of the gradients variance estimates*, and *Huber/White/sandwich robust variance estimates*, and you can understand not only what the terms mean but also some of the theory behind them.

As for practice, a little understanding of how numerical optimizers work goes a long way toward reducing the frustration of programming ML estimators. A knowledgeable person can glance at output and conclude when better starting values are needed, when more iterations are needed, or when—even though the software reported convergence—the process has not converged.

1.1   The likelihood-maximization problem

The foundation for the theory and practice of ML estimation is a probability model

$$\Pr(Z \leq z) = F(z; \boldsymbol{\theta})$$

where Z is the random variable distributed according to a cumulative probability distribution function $F()$ with parameter vector $\boldsymbol{\theta}' = (\theta_1, \theta_2, \ldots, \theta_E)$ from $\boldsymbol{\Theta}$, which is the parameter space for $F()$. Typically, there is more than one variable of interest, so the model

$$\Pr(Z_1 \leq z_1, Z_2 \leq z_2, \ldots, Z_k \leq z_k) = F(\mathbf{z}; \boldsymbol{\theta}) \tag{1.1}$$

describes the joint distribution of the random variables, with $\mathbf{z} = (z_1, z_2, \ldots, z_k)$. Using $F()$, we can compute probabilities for values of the Zs given values of the parameters $\boldsymbol{\theta}$.

In likelihood theory, we turn things around. Given observed values $\mathbf{z}$ of the variables, the likelihood function is

$$\ell(\boldsymbol{\theta}; \mathbf{z}) = f(\mathbf{z}; \boldsymbol{\theta})$$

where $f()$ is the probability density function corresponding to $F()$. The point is that we are interested in the element (vector) of $\boldsymbol{\Theta}$ that was used to generate $\mathbf{z}$. We denote this vector by $\boldsymbol{\theta}_T$.

Data typically consist of multiple observations on relevant variables, so we will denote a dataset with the matrix $\mathbf{Z}$. Each of N rows, $\mathbf{z}_j$, of $\mathbf{Z}$ consists of jointly observed values of the relevant variables. In this case, we write

$$L(\boldsymbol{\theta}; \mathbf{Z}) = f(\mathbf{Z}; \boldsymbol{\theta}) \tag{1.2}$$

and acknowledge that $f()$ is now the joint-distribution function of the data-generating process. This means that the method by which the data were collected now plays a role in the functional form of $f()$; at this point, if this were a textbook, we would introduce the assumption that "observations" are independent and identically distributed (i.i.d.) and rewrite the likelihood as

$$L(\boldsymbol{\theta}; \mathbf{Z}) = \ell(\boldsymbol{\theta}; \mathbf{z}_1) \times \ell(\boldsymbol{\theta}; \mathbf{z}_2) \times \cdots \times \ell(\boldsymbol{\theta}; \mathbf{z}_N)$$

The ML estimates for $\boldsymbol{\theta}$ are the values $\widehat{\boldsymbol{\theta}}$ such that

$$L(\widehat{\boldsymbol{\theta}}; \mathbf{Z}) = \max_{\mathbf{t} \in \boldsymbol{\Theta}} \ L(\mathbf{t}; \mathbf{Z})$$

Most texts will note that the above is equivalent to finding $\widehat{\boldsymbol{\theta}}$ such that

$$\ln L(\widehat{\boldsymbol{\theta}}; \mathbf{Z}) = \max_{\mathbf{t} \in \boldsymbol{\Theta}} \ \ln L(\mathbf{t}; \mathbf{Z})$$

This is true because $L()$ is a positive function and $\ln()$ is a monotone increasing transformation. Under the i.i.d. assumption, we can rewrite the log likelihood as

$$\ln L(\boldsymbol{\theta}; \mathbf{Z}) = \ln \ell(\boldsymbol{\theta}; \mathbf{z}_1) + \ln \ell(\boldsymbol{\theta}; \mathbf{z}_2) + \cdots + \ln \ell(\boldsymbol{\theta}; \mathbf{z}_N)$$

Why do we take logarithms?

1. Speaking statistically, we know how to take expectations (and variances) of sums, and it is particularly easy when the individual terms are independent.

2. Speaking numerically, some models would be impossible to fit if we did not take logs. That is, we would want to take logs even if logs were not, in the statistical sense, convenient.

To better understand the second point, consider a likelihood function for discrete data, meaning that the likelihoods correspond to probabilities; logit and probit models are examples. In such cases,

$$\ell(\boldsymbol{\theta}; \mathbf{z}_j) = \Pr(\text{we would observe } \mathbf{z}_j)$$

where $\mathbf{z}_j$ is a vector of observed values of one or more response (dependent) variables. All predictor (independent) variables, $\mathbf{x} = (x_1, \ldots, x_p)$, along with their coefficients, $\boldsymbol{\beta}' = (\beta_1, \ldots, \beta_p)$, are part of the model parameterization, so we can refer to the parameter values for the jth observation as $\boldsymbol{\theta}_j$. For instance, $\ell(\boldsymbol{\theta}_j; \mathbf{z}_j)$ might be the probability that $y_j = 1$, conditional on $\mathbf{x}_j$; thus $\mathbf{z}_j = y_j$ and $\boldsymbol{\theta}_j = \mathbf{x}_j \boldsymbol{\beta}$. The overall likelihood function is then the probability that we would observe the y values given the $\mathbf{x}$ values and

$$L(\boldsymbol{\theta}; \mathbf{Z}) = \ell(\boldsymbol{\theta}; \mathbf{z}_1) \times \ell(\boldsymbol{\theta}; \mathbf{z}_2) \times \cdots \times \ell(\boldsymbol{\theta}; \mathbf{z}_N)$$

because the N observations are assumed to be independent. Said differently,

$$\Pr(\text{dataset}) = \Pr(\text{datum } 1) \times \Pr(\text{datum } 2) \times \cdots \times \Pr(\text{datum } N)$$

Probabilities are bound by 0 and 1. In the simple probit or logit case, we can hope that $\ell(\boldsymbol{\theta}; \mathbf{z}_j) > 0.5$ for almost all j, but that may not be true. If there were many possible outcomes, such as multinomial logit, it is unlikely that $\ell(\boldsymbol{\theta}; \mathbf{z}_j)$ would be greater than 0.5. Anyway, suppose that we are lucky and $\ell(\boldsymbol{\theta}; \mathbf{z}_j)$ is right around 0.5 for all N observations. What would be the value of $L()$ if we had, say, 500 observations? It would be

$$0.5^{500} \approx 3 \times 10^{-151}$$

That is a very small number. What if we had 1,000 observations? The likelihood would be

$$0.5^{1000} \approx 9 \times 10^{-302}$$

What if we had 2,000 observations? The likelihood would be

$$0.5^{2000} \approx \texttt{<COMPUTER UNDERFLOW>}$$

Mathematically, we can calculate it to be roughly 2×10^{-603}, but that number is too small for most digital computers. Modern computers can process a range of roughly 10^{-301} to 10^{301}.

Therefore, if we were considering ML estimators for the logit or probit models and if we implemented our likelihood function in natural units, we could not deal with more than about 1,000 observations! Taking logs is how programmers solve such problems because logs remap small positive numbers to the entire range of negative numbers. In logs,

$$\ln\left(0.5^{1000}\right) = 1000 \times \ln(0.5) \approx 1000 \times -0.6931 = -693.1$$

a number well within computational range. Similarly, 2,000 observations is not a problem:

$$\ln\left(0.5^{2000}\right) = 2000 \times \ln(0.5) \approx 2000 \times -0.6931 = -1386.2$$

1.2   Likelihood theory

In this section, we examine the asymptotic properties of $\widehat{\boldsymbol{\theta}}$, namely, its probability limit, variance estimators, and distribution. We will reference several theoretical results—the Mean Value Theorem, the Central Limit Theorem, Slutsky's Theorem, the Law of Large Numbers—and provide some details on how they relate to ML estimation. We refer you to Stuart and Ord (1991, 649–706) and Welsh (1996, chap. 4) for more thorough reviews of this and related topics.

Because we are dealing with likelihood functions that are continuous in their parameters, let's define some differential operators to simplify the notation. For any real valued function $a(\mathbf{t})$, we define D by

$$D\, a(\boldsymbol{\theta}) = \left. \frac{\partial a(\mathbf{t})}{\partial \mathbf{t}} \right|_{\mathbf{t} = \boldsymbol{\theta}}$$

and D^2 by

$$D^2 a(\boldsymbol{\theta}) = \left. \frac{\partial^2 a(\mathbf{t})}{\partial \mathbf{t} \partial \mathbf{t}'} \right|_{\mathbf{t}=\boldsymbol{\theta}}$$

The first derivative of the log-likelihood function with respect to its parameters is commonly referred to as the *gradient vector*, or *score vector*. We denote the gradient vector by

$$\mathbf{g}(\boldsymbol{\theta}) = D \ln L(\boldsymbol{\theta}) = \left. \frac{\partial \ln L(\mathbf{t}; \mathbf{Z})}{\partial \mathbf{t}} \right|_{\mathbf{t}=\boldsymbol{\theta}}$$

The second derivative of the log-likelihood function with respect to its parameters is commonly referred to as the *Hessian matrix*. We denote the Hessian matrix by

$$\mathbf{H}(\boldsymbol{\theta}) = D \, \mathbf{g}(\boldsymbol{\theta}) = D^2 \ln L(\boldsymbol{\theta}) = \left. \frac{\partial^2 \ln L(\mathbf{t}; \mathbf{Z})}{\partial \mathbf{t} \partial \mathbf{t}'} \right|_{\mathbf{t}=\boldsymbol{\theta}} \tag{1.3}$$

Let's assume that through some miracle, we obtain $\widehat{\boldsymbol{\theta}}$ satisfying

$$\ln L(\widehat{\boldsymbol{\theta}}; \mathbf{Z}) = \max_{\mathbf{t} \in \Theta} \{ \ln \ell(\mathbf{t}; \mathbf{z}_1) + \ln \ell(\mathbf{t}; \mathbf{z}_2) + \cdots + \ln \ell(\mathbf{t}; \mathbf{z}_N) \}$$

Now $\widehat{\boldsymbol{\theta}}$ identifies the peak of a continuous surface; thus the slope at this point is zero:

$$\mathbf{g}(\widehat{\boldsymbol{\theta}}) = \mathbf{0} \tag{1.4}$$

By the Mean Value Theorem, that is, a Taylor expansion with first-order remainder, there exists a $\boldsymbol{\theta}^*$ between $\widehat{\boldsymbol{\theta}}$ (the estimated values) and $\boldsymbol{\theta}_T$ (the true values) such that

$$\mathbf{g}(\widehat{\boldsymbol{\theta}}) - \mathbf{g}(\boldsymbol{\theta}_T) = D\mathbf{g}(\boldsymbol{\theta}^*) (\widehat{\boldsymbol{\theta}} - \boldsymbol{\theta}_T)$$

From (1.3) and (1.4), we have

$$-\mathbf{g}(\boldsymbol{\theta}_T) = \mathbf{H}(\boldsymbol{\theta}^*) (\widehat{\boldsymbol{\theta}} - \boldsymbol{\theta}_T)$$

and assuming that $\mathbf{H}(\boldsymbol{\theta}^*)$ is nonsingular, we can rewrite this as

$$\widehat{\boldsymbol{\theta}} = \boldsymbol{\theta}_T + \{-\mathbf{H}(\boldsymbol{\theta}^*)\}^{-1} \mathbf{g}(\boldsymbol{\theta}_T)$$

which, as we will see in section 1.3.1, is motivation for the update step in the Newton–Raphson algorithm.

We are assuming that the $\mathbf{z}_j$ are i.i.d., so

$$\mathbf{g}(\boldsymbol{\theta}) = D \ln \ell(\boldsymbol{\theta}; \mathbf{z}_1) + \cdots + D \ln \ell(\boldsymbol{\theta}; \mathbf{z}_N)$$

and $\mathbf{g}(\boldsymbol{\theta})$ is the sum of N i.i.d. random variables. By the Central Limit Theorem, the asymptotic distribution of $\mathbf{g}(\boldsymbol{\theta})$ is multivariate normal with mean vector $E\{\mathbf{g}(\boldsymbol{\theta})\}$ and variance matrix $\mathrm{Var}\{\mathbf{g}(\boldsymbol{\theta})\}$. Note that, for clustered observations, we can use similar arguments by identifying the independent cluster groups.

Lemma 1. *Let $E()$ denote expectation with respect to the probability measure defined by $F()$ in (1.1). Then, given the previous notation (and under the usual regularity conditions),*

$$E\{\mathbf{g}(\boldsymbol{\theta}_T)\} = \mathbf{0} \tag{1.5}$$

$$\text{Var}\{\mathbf{g}(\boldsymbol{\theta}_T)\} = E[\{\mathbf{g}(\boldsymbol{\theta}_T)\}\{\mathbf{g}(\boldsymbol{\theta}_T)\}'] \tag{1.6}$$

❑ **Proof**

Let $\mathcal{Z}$ denote the sample space associated with the model defined in (1.2). Note that

$$\int_{\mathcal{Z}} L(\boldsymbol{\theta}_T; \mathbf{Z}) \, d\mathbf{Z} = \int_{\mathcal{Z}} f(\mathbf{Z}; \boldsymbol{\theta}_T) \, d\mathbf{Z} = \int_{\mathcal{Z}} dF = 1$$

because $f()$ is the density function corresponding to $F()$. The standard line at this point being "under appropriate regularity conditions", we can move the derivative under the integral sign to get

$$\mathbf{0} = D \left\{ \int_{\mathcal{Z}} L(\boldsymbol{\theta}_T; \mathbf{Z}) \, d\mathbf{Z} \right\} = \int_{\mathcal{Z}} \{ D \, L(\boldsymbol{\theta}_T; \mathbf{Z}) \} \, d\mathbf{Z}$$

You might think that these regularity conditions are inconsequential for practical problems, but one of the conditions is that the sample space $\mathcal{Z}$ does not depend on $\boldsymbol{\theta}_T$. If it does, all the following likelihood theory falls apart and the following estimation techniques will not work. Thus if the range of the values in the data $\mathbf{Z}$ depends on $\boldsymbol{\theta}_T$, you have to start from scratch.

In any case,

$$\mathbf{0} = \int_{\mathcal{Z}} \{ D \, L(\boldsymbol{\theta}_T; \mathbf{Z}) \} \, d\mathbf{Z} = \int_{\mathcal{Z}} \{ 1/L(\boldsymbol{\theta}_T; \mathbf{Z}) \} \{ D \, L(\boldsymbol{\theta}_T; \mathbf{Z}) \} \, L(\boldsymbol{\theta}_T; \mathbf{Z}) \, d\mathbf{Z}$$

$$= \int_{\mathcal{Z}} \{ D \ln L(\boldsymbol{\theta}_T; \mathbf{Z}) \} \, f(\mathbf{Z}; \boldsymbol{\theta}_T) \, d\mathbf{Z}$$

$$= \int_{\mathcal{Z}} \mathbf{g}(\boldsymbol{\theta}_T) \, f(\mathbf{Z}; \boldsymbol{\theta}_T) \, d\mathbf{Z}$$

$$= E\{\mathbf{g}(\boldsymbol{\theta}_T)\}$$

which concludes the proof of (1.5).

Note that (1.6) follows from (1.5) and the definition of the variance.

❑

The following large-sample arguments may be made once it is established that $\widehat{\boldsymbol{\theta}}$ is consistent for $\boldsymbol{\theta}_T$. By consistent, we mean that $\widehat{\boldsymbol{\theta}}$ converges to $\boldsymbol{\theta}_T$ in probability. Formally, $\widehat{\boldsymbol{\theta}}$ converges to $\boldsymbol{\theta}_T$ in probability if for every $\varepsilon > 0$

$$\lim_{N \to \infty} \Pr(|\widehat{\boldsymbol{\theta}} - \boldsymbol{\theta}_T| > \varepsilon) = 0$$

We will write this as

$$\operatorname*{plim}_{N\to\infty} \widehat{\boldsymbol{\theta}} = \boldsymbol{\theta}_T$$

or

$$\widehat{\boldsymbol{\theta}} \xrightarrow{p} \boldsymbol{\theta}_T \quad \text{as} \quad N \to \infty$$

There are multiple papers in the statistics literature that prove this—some are referenced in Welsh (1996). We will accept this without providing the outline for a proof.

Given that

$$\mathbf{H}(\boldsymbol{\theta}) = D^2 \ln \ell(\boldsymbol{\theta}; \mathbf{z}_1) + \cdots D^2 \ln \ell(\boldsymbol{\theta}; \mathbf{z}_N)$$

we now note that consistency of $\widehat{\boldsymbol{\theta}}$ and the Law of Large Numbers imply

$$\operatorname*{plim}_{N\to\infty} \{-\mathbf{H}(\widehat{\boldsymbol{\theta}})\} = \operatorname*{plim}_{N\to\infty} \{-\mathbf{H}(\boldsymbol{\theta}_T)\} = -E\{\mathbf{H}(\boldsymbol{\theta}_T)\} \qquad (1.7)$$

which implies

$$\operatorname*{plim}_{N\to\infty} \{-\mathbf{H}(\boldsymbol{\theta}^*)\} = -E\{\mathbf{H}(\boldsymbol{\theta}_T)\}$$

If we can assume that $-E\{\mathbf{H}(\boldsymbol{\theta}_T)\}$ is a nonsingular matrix, then by Slutsky's Theorem, the asymptotic distribution of $\widehat{\boldsymbol{\theta}}$ is multivariate normal with mean vector $\boldsymbol{\theta}_T$ and variance matrix

$$\mathbf{V}_1 = [-E\{\mathbf{H}(\boldsymbol{\theta}_T)\}]^{-1} \operatorname{Var}\{\mathbf{g}(\boldsymbol{\theta}_T)\} [-E\{\mathbf{H}(\boldsymbol{\theta}_T)\}]^{-1} \qquad (1.8)$$

which you may recognize as the form of the sandwich (robust variance) estimator; see section 1.2.4. Note that although we defined $\mathbf{g}(\boldsymbol{\theta})$ and $\mathbf{H}(\boldsymbol{\theta})$ in terms of a log-likelihood function, it was only in proving lemma 1 that we used this distributional assumption. There are other, more technical proofs that do not require this assumption.

If, as in the proof of lemma 1, we can take the derivative under the integral a second time, we are left with a simpler formula for the asymptotic variance of $\widehat{\boldsymbol{\theta}}$.

Lemma 2. *Given the assumptions of lemma 1,*

$$\operatorname{Var}\{\mathbf{g}(\boldsymbol{\theta}_T)\} = -E\{\mathbf{H}(\boldsymbol{\theta}_T)\} \qquad (1.9)$$

❑ **Proof**

Taking the derivative of $E\{\mathbf{g}(\boldsymbol{\theta}_T)\} = \mathbf{0}$ yields

$$
\begin{aligned}
\mathbf{0} &= D\left[E\{\mathbf{g}(\boldsymbol{\theta}_T)\}\right] \\
&= D\int_{\mathcal{X}} \{D\ln L(\boldsymbol{\theta}_T; \mathbf{Z})\}\, L(\boldsymbol{\theta}_T; \mathbf{Z})\, d\mathbf{Z} \\
&= \int_{\mathcal{X}} D\left[\{D\ln L(\boldsymbol{\theta}_T; \mathbf{Z})\}\, L(\boldsymbol{\theta}_T; \mathbf{Z})\right] d\mathbf{Z} \\
&= \int_{\mathcal{X}} \left[\{D^2\ln L(\boldsymbol{\theta}_T; \mathbf{Z})\}L(\boldsymbol{\theta}_T; \mathbf{Z}) + \{D\ln L(\boldsymbol{\theta}_T; \mathbf{Z})\}\{D\,L(\boldsymbol{\theta}_T; \mathbf{Z})\}'\right] d\mathbf{Z} \\
&= \int_{\mathcal{X}} \left[\{D^2\ln L(\boldsymbol{\theta}_T; \mathbf{Z})\} + \{D\ln L(\boldsymbol{\theta}_T; \mathbf{Z})\}\{D\ln L(\boldsymbol{\theta}_T; \mathbf{Z})\}'\right] L(\boldsymbol{\theta}_T; \mathbf{Z})\, d\mathbf{Z} \\
&= E\{D^2\ln L(\boldsymbol{\theta}_T; \mathbf{Z})\} + E[\{D\ln L(\boldsymbol{\theta}_T; \mathbf{Z})\}\{D\ln L(\boldsymbol{\theta}_T; \mathbf{Z})\}'] \\
&= E\{\mathbf{H}(\boldsymbol{\theta}_T)\} + E\{\mathbf{g}(\boldsymbol{\theta}_T)\mathbf{g}(\boldsymbol{\theta}_T)'\}
\end{aligned}
$$

Thus

$$
\text{Var}\{\mathbf{g}(\boldsymbol{\theta}_T)\} = E\{\mathbf{g}(\boldsymbol{\theta}_T)\mathbf{g}(\boldsymbol{\theta}_T)'\} = -E\{\mathbf{H}(\boldsymbol{\theta}_T)\}
$$

which concludes the proof.

❑

Thus $\widehat{\boldsymbol{\theta}}$ is asymptotically multivariate normal with mean $\boldsymbol{\theta}_T$ and variance

$$
\mathbf{V}_2 = [-E\{\mathbf{H}(\boldsymbol{\theta}_T)\}]^{-1}
$$

which follows from (1.8) and (1.9).

It is common to use

$$
\widehat{\mathbf{V}}_2 = \{-\mathbf{H}(\widehat{\boldsymbol{\theta}})\}^{-1}
$$

as a variance estimator for $\widehat{\boldsymbol{\theta}}$. This is justified by (1.7).

1.2.1   All results are asymptotic

The first important consequence of the above discussion is that all results are asymptotic.

1. $\widehat{\boldsymbol{\theta}} \xrightarrow{p} \boldsymbol{\theta}_T$ is guaranteed only as $N \to \infty$.
2. It is not true in general that $E(\widehat{\boldsymbol{\theta}}) = \boldsymbol{\theta}_T$ for finite N. $\widehat{\boldsymbol{\theta}}$ may be biased, and the bias may be substantial for small N.
3. The variance of $\widehat{\boldsymbol{\theta}}$ is $\{-\mathbf{H}(\widehat{\boldsymbol{\theta}})\}^{-1}$ asymptotically. For finite N, we are not sure how good this variance estimator is.

For these reasons, you should not fit an ML model with only a handful of observations.

1.2.2 Likelihood-ratio tests and Wald tests

Stata's `test` command performs a Wald test, which is a statistical test of the coefficients based on the estimated variance $\{-\mathbf{H}(\widehat{\boldsymbol{\theta}})\}^{-1}$. Likelihood-ratio (LR) tests, on the other hand, compare the heights of the likelihood function at $\widehat{\boldsymbol{\theta}}$ and $\boldsymbol{\theta}_0$, where $\boldsymbol{\theta}_0$ is the vector of hypothesized values. Stata's `lrtest` command performs LR tests.

The LR is defined as

$$\mathrm{LR} = \frac{\max_{\mathbf{t}=\boldsymbol{\theta}_0} L(\mathbf{t}; \mathbf{Z})}{\max_{\mathbf{t}\in\boldsymbol{\Theta}} L(\mathbf{t}; \mathbf{Z})} = \frac{\max_{\boldsymbol{\theta}=\boldsymbol{\theta}_0} L(\boldsymbol{\theta}; \mathbf{Z})}{L(\widehat{\boldsymbol{\theta}}; \mathbf{Z})}$$

The null hypothesis, $\mathrm{H}_0 : \boldsymbol{\theta} = \boldsymbol{\theta}_0$, may be simple—all values of $\boldsymbol{\theta}$ are hypothesized to be some set of values, such as $\boldsymbol{\theta}_0' = (0, 0, \ldots, 0)$—or it may be a composite hypothesis—only some values of $\boldsymbol{\theta}$ are hypothesized, such as $\boldsymbol{\theta}_0' = (0, ?, ?, \ldots, ?)$, where ? means that the value can be anything feasible.

In general, we can write $\boldsymbol{\theta}_0' = (\boldsymbol{\theta}_r', \boldsymbol{\theta}_u')$, where $\boldsymbol{\theta}_r$ is fixed and $\boldsymbol{\theta}_u$ is not. Thus

$$\max_{\boldsymbol{\theta}=\boldsymbol{\theta}_0} L(\boldsymbol{\theta}; \mathbf{Z}) = \max_{\boldsymbol{\theta}_u} L(\boldsymbol{\theta}_r, \boldsymbol{\theta}_u; \mathbf{Z}) = L(\boldsymbol{\theta}_r, \widehat{\boldsymbol{\theta}}_u; \mathbf{Z})$$

and the LR becomes

$$\mathrm{LR} = \frac{L(\boldsymbol{\theta}_r, \widehat{\boldsymbol{\theta}}_u; \mathbf{Z})}{L(\widehat{\boldsymbol{\theta}}; \mathbf{Z})}$$

It can be shown that $-2 \ln(\mathrm{LR})$ is asymptotically distributed as chi-squared with r degrees of freedom, where r is the dimension of $\boldsymbol{\theta}_r$. This theory holds only for small to moderate values of r. If r is very large (say, $r > 100$), the chi-squared distribution result becomes questionable.

To compute LR, we must do two maximizations: one to get $\widehat{\boldsymbol{\theta}}$ and another to compute $\widehat{\boldsymbol{\theta}}_u$. Hence, LR tests can be time consuming because you must do an "extra" maximization for each hypothesis test.

The Wald test, on the other hand, simply uses $\mathrm{Var}(\widehat{\boldsymbol{\theta}})$, which is estimated assuming the true values of $\boldsymbol{\theta}$; that is, it uses $\{-\mathbf{H}(\widehat{\boldsymbol{\theta}})\}^{-1}$ for the variance. For a linear hypothesis $\mathbf{R}\boldsymbol{\theta} = \mathbf{r}$, the Wald test statistic is

$$W = (\mathbf{R}\widehat{\boldsymbol{\theta}} - \mathbf{r})'\{\mathbf{R}\mathrm{Var}(\widehat{\boldsymbol{\theta}})\mathbf{R}'\}^{-1}(\mathbf{R}\widehat{\boldsymbol{\theta}} - \mathbf{r})$$

If $\widehat{\boldsymbol{\theta}}$ is normally distributed, W has a chi-squared distribution with $r = \mathrm{rank}(\mathbf{R}\mathbf{R}')$ degrees of freedom. Obviously, this is an easy computation to make once we have $\mathrm{Var}(\widehat{\boldsymbol{\theta}})$.

As Davidson and MacKinnon (1993, 278) discuss, when the same size is not large, these tests may have very different finite-sample properties, so in general, we cannot claim that one test is better than the other in all cases.

One advantage that the LR test does have over the Wald test is the so-called invariance property. Suppose we want to test the hypothesis $\mathrm{H}_0 : \beta_2 = 2$. Saying $\beta_2 = 2$ is clearly equivalent to saying $1/\beta_2 = 1/2$. With the LR test, whether we state our null hypothesis as $H_0 : \beta_2 = 2$ or $H_0 : 1/\beta_2 = 1/2$ makes absolutely no difference; we will reach the same conclusion either way. On the other hand, you may be surprised to know that we will obtain different Wald test statistics for tests of these two hypotheses. In fact, in some cases, the Wald test may lead to the rejection of one null hypothesis but does not allow the rejection of a mathematically equivalent null hypothesis formulated differently. The LR test is said to be "invariant to nonlinear transformations", and the Wald test is said to be "manipulable."

1.2.3   The outer product of gradients variance estimator

In discussing the large-sample properties of $\widehat{\boldsymbol{\theta}}$, we established

$$\text{Var}(\widehat{\boldsymbol{\theta}}) \approx [-E\{\mathbf{H}(\boldsymbol{\theta}_T)\}]^{-1}$$

and indicated that $\{-\mathbf{H}(\widehat{\boldsymbol{\theta}})\}^{-1}$ is consistent for the variance of $\widehat{\boldsymbol{\theta}}$.

There are other ways we can obtain estimates of the variance. From (1.9), we may rewrite the above as

$$\text{Var}(\widehat{\boldsymbol{\theta}}) \approx [\text{Var}\{\mathbf{g}(\boldsymbol{\theta}_T)\}]^{-1}$$

Note that

$$\mathbf{g}(\boldsymbol{\theta}) = \mathbf{g}_1(\boldsymbol{\theta}) + \cdots + \mathbf{g}_N(\boldsymbol{\theta})$$

where

$$\mathbf{g}_j(\boldsymbol{\theta}) = D \ln \ell(\boldsymbol{\theta}; \mathbf{z}_j)$$

which is to say, $\mathbf{g}(\boldsymbol{\theta})$ is the sum of N random variables. If the observations are independent, we can further say that $\mathbf{g}(\boldsymbol{\theta})$ is the sum of N i.i.d. random variables. What is a good estimator for the variance of the sum of N i.i.d. random variables? Do not let the fact that we are summing evaluations of functions fool you. A good estimator for the variance of the mean of N i.i.d. random variates $y_1, y_2, \ldots, y_N$ is

$$\frac{s^2}{N} = \frac{1}{N(N-1)} \sum_{j=1}^{N} (y_j - \overline{y})^2$$

Therefore, a good estimator for the variance of the total is simply N^2 times the variance estimator of the mean:

$$N s^2 = \frac{N}{N-1} \sum_{j=1}^{N} (y_j - \overline{y})^2 \tag{1.10}$$

We estimate $\text{Var}\{\mathbf{g}(\boldsymbol{\theta}_T)\}$ by simply substituting $y_j = \mathbf{g}_j(\boldsymbol{\theta}_T)$ in (1.10), using the respective matrix operators. Let $\overline{\mathbf{g}}(\boldsymbol{\theta}) = \mathbf{g}(\boldsymbol{\theta})/N$, and then we have

$$\text{Var}\{\mathbf{g}(\boldsymbol{\theta}_T)\} \approx \frac{N}{N-1} \sum_{j=1}^{N} \{\mathbf{g}_j(\boldsymbol{\theta}_T) - \overline{\mathbf{g}}(\boldsymbol{\theta}_T)\} \{\mathbf{g}_j(\boldsymbol{\theta}_T) - \overline{\mathbf{g}}(\boldsymbol{\theta}_T)\}' \tag{1.11}$$

and thus another variance estimator for $\widehat{\boldsymbol{\theta}}$ is

$$
\text{Var}(\widehat{\boldsymbol{\theta}}) \approx \left\{ \sum_{j=1}^{N} \mathbf{g}_j(\widehat{\boldsymbol{\theta}}) \mathbf{g}_j(\widehat{\boldsymbol{\theta}})' \right\}^{-1} \tag{1.12}
$$

where we substituted $\widehat{\boldsymbol{\theta}}$ for $\boldsymbol{\theta}_T$ and simplified by remembering that $\mathbf{g}(\widehat{\boldsymbol{\theta}}) = \mathbf{0}$ (implying $\overline{\mathbf{g}}(\widehat{\boldsymbol{\theta}})$ is also $\mathbf{0}$) and by substituting 1 for $N/(N-1)$ (which it is for large N). In any case, (1.12) is known as the outer product of the gradients variance estimator—often called the OPG—and, in fact, the OPG has the same asymptotic properties as $\{-\mathbf{H}(\widehat{\boldsymbol{\theta}})\}^{-1}$, the "conventional variance estimator".

We refer to $\{-\mathbf{H}(\widehat{\boldsymbol{\theta}})\}^{-1}$ as the conventional variance estimator not because it is better but because it is more commonly reported. As we will discuss later, if we use an optimization method like Newton–Raphson for maximizing the likelihood function, we need to compute the Hessian matrix anyway, and thus at the end of the process, we will have $\{-\mathbf{H}(\widehat{\boldsymbol{\theta}})\}^{-1}$ at our fingertips.

We will also discuss ways in which functions can be maximized that do not require calculation of $\{-\mathbf{H}(\widehat{\boldsymbol{\theta}})\}^{-1}$, such as the Davidon–Fletcher–Powell algorithm. These algorithms are often used on functions for which calculating the second derivatives would be computationally expensive, and given that expense, it is common to report the OPG variance estimate once the maximum is found.

The OPG variance estimator has much to recommend it. In fact, it is more empirical than the conventional calculation because it is like using the sample variance of a random variable instead of plugging the ML parameter estimates into the variance expression for the assumed probability distribution. Note, however, that the above development requires that the data come from a simple random sample.

1.2.4   Robust variance estimates

Our result that $\text{Var}(\widehat{\boldsymbol{\theta}})$ is asymptotically $\{-\mathbf{H}(\widehat{\boldsymbol{\theta}})\}^{-1}$ hinges on lemma 1 and lemma 2. In proving these lemmas, and only in proving them, we assumed the likelihood function $L(\boldsymbol{\theta}_T; \mathbf{Z})$ was the density function for $\mathbf{Z}$. If $L(\boldsymbol{\theta}_T; \mathbf{Z})$ is not the true density function, our lemmas do not apply and all our subsequent results do not necessarily hold. In practice, the ML estimator and this variance estimator still work reasonably well if $L(\boldsymbol{\theta}_T; \mathbf{Z})$ is a little off from the true density function. For example, if the true density is probit and you fit a logit model (or vice versa), the results will still be accurate.

You can also derive a variance estimator that does not require $L(\boldsymbol{\theta}_T; \mathbf{Z})$ to be the density function for $\mathbf{Z}$. This is the *robust variance estimator*, which is implemented in many Stata estimation commands and in Stata's survey (svy) commands.

The robust variance estimator was first published, we believe, by Peter Huber, a mathematical statistician, in 1967 in conference proceedings (Huber 1967). Survey statisticians were thinking about the same things around this time, at least for linear regression. In the 1970s, the survey statisticians wrote up their work, including

Kish and Frankel (1974), Fuller (1975), and others, all of which was summarized and generalized in an excellent paper by Binder (1983). White, an economist, independently derived the estimator and published it in 1980 for linear regression and in 1982 for ML estimates, both in economics literature. Many others have extended its development, including Kent (1982); Royall (1986); Gail, Tan, and Piantadosi (1988); and Lin and Wei (1989).

The robust variance estimator is called different things by different people. At Stata, we originally called it the Huber variance estimator (Bill Rogers, who first implemented it here, was a student of Huber). Some people call it the sandwich estimator. Survey statisticians call it the Taylor-series linearization method, linearization method, or design-based variance estimate. Economists often call it the White estimator. Statisticians often refer to it as the empirical variance estimator. In any case, they all mean the same variance estimator. We will sketch the derivation here.

The starting point is

$$\mathrm{Var}(\widehat{\boldsymbol{\theta}}) \approx \{-\mathbf{H}(\widehat{\boldsymbol{\theta}})\}^{-1} \mathrm{Var}\{\mathbf{g}(\widehat{\boldsymbol{\theta}})\} \{-\mathbf{H}(\widehat{\boldsymbol{\theta}})\}^{-1} \tag{1.13}$$

Because of this starting point, some people refer to robust variance estimates as the Taylor-series linearization method; we obtained (1.13) from the delta method, which is based on a first-order (linear term only) Taylor series.

The next step causes other people to refer to this as the empirical variance estimator. Using the empirical variance estimator of $\mathrm{Var}\{\mathbf{g}(\boldsymbol{\theta}_T)\}$ from (1.11) for $\mathrm{Var}\{\mathbf{g}(\widehat{\boldsymbol{\theta}})\}$, we have that

$$\mathrm{Var}(\widehat{\boldsymbol{\theta}}) \approx \{-\mathbf{H}(\widehat{\boldsymbol{\theta}})\}^{-1} \left\{ \frac{N}{N-1} \sum_{j=1}^{N} \mathbf{g}_j(\widehat{\boldsymbol{\theta}}) \mathbf{g}_j(\widehat{\boldsymbol{\theta}})' \right\} \{-\mathbf{H}(\widehat{\boldsymbol{\theta}})\}^{-1}$$

This is the robust variance estimator. It is robust for the same reasons that the estimator for the variance of a mean is robust; note our lack of assumptions about $\ln \ell()$ and hence $L()$. That is, (1.13) arises from the fact that $\widehat{\boldsymbol{\theta}}$ solves $\mathbf{g}(\widehat{\boldsymbol{\theta}}) = \mathbf{0}$, regardless of whether $\mathbf{g}()$ is obtained from the true data-generating process.

The estimator for the variance of the total of the $\mathbf{g}_j(\widehat{\boldsymbol{\theta}})$ values relies only on our data's coming from simple random sampling (the observations are i.i.d.). Thus the essential assumption of the robust variance estimator is that the observations are independent selections from the same population.

For cluster sampling, we merely change our estimator for the variance of the total of the $\mathbf{g}_j(\widehat{\boldsymbol{\theta}})$ values to reflect this sampling scheme. Consider superobservations made up of the sum of $\mathbf{g}_j(\widehat{\boldsymbol{\theta}})$ for a cluster: these superobservations are independent, and the above formulas hold with $\mathbf{g}_j(\widehat{\boldsymbol{\theta}})$ replaced by the cluster sums (and N replaced by the number of clusters). See [P] _robust, [SVY] **variance estimation**, Rogers (1993), Williams (2000), and Wooldridge (2002) for more details.

1.3   The maximization problem

Given the problem

$$L(\widehat{\boldsymbol{\theta}}; \mathbf{Z}) = \max_{\mathbf{t} \in \boldsymbol{\Theta}} \ L(\mathbf{t}; \mathbf{Z})$$

how do we obtain the solution? One way we might solve this analytically is by taking derivatives and setting them to zero:

$$\text{Solve for } \widehat{\boldsymbol{\theta}}: \qquad \frac{\partial \ln L(\mathbf{t}; \mathbf{Z})}{\partial \mathbf{t}} \bigg|_{\mathbf{t} = \widehat{\boldsymbol{\theta}}} = \mathbf{0} \qquad (1.14)$$

Let's use a simplified example. If $\ln L(\theta) = -\theta^2 + \theta$, then

$$\frac{d \ln L(\theta)}{d\theta} = -2\theta + 1 = 0 \qquad \Rightarrow \qquad \theta = 1/2$$

In general, however, $\partial \ln L(\mathbf{t}; \mathbf{Z})/\partial \mathbf{t} = \mathbf{0}$ is too complicated to admit an analytic solution, and we are forced to seek a numerical one. We start with (1.14) but use the gradient vector, so our problem is

$$\text{Solve for } \widehat{\boldsymbol{\theta}}: \qquad \mathbf{g}(\widehat{\boldsymbol{\theta}}; \mathbf{Z}) = \mathbf{0}$$

We have converted the maximization problem into a root-finding problem. Thus ML estimates $\widehat{\boldsymbol{\theta}}$ are obtained by finding the roots of $\mathbf{g}(\widehat{\boldsymbol{\theta}}; \mathbf{Z})$. Many computer optimizers (the generic name for maximizers and minimizers) are, at heart, root finders.

1.3.1   Numerical root finding

For one dimension, so that θ and $g(\theta)$ are both scalars, one method of finding roots is Newton's method. This is an iterative technique. You have a guess θ_i (called the initial value), and you update it as follows.

Newton's method

1. The current guess is θ_i.
2. Calculate $g(\theta_i)$.
3. Calculate $g'(\theta_i)$, the slope of $g()$ at the current guess.
4. Draw the tangent line at the current guess: $y = g'(\theta_i)(\theta - \theta_i) + g(\theta_i)$.
5. The new guess is θ_{i+1}, the root of the above tangent line:

$$0 = g'(\theta_i)(\theta_{i+1} - \theta_i) + g(\theta_i) \qquad \Rightarrow \qquad \theta_{i+1} = \theta_i - \frac{g(\theta_i)}{g'(\theta_i)}$$

6. Repeat.

The above sequence is not guaranteed to converge, but it generally does. Figure 1.1 illustrates a step in Newton's method.

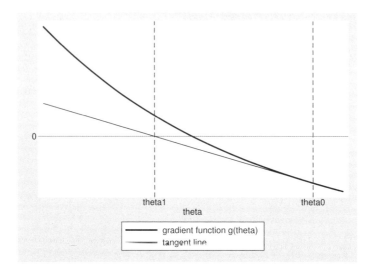

Figure 1.1. Newton's method. The next guess, `theta1`, is the point where the tangent line crosses the horizontal axis.

Newton's method for finding roots can be converted to finding minimums and maximums of a function—say, $f()$—by substituting $f'()$ for $g()$ and $f''()$ for $g'()$:

To find θ such that $f(\theta)$ is maximized,

1. Start with guess θ_i.
2. Calculate new guess $\theta_{i+1} = \theta_i - f'(\theta_i)/f''(\theta_i)$.
3. Repeat.

We can generalize this to allow θ to be a vector, $\boldsymbol{\theta}$.

To find vector $\widehat{\boldsymbol{\theta}}$ such that $f(\widehat{\boldsymbol{\theta}})$ is maximized,

1. Start with guess $\boldsymbol{\theta}_i$.
2. Calculate new guess $\boldsymbol{\theta}_{i+1} = \boldsymbol{\theta}_i - \{\mathbf{H}(\boldsymbol{\theta}_i)\}^{-1}\mathbf{g}(\boldsymbol{\theta}_i)$, where $\mathbf{g}(\boldsymbol{\theta}_i)$ is the gradient vector and $\mathbf{H}(\boldsymbol{\theta}_i)$ is the matrix of second derivatives. (In the future, we will write this as $\boldsymbol{\theta}_{i+1} = \boldsymbol{\theta}_i + \{-\mathbf{H}(\boldsymbol{\theta}_i)\}^{-1}\mathbf{g}(\boldsymbol{\theta}_i)$ because the matrix $-\mathbf{H}(\boldsymbol{\theta}_i)$ is positive definite and calculating the inverse of a positive-definite matrix is easier.)
3. Repeat.

This is the Newton–Raphson algorithm, which is the method Stata and most other statistical packages use by default, although a few details are commonly changed. In fact, any package that claims to use Newton–Raphson is probably oversimplifying the story. These details are changed because

1. Calculating $f(\boldsymbol{\theta}_i)$ is computationally expensive.
2. Calculating $\mathbf{g}(\boldsymbol{\theta}_i)$ is more computationally expensive.
3. Calculating $\mathbf{H}(\boldsymbol{\theta}_i)$ is even more computationally expensive.

The Newton–Raphson algorithm

Relatively speaking, we can calculate $f(\boldsymbol{\theta}_i)$ many times for the same computer time required to calculate $\mathbf{H}(\boldsymbol{\theta}_i)$ just once. This leads to separating the direction calculation from the stepsize:

To find vector $\widehat{\boldsymbol{\theta}}$ such that $f(\widehat{\boldsymbol{\theta}})$ is maximized,

1. Start with a guess $\boldsymbol{\theta}_i$.
2. Calculate a direction vector $\mathbf{d} = \{-\mathbf{H}(\boldsymbol{\theta}_i)\}^{-1}\mathbf{g}(\boldsymbol{\theta}_i)$.
3. Calculate a new guess $\boldsymbol{\theta}_{i+1} = \boldsymbol{\theta}_i + s\mathbf{d}$, where s is a scalar, for instance:
   a. Start with $s = 1$.
   b. If $f(\boldsymbol{\theta}_i + \mathbf{d}) > f(\boldsymbol{\theta}_i)$, try $s = 2$. If $f(\boldsymbol{\theta}_i + 2\mathbf{d}) > f(\boldsymbol{\theta}_i + \mathbf{d})$, try $s = 3$ or even $s = 4$, and so on.
   c. If $f(\boldsymbol{\theta}_i + \mathbf{d}) \leq f(\boldsymbol{\theta}_i)$, back up and try $s = 0.5$ or $s = 0.25$, etc.
4. Repeat.

The way that s is obtained varies from package to package, but the idea is simple enough: go in the direction $\mathbf{d} = \{-\mathbf{H}(\boldsymbol{\theta}_i)\}^{-1}\mathbf{g}(\boldsymbol{\theta}_i)$ as far as you possibly can before spending computer cycles to compute a new direction. This all goes under the rubric "stepsize calculation".

The other important detail goes under the rubric "nonconcavity", meaning that if $\{-\mathbf{H}(\boldsymbol{\theta}_i)\}^{-1}$ does not exist—if $\mathbf{H}(\boldsymbol{\theta}_i)$ is singular—then the direction calculation $\mathbf{d} = \{-\mathbf{H}(\boldsymbol{\theta}_i)\}^{-1}\mathbf{g}(\boldsymbol{\theta}_i)$ cannot be made. It is worth thinking about this in the single-dimension case: in the maximization problem, it means that the second derivative of $f()$ is zero. In the root-finding problem, it means that the derivative of $g()$ is zero. The gradient is flat, so there is no root, as illustrated in figure 1.2.

(Continued on next page)

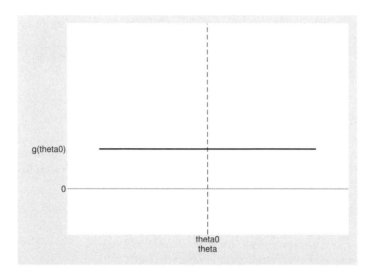

Figure 1.2. Nonconcavity. The slope of the tangent line at `theta` is zero, so where is the root?

Remember, all you really know is that $g'()$ evaluated at θ_i is 0. Even though we have drawn a line, you do not really know that $g'(\theta) = 0$ for all θ. In fact, you are reasonably certain that "$g'(\theta) = 0$ for all θ" is not true because $g()$ is the gradient of a likelihood function and the likelihood certainly cannot be increasing or decreasing without bound. Probably, θ_i is just a poor guess and far from the heart of the function.

The tangent line in figure 1.1 is how you would usually project to find the new guess θ_{i+1}—it goes through the point $\{\theta_i, g(\theta_i)\}$ and has slope $g'(\theta_i)$. Figure 1.2 emphasizes why this rule is not helpful in regions of nonconcavity.

So, pretend that you are the optimizer charged with finding the maximum of $f(\theta)$; θ_i is the current guess and $g'(\theta_i) = 0$. What is your next guess? Put that aside. Is your next guess θ_{i+1} to the left or right of θ_i?

This is where knowing that this is a maximization problem rather than merely a root-finding problem is useful. If you told us just to find the roots, we would have no reason to select $\theta_{i+1} > \theta_i$ or $\theta_{i+1} < \theta_i$; we would have to try both. But if you tell us this is a maximization problem, we observe that in the graph drawn, $g(\theta_i) > 0$, so increasing θ_i increases the value of $f()$. We would try something to the right of θ_i. If $g(\theta_i) < 0$, we would try something to the left of θ_i. And if this were a minimization rather than a maximization problem, we would reverse the rules.

The important point is that at least we have some idea of which direction to go.

When we generalize to allow θ to be a vector, our choices proliferate. Corresponding to our scalar go-with-the-sign-of-$g()$ rule, the vector equivalent is the go-with-the-vector-$\mathbf{g}()$ rule, which amounts to treating the noninvertible matrix $\mathbf{H}(\boldsymbol{\theta}_i)$ as if $\{-\mathbf{H}(\boldsymbol{\theta}_i)\}^{-1} = \mathbf{I}$

because our general rule is direction $\mathbf{d} = \{-\mathbf{H}(\boldsymbol{\theta}_i)\}^{-1}\mathbf{g}(\boldsymbol{\theta}_i)$. That is called *steepest ascent*. It is called ascent because, just as in the scalar case, we choose the direction based on the sign of the gradient and the knowledge that we want to maximize $f()$. It is called steepest because, mathematically, it works out that among all directions, we could go with $f()$ increasing, the direction $\mathbf{g}(\boldsymbol{\theta}_i)$ is steepest. By the way, the steepest-ascent rule is often called such because most of the numerical optimization literature is written in terms of function minimization rather than maximization.

This rule can be improved upon. Rather than treating $\{-\mathbf{H}(\boldsymbol{\theta}_i)\}^{-1}$ as $\mathbf{I}$, we can add a constant to the diagonal elements of $\{-\mathbf{H}(\boldsymbol{\theta}_i)\}^{-1}$ until it is invertible. This is like mixing in a little steepest ascent to the otherwise "best" direction and is called the *modified Marquardt algorithm* (Marquardt 1963).

Stata follows one of two variations on this scheme. In the first variation, rather than adding a constant, c, to the diagonal terms, `ml` adds a proportional term $c|h_{ii}|$, where h_{ii} is the ith diagonal element of $\{-\mathbf{H}(\boldsymbol{\theta})\}^{-1}$. `ml` then renormalizes the resulting matrix so that it has the original trace, thus attempting to keep the scale of the matrix the same.

If you specify `ml`'s `difficult` option, it does more work, computing the eigenvalues of $\{-\mathbf{H}(\boldsymbol{\theta})\}^{-1}$. Then for the part of the orthogonal subspace where the eigenvalues are negative or small positive numbers, `difficult` uses steepest ascent; in the other subspace, it uses a regular Newton–Raphson step.

1.3.2   Quasi-Newton methods

The Newton–Raphson algorithm is not the only method for finding maximums. Related techniques are known as quasi-Newton methods. These methods follow the same steps as the Newton–Raphson algorithm (see page 15) but make a substitution for the Hessian matrix $\mathbf{H}(\boldsymbol{\theta}_i)$ in computing the direction to go, $\mathbf{d} = \{-\mathbf{H}(\boldsymbol{\theta}_i)\}^{-1}\mathbf{g}(\boldsymbol{\theta}_i)$.

Often calculation of $\mathbf{H}(\boldsymbol{\theta}_i)$ can be computationally expensive—very computationally expensive—and these quasi-Newton methods seek to avoid that expense by substituting something cheaper to calculate and (it is advertised) nearly as good.

You may be bothered by having choices with no criterion with which to choose among them. Certainly, you will say to yourself, there must be a best way to maximize a function. Were computation free and computers infinitely fast, then we could seek a method certain to find the maximum regardless of the amount of computation it took to reach our goal. That is not a difficult problem, and you can probably think of a few solutions yourself. Perhaps the easiest would be just to calculate the function at every point in $\boldsymbol{\Theta}$ and choose the maximum, not worrying that an infinite amount of calculations would be required.

In the real world, the problem is to find the maximum with as little calculation as possible. What you must remember is that there are many paths to the top of a hill and the particular path you choose does not matter, assuming you reach the top. The paths differ only in the amount of computation you expend in arriving at the solution. These quasi-Newton methods in general choose poorer paths, but they expend less computation on the path. They are of particular interest when the calculation of $\mathbf{H}(\boldsymbol{\theta}_i)$ is known to be exorbitantly expensive.

Even Newton–Raphson is not guaranteed to find the maximum. In general, Newton–Raphson works well, but if it does fail, it is often because of some very fine detail in the likelihood function that interacts with the Newton–Raphson rules, and the substitution of another set of rules—any other method—will be sufficient to work around the problem.

The BHHH algorithm

BHHH stands for Berndt, Hall, Hall, and Hausman—its authors—and is often pronounced B-H cubed.

The BHHH algorithm substitutes the outer product of the gradients for the observed negative Hessian matrix. Thus if $\mathbf{g}_j(\boldsymbol{\theta}_i)$ is the gradient vector from the jth observation using the current guess $\boldsymbol{\theta}_i$, then the BHHH algorithm uses

$$-\mathbf{H}_{\mathrm{BHHH}}(\boldsymbol{\theta}_i) = \sum_{j=1}^{N} \mathbf{g}_j(\boldsymbol{\theta}_i)\mathbf{g}_j(\boldsymbol{\theta}_i)'$$

instead of the observed negative Hessian matrix. For further details about this algorithm, see Berndt et al. (1974).

The DFP and BFGS algorithms

The DFP and BFGS methods substitute an ever-improving estimate for $\mathbf{H}(\boldsymbol{\theta}_i)$ at every step. DFP stands for Davidon (1959), Fletcher (1970), and Powell (Fletcher and Powell 1963), and BFGS stands for Broyden (1967), Fletcher (1970), Goldfarb (1970), and Shanno (1970).

As above, let i be the index for the current iteration; $\boldsymbol{\theta}_i$ be the guess for the ML estimate of $\boldsymbol{\theta}_T$ at the ith iteration; $\mathbf{g}_i = \mathbf{g}(\boldsymbol{\theta}_i)$, the associated gradient vector; and $\mathbf{H}_i = \mathbf{H}(\boldsymbol{\theta}_i)$, the associated Hessian matrix. According to the Newton–Raphson algorithm, the next guess is

$$\boldsymbol{\theta}_{i+1} = \boldsymbol{\theta}_i + (-\mathbf{H}_i)^{-1}\mathbf{g}_i$$

Suppose that $\mathbf{A}_i$ is a substitute for $(-\mathbf{H}_i)^{-1}$. For both DFP and BFGS, $\mathbf{A}_{i+1}$ is calculated by adding in a correction to $\mathbf{A}_i$ so that

$$\Delta\boldsymbol{\theta}_{i+1} = \mathbf{A}_{i+1}\Delta\mathbf{g}_{i+1}$$

where $\Delta\boldsymbol{\theta}_{i+1} = \boldsymbol{\theta}_{i+1} - \boldsymbol{\theta}_i$ and $\Delta\mathbf{g}_{i+1} = \mathbf{g}_{i+1} - \mathbf{g}_i$.

In the DFP algorithm, $\mathbf{A}_{i+1}$ is updated by

$$\mathbf{A}_{i+1} = \mathbf{A}_i + \frac{\Delta\boldsymbol{\theta}_{i+1}\Delta\boldsymbol{\theta}'_{i+1}}{\Delta\boldsymbol{\theta}'_{i+1}\Delta\mathbf{g}_{i+1}} - \frac{\mathbf{A}_i\Delta\mathbf{g}_{i+1}\Delta\mathbf{g}'_{i+1}\mathbf{A}_i}{\Delta\mathbf{g}'_{i+1}\mathbf{A}_i\Delta\mathbf{g}_{i+1}}$$

In the BFGS algorithm, $\mathbf{A}_{i+1}$ is updated by

$$\mathbf{A}_{i+1} = \mathbf{A}_i + \frac{\Delta\boldsymbol{\theta}_{i+1}\Delta\boldsymbol{\theta}'_{i+1}}{\Delta\boldsymbol{\theta}'_{i+1}\Delta\mathbf{g}_{i+1}} - \frac{\mathbf{A}_i\Delta\mathbf{g}_{i+1}\Delta\mathbf{g}'_{i+1}\mathbf{A}_i}{\Delta\mathbf{g}'_{i+1}\mathbf{A}_i\Delta\mathbf{g}_{i+1}} + \Delta\mathbf{g}'_{i+1}\mathbf{A}_i\Delta\mathbf{g}_{i+1}\mathbf{u}\mathbf{u}'$$

where

$$\mathbf{u} = \frac{\Delta\boldsymbol{\theta}_{i+1}}{\Delta\boldsymbol{\theta}'_{i+1}\Delta\mathbf{g}_{i+1}} - \frac{\mathbf{A}_i\Delta\mathbf{g}_{i+1}}{\Delta\mathbf{g}'_{i+1}\mathbf{A}_i\Delta\mathbf{g}_{i+1}}$$

Thus the BFGS algorithm is essentially the same as DFP with an added correction to $\mathbf{A}_i$ when calculating $\mathbf{A}_{i+1}$. For both algorithms, `ml` uses $\mathbf{g}_0 = \mathbf{0}$ (vector of zeros) and $\mathbf{A}_0 = I$ (identity matrix) as starting values. For further details about these algorithms, see Fletcher (1987) and Press et al. (2007).

1.3.3   Numerical maximization

To summarize, to find $\widehat{\boldsymbol{\theta}}$ such that $f(\widehat{\boldsymbol{\theta}})$ is maximized,

1. Start with a guess $\boldsymbol{\theta}_i$.

2. Calculate a direction vector $\mathbf{d} = \{-\mathbf{H}(\boldsymbol{\theta}_i)\}^{-1}\mathbf{g}(\boldsymbol{\theta}_i)$, if you can. $\mathbf{H}(\boldsymbol{\theta}_i)$ can be calculated several ways: by direct calculation of the second derivatives, by the outer product of the gradients, or by an iterative technique. However you get $\mathbf{H}(\boldsymbol{\theta}_i)$, if $-\mathbf{H}(\boldsymbol{\theta}_i)$ is not invertible, then substitute for $-\mathbf{H}(\boldsymbol{\theta}_i)$ the matrix $\mathbf{I}$, or $-\mathbf{H}(\boldsymbol{\theta}_i) + c\mathbf{I}$ for some small, positive scalar c, or some other matrix related to $-\mathbf{H}(\boldsymbol{\theta}_i)$.

3. Calculate a new guess $\boldsymbol{\theta}_{i+1} = \boldsymbol{\theta}_i + s\mathbf{d}$, where s is a scalar.

    a. For instance, start with $s = 1$.

    b. If $f(\boldsymbol{\theta}_i + \mathbf{d}) > f(\boldsymbol{\theta}_i)$, try $s = 2$. If $f(\boldsymbol{\theta}_i + 2\mathbf{d}) > f(\boldsymbol{\theta}_i + \mathbf{d})$, try $s = 3$ or even $s = 4$, and so on.

    c. If $f(\boldsymbol{\theta}_i + \mathbf{d}) \leq f(\boldsymbol{\theta}_i)$, back up and try $s = 0.5$ or $s = 0.25$, etc.

4. Repeat.

If you have ever used Stata's `ml` command or any of Stata's preprogrammed ML estimators, you have probably seen remnants of these rules.

When Stata says

```
Iteration #:  log likelihood = ...
```

that means it is at step 2 and calculating a direction. The log likelihood reported is $f(\boldsymbol{\theta}_i)$.

You may have seen

```
Iteration #:  log likelihood = ...   (not concave)
```

That arose during step 2. $-\mathbf{H}(\boldsymbol{\theta}_i)$ was not invertible, and Stata had to use something else to choose a direction. So what does it mean? If the message occurs in early iterations, you can safely ignore it. It just means that the current guess for $\widehat{\boldsymbol{\theta}}$ is rather poor and that the matrix of second derivatives calculated at that point is noninvertible—corresponding in the scalar case to the gradient being flat.

In the multidimensional case, a noninvertible $-\mathbf{H}(\boldsymbol{\theta}_i)$ means that the likelihood function $f()$ locally has a ridge, a flat section, or a saddle point. Real likelihood functions have few points like this, and when such arise, it is typically because the gradient is almost, but not perfectly, flat. Roundoff error makes the nearly flat gradient appear absolutely flat. In any case, the optimizer encountered the problem and worked around it.

If, on the other hand, this message were to arise at the final iteration, just before results are presented, it would be of great concern.

If you have seen Stata say

```
Iteration #:  log likelihood = ...   (backed up)
```

that arose in step 3. Stata started with $s = 1$. With $s = 1$, Stata expected $f(\boldsymbol{\theta}_i + s\mathbf{d}) = f(\boldsymbol{\theta}_i + \mathbf{d}) > f(\boldsymbol{\theta}_i)$, but that did not happen. Actually, Stata mentions that it backed up only if it halved the original stepsize six or more times. As with the not-concave message, this should not concern you unless it occurs at the last iteration, in which case the declaration of convergence is questionable. (You will learn to specify ml's `gradient` option to see the gradient vector in cases such as this so that you can determine whether its elements really are close to zero.)

1.3.4   Numerical derivatives

The maximizer that we have sketched has substantial analytic requirements. In addition to specifying $f(\mathbf{t}) = \ln L(\mathbf{t}; \mathbf{Z})$, the routine must be able to calculate $\mathbf{g}(\mathbf{t}) = \partial f(\mathbf{t})/\partial \mathbf{t}$ and, in the Newton–Raphson case, $\mathbf{H}(\mathbf{t}) = \partial^2 f(\mathbf{t})/\partial \mathbf{t}\partial \mathbf{t}'$. It needs the function, its first derivatives, and its second derivatives.

Stata provides facilities so that you do not actually have to program the derivative calculations, but really, Stata has only one optimizer. When you do not provide the derivatives, Stata calculates them numerically.

The definition of an analytic derivative is

$$f'(z) = \frac{df(z)}{dz} = \lim_{h \to 0} \frac{f(z+h) - f(z)}{h}$$

and that leads to a simple approximation formula,

$$f'(z) \approx \frac{f(z+h) - f(z)}{h}$$

for a suitably small but large enough h.

We have a lot to say about how Stata finds a suitably small but large enough h. Stata's use of a formula like $\{f(z+h) - f(z)\}/h$ has an important implication for your program that calculates $f()$, the log-likelihood function.

If Stata chooses h just right, in general this formula will be only about half as accurate as the routine that calculates $f()$. Because of the way Stata chooses h, it will not be less accurate than that, and it might be more accurate. The loss of accuracy comes about because of the subtraction in the numerator.

Let's show this by calculating the derivative of $\exp(z)$ at $z = 2$. The true answer is $\exp(2)$ because $d\{\exp(z)\}/dz = \exp(z)$. Let's try this formula with $h = 10^{-8}$ ($h = 1\text{e-}8$ in computer speak). We will carefully do this calculation with 16 digits of accuracy:

```
exp(2 + 1e-8)       =  7.389056172821211   (accuracy is 16 digits)
exp(2)              =  7.389056098930650   (accuracy is 16 digits)
                       ─────────────────
difference          =   .000000073890561

exp(2)              =  7.389056098930650   (true answer)
difference / 1e-8   =  7.389056033701991   (approximation formula)
                       ─────────────────
error                   .000000065228659   approximation is correct to
                       1 2345678           8 digits
```

The major source of error was introduced when we calculated the difference $\exp(2 + 1\text{e-}8) - \exp(2)$.

```
exp(2 + 1e-8)       =  7.389056172821211   (accuracy is 16 digits)
exp(2)              =  7.389056098930650   (accuracy is 16 digits)
                       ─────────────────
difference          =   .000000073890561
                              12345678    (accuracy is 8 digits)
```

Our full 16 digits of accuracy were lost because half the digits of $\exp(2 + 1\text{e-}8)$ and $\exp(2)$ were in common and so canceled each other.

This is an unpleasant feature of numerical derivatives. Given how Stata chooses h, if you start with k digits of accuracy, you will fall to $k/2$ digits at worst. You can get lucky, the best case being when $f(z) = 0$, in which case you can get all k digits back, but that is ignoring the inaccuracy introduced by the algorithm itself.

Thus as a programmer of likelihood functions, if you are going to use numerical derivatives, it is vitally important that you supply $\ln L()$ as accurately as you can. This means that all your `generate` statements should explicitly specify `double`:

```
. gen double ... = ...
```

If you do not specify `double`, Stata will store the result as `float`, which has about 7 digits of accuracy, meaning that the numerical first derivatives will be accurate to only 3.5 digits. (Even if you specify `double`, your routine is probably not accurate to 16 digits; it is merely carrying around 16 digits, and the roundoff error inherent in your calculation probably means you will supply only 13 or 14 accurate digits and possibly fewer. The minute you include one `generate` without `double`, however, accuracy falls to 7 digits, and then the roundoff error further reduces it from there.)

The issue in calculating numerical derivatives is choosing h, which Stata does for you. Just so that you are properly appreciative, let's consider that problem:

1. If Stata chooses a value for h that is too small, $f(z + h)$ will numerically equal $f(z)$, so the approximation to $f'(z)$ will be zero.

2. Values that are larger than that but still too small result in poor accuracy because of the numerical roundoff error associated with subtraction that we have just seen. If $f(z + h)$ is very nearly equal to $f(z)$, nearly all the digits are in common, and the difference has few significant digits. For instance, say $f(z)$ and $f(z+h)$ differ only in the 15th and 16th digits:

   ```
   f(x0+h)             =   y.xxxxxxxxxxxxxab
   f(x0)               =   y.xxxxxxxxxxxxxcd
   ─────────────────────────────────────────
   difference          =   0.0000000000000ef    (2 digits of accuracy)
   ```

   Even worse, the 16-digit numbers shown might be accurate to only 14 digits themselves. In that case, the difference $f(z + h) - f(z)$ would be purely noise.

3. On the other hand, if Stata chooses a value for h that is too large, the approximation formula will not estimate the derivative accurately.

Choosing the right value for h is important. Above we considered numerically evaluating the derivative of $\exp(z)$ at $z = 2$. Below we calculate the error for a variety of h's:

```
. drop _all
. set obs 21
obs was 0, now 21
. gen double h = 10^(-(_N-_n))
. gen double approx = (exp(2+h)-exp(2))/h
. gen double error = exp(2) - approx
. gen double relerr = abs(error/exp(2))
```

```
. list, sep(0) divider
```

	h	approx	error	relerr
1.	1.000e-20	0	7.3890561	1
2.	1.000e-19	0	7.3890561	1
3.	1.000e-18	0	7.3890561	1
4.	1.000e-17	0	7.3890561	1
5.	1.000e-16	0	7.3890561	1
6.	1.000e-15	6.2172489	1.1718072	.15858685
7.	1.000e-14	7.5495166	-.16046047	.02171596
8.	1.000e-13	7.3807627	.00829343	.00112239
9.	1.000e-12	7.3896445	-.00058835	.00007962
10.	1.000e-11	7.3890227	.00003337	4.516e-06
11.	1.000e-10	7.3890583	-2.155e-06	2.917e-07
12.	1.000e-09	7.3890565	-3.789e-07	5.127e-08
13.	1.000e-08	7.389056	6.523e-08	8.828e-09
14.	1.000e-07	7.3890565	-3.522e-07	4.767e-08
15.	1.000e-06	7.3890598	-3.695e-06	5.001e-07
16.	.00001	7.389093	-.00003695	5.000e-06
17.	.0001	7.3894256	-.00036947	.00005
18.	.001	7.3927519	-.00369576	.00050017
19.	.01	7.4261248	-.03706874	.00501671
20.	.1	7.7711381	-.38208204	.05170918
21.	1	12.696481	-5.3074247	.71828183

In comparing these alternative h's, you should focus on the `relerr` column. The minimum relative error in this experiment is recorded in observation 13, $h = $ 1e–8.

So should h be 1e–8? Not in general: 1e–8 is a good choice for calculating the derivative of exp() at 2, but at some other location, a different value of h would be best. Even if you map out the entire exp() function, you will not have solved the problem. Change the function, and you change the best value of h.

Nash (1990, 219) suggested that we choose h so that $z + h$ differs from z in at least half its digits (the least significant ones). In particular, Nash suggested the formula

$$h(z) = \epsilon \left(|z| + \epsilon \right)$$

where ϵ is the square root of machine precision. Machine precision is roughly 1e–16 for double precision, and therefore, $h(2) \approx$ 2e–8. This way of dynamically adjusting h to the value of z works pretty well.

Nash's suggestion can be improved upon if we are willing to spend computer time. Two issues require balancing:

1. If the calculation were carried out in infinite precision, then the smaller h is, the more accurately $\{f(z + h) - f(z)\}/h$ would approximate the derivative.

2. Taking into consideration finite precision, the closer the values $f(z + h)$ and $f(z)$ are, the fewer the significant digits in the calculation $f(z + h) - f(z)$ are.

Thus we use a variation on Nash's suggestion and control the numerator:

$$f(z+h) \text{ and } f(z) \text{ should differ in about half their digits.}$$

That is, we adopt the rule of setting h as small as possible, subject to the constraint that $f(z+h) - f(z)$ will be calculated to at least half accuracy.

This is a computationally expensive rule to implement because, each time a numerical derivative is calculated, the program must search for the optimal value of h, meaning that $f(z+h)$ must be calculated for trial values. The payoff, however, is that `ml` is remarkably robust.

`ml` also uses a centered derivative calculation. Rather than using

$$f'(z) \approx \frac{f(z+h) - f(z)}{h}$$

`ml` uses

$$f'(z) \approx \frac{f(z+h/2) - f(z-h/2)}{h}$$

This also increases the computational time required but results in an important improvement, reducing the order of magnitude of the error (ignoring numerically induced error) from $O(h)$ to $O(h^2)$.

1.3.5 Numerical second derivatives

Newton–Raphson requires not only $f'(z)$ but also $f''(z)$, the second derivatives. We use the same method to calculate the second derivatives that we use to calculate the first derivatives:

$$f''(z) \approx \frac{f'(z+h/2) - f'(z-h/2)}{h}$$

If you make the substitution for $f'(z)$, you obtain

$$f''(z) \approx \frac{f(z+h) - 2f(z) + f(z-h)}{h^2}$$

Think about the accuracy of this. You might first be tempted to reason that if $f()$ is calculated to 16 decimal places, $f'()$ will be accurate to 16/2 decimal places at worst, assuming optimal h. If $f'()$ is accurate to 8 decimal places, $f''()$ will be accurate to at least 8/2 decimal places, at worst. Thus $f''()$ is accurate to at least 4 decimal places, at worst.

That is not quite correct: you must look at the numerical calculation formula. The numerator can be rewritten as

$$\{f(z+h) - f(z)\} + \{f(z-h) - f(z)\}$$

The first term, $f(z+h) - f(z)$, can be calculated to half accuracy. The second term can similarly be calculated to the same accuracy if we assume that $f(z+h)$ and $f(z-h)$

are of the same order of magnitude (binary magnitude for modern digital computers). Given that roughly half the digits have already been lost, the numerical summation of the two half-accurate results can be done without further loss of precision, assuming that they are of the same order of (binary) magnitude. The net result is that the numerator can be calculated nearly as accurately as the numerator for the first derivatives.

1.4   Monitoring convergence

When you perform ML estimation, before presenting the estimates, Stata displays something like this:

```
Iteration 0:   log likelihood =  -45.03321
Iteration 1:   log likelihood = -27.990675
Iteration 2:   log likelihood = -23.529093
Iteration 3:   log likelihood = -21.692004
Iteration 4:   log likelihood = -20.785625
Iteration 5:   log likelihood = -20.510315
Iteration 6:   log likelihood = -20.483776
Iteration 7:   log likelihood = -20.483495
```

Stata displays this partly for the entertainment value. Placing something on the screen periodically convinces you that your computer is still working.

The iteration log, however, contains reassuring information if you know how to read it. Now that you know how Stata's modified Newton–Raphson works—it is based on Newton's method for finding roots—it should be obvious to you that, for smooth functions, it may jump around at first; however, once it gets close to the maximum (root), it should move smoothly toward it, taking smaller and smaller steps. That is what happened above, and the first graph in figure 1.3 plots these log-likelihood values.

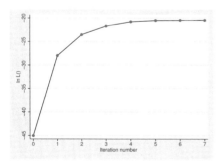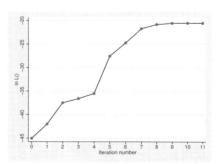

Figure 1.3. Monitoring convergence. Left: the log-likelihood values displayed earlier; right: the log-likelihood values jumping around early on but smoothing out as we get closer to convergence.

When we see this, we know that the numerical routines have really converged to a maximum. Every step changes the log likelihood by less than the previous step. It would not bother us too much if the early iterations jumped around, as long as the later iterations followed the expected pattern. An example of this is the second graph in figure 1.3. All this means is that, wherever we were at the second through fourth iterations, the likelihood function flattened out.

The graphs in figure 1.4, however, would concern us because we are not seeing the expected slowing down of increments in the likelihood function as we get close to the solution.

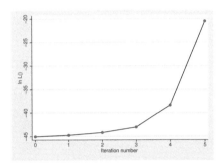 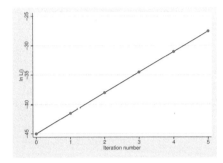

Figure 1.4. Disconcerting "convergence". Left: larger than the previous, indicating that there is still a long way to go to get to the maximum; right: the steps have the same amount of progress, also indicating that there is still a long way to the maximum.

Looking at the second graph of figure 1.4, after thinking about the possibility of a programming bug and dismissing it, we would force more iterations and hope to see something like the first graph in figure 1.5; however, we would not be shocked to see something like the second graph. Either way, it would mean that we just had not done enough iterations previously.

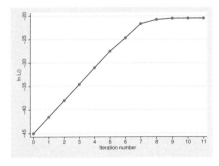 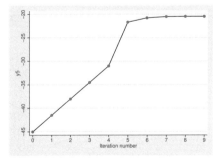

Figure 1.5. Achieving convergence: two possible results from forcing more iterations from the problem identified by the graph on the right in figure 1.4.

We suggest that, after estimation, you type `ml graph` to obtain graphs like the ones above.

Stata is reasonably conservative about declaring convergence, and in fact, it is not uncommon to see irritating evidence of that:

```
Iteration 0:   log likelihood =   -45.03321
Iteration 1:   log likelihood = -27.990675
Iteration 2:   log likelihood = -23.529093
Iteration 3:   log likelihood = -21.692004
Iteration 4:   log likelihood = -20.785625
Iteration 5:   log likelihood = -20.510315
Iteration 6:   log likelihood = -20.483776
Iteration 7:   log likelihood = -20.483495
Iteration 8:   log likelihood = -20.483495
```

Iterations 7 and 8 both present the same value (to the sixth decimal place) of the log-likelihood function. This extra step can be irritating when iterations take 10 or 15 seconds, but the equal value at the end is very reassuring. Stata does not always do that, but we like to see it the first time we code a likelihood function, and whatever Stata does, we invariably force a few extra iterations just for the reassurance.

2 Introduction to ml

`ml` is the Stata command to maximize user-defined likelihoods. Obtaining maximum likelihood (ML) estimates requires the following steps:

1. Derive the log-likelihood function from your probability model.

2. Write a program that calculates the log-likelihood values and, optionally, its derivatives. This program is known as a likelihood evaluator.

3. Identify a particular model to fit using your data variables and the `ml model` statement.

4. Fit the model using `ml maximize`.

This chapter illustrates steps 2, 3, and 4 using the probit model for dichotomous (0/1) variables and the linear regression model assuming normally distributed errors.

In this chapter, we fit our models explicitly, handling each coefficient and variable individually. New users of `ml` will appreciate this approach because it closely reflects how you would write down the model you wish to fit on paper; and it allows us to focus on some of the basic features of `ml` without becoming overly encumbered with programming details. We will also illustrate this strategy's shortcomings so that once you become familiar with the basics of `ml` by reading this chapter, you will want to think of your model in a slightly more abstract form, providing much more flexibility.

In the next chapter, we discuss `ml`'s probability model parameter notation, which is particularly useful when, as is inevitably the case, you decide to change some of the variables appearing in your model. If you are already familiar with `ml`'s θ-parameter notation, you can skip this chapter with virtually no loss of continuity with the rest of the book.

Chapter 15 contains the derivations of log-likelihood functions (step 1) for models discussed in this book.

2.1 The probit model

Say that we want to fit a probit model to predict whether a car is foreign or domestic based on its weight and price using the venerable `auto.dta` that comes with Stata. Our statistical model is

$$\begin{aligned}
\pi_j &= \Pr(\texttt{foreign}_j \mid \texttt{weight}_j, \texttt{price}_j) \\
&= \Phi(\beta_1 \texttt{weight}_j + \beta_2 \texttt{price}_j + \beta_0)
\end{aligned}$$

where we use the subscript j to denote observations and $\Phi(\cdot)$ denotes the standard normal distribution function. The log likelihood for the jth observation is

$$\ln \ell_j = \begin{cases} \ln \Phi(\beta_1 \texttt{weight}_j + \beta_2 \texttt{price}_j + \beta_0) & \text{if } \texttt{foreign}_j = 1 \\ 1 - \ln \Phi(\beta_1 \texttt{weight}_j + \beta_2 \texttt{price}_j + \beta_0) & \text{if } \texttt{foreign}_j = 0 \end{cases}$$

Because the normal density function is symmetric about zero, $1 - \Phi(w) = \Phi(-w)$, and computers can more accurately calculate the latter than the former. Therefore, we are better off writing the log likelihood as

$$\ln \ell_j = \begin{cases} \ln \Phi(\beta_1 \texttt{weight}_j + \beta_2 \texttt{price}_j + \beta_0) & \text{if } \texttt{foreign}_j = 1 \\ \ln \Phi(-\beta_1 \texttt{weight}_j - \beta_2 \texttt{price}_j - \beta_0) & \text{if } \texttt{foreign}_j = 0 \end{cases} \tag{2.1}$$

With our log-likelihood function in hand, we write a program to evaluate it:

```
───────────────────────────────────────── begin myprobit_gf0.ado ─────────
program myprobit_gf0
        args todo b lnfj
        tempvar xb
        quietly generate double `xb' = `b'[1,1]*weight + `b'[1,2]*price + ///
                               `b'[1,3]
        quietly replace `lnfj' = ln(normal(`xb')) if foreign == 1
        quietly replace `lnfj' = ln(normal(-1*`xb')) if foreign == 0
end
─────────────────────────────────────────── end myprobit_gf0.ado ─────────
```

We named our program `myprobit_gf0.ado`, but you could name it anything you want as long as it has the extension `.ado`. The name without the `.ado` extension is what we use to tell `ml model` about our likelihood function. We added `gf0` to our name to emphasize that our evaluator is a general-form problem and that we are going to specify no (0) derivatives. We will return to this issue when we use the `ml model` statement.

Our program accepts three arguments. The first, `todo`, we can safely ignore for now. In later chapters, when we discuss other types of likelihood-evaluator programs, we will need that argument. The second, `b`, contains a row vector containing the parameters of our model (β_0, β_1, and β_2). The third argument, `lnfj`, is the name of a temporary variable that we are to fill in with the values of the log-likelihood function evaluated at the coefficient vector b. Our program then created a temporary variable to hold the values of $\beta_1 \texttt{weight}_j + \beta_2 \texttt{price}_j + \beta_0$. We created that variable to have storage type `double`; we will discuss this point in greater detail in the next chapter, but for now you should remember that when coding your likelihood evaluator, you must create

temporary variables as `doubles`. The last two lines replace `lnfj` with the values of
the log likelihood for $foreign_j$ equal to 0 and 1, respectively. Because `b` and `lnfj` are
arguments passed to our program and `xb` is a temporary variable we created with the
`tempvar` commands, their names are local macros that must be dereferenced using left-
and right-hand single quote marks to use them; see [U] **18.7 Temporary objects**.

The next step is to identify our model using `ml model`. To that end, we type

```
. sysuse auto
(1978 Automobile Data)
. ml model gf0 myprobit_gf0 (foreign = weight price)
```

We first loaded in our dataset, because `ml model` will not work without the dataset
in memory. Next we told `ml` that we have a method-`gf0` likelihood evaluator named
`myprobit_gf0`, our dependent variable is `foreign`, and our independent variables are
`weight` and `price`. In subsequent chapters, we examine all the likelihood-evaluator
types; method-`gf0` (general form) evaluator programs most closely follow the mathematical notation we used in (2.1) and are therefore perhaps easiest for new users of `ml`
to grasp, but we will see that they have disadvantages as well. General-form evaluators simply receive a vector of parameters and a variable into which the observations'
log-likelihood values are to be stored.

The final step is to tell `ml` to maximize the likelihood function and report the coefficients:

```
. ml maximize
initial:       log likelihood = -51.292891
alternative:   log likelihood = -45.055272
rescale:       log likelihood = -45.055272
Iteration 0:   log likelihood = -45.055272
Iteration 1:   log likelihood = -20.770386
Iteration 2:   log likelihood = -18.023563
Iteration 3:   log likelihood = -18.006584
Iteration 4:   log likelihood = -18.006571
Iteration 5:   log likelihood = -18.006571

                                       Number of obs   =         74
                                       Wald chi2(2)    =      14.09
Log likelihood = -18.006571            Prob > chi2     =     0.0009
```

foreign	Coef.	Std. Err.	z	P>|z|	[95% Conf. Interval]
weight	-.003238	.0008643	-3.75	0.000	-.004932	-.0015441
price	.000517	.0001591	3.25	0.001	.0002052	.0008287
_cons	4.921935	1.330066	3.70	0.000	2.315054	7.528816

You can verify that we would obtain identical results using `probit`:

```
. probit foreign weight price
```

This example was straightforward because we had only one equation and no auxiliary parameters. Next we consider linear regression with normally distributed errors.

2.2   Normal linear regression

Now suppose we want to fit a linear regression of `turn` on `length` and `headroom`:

$$\text{turn}_j = \beta_1 \text{length}_j + \beta_2 \text{headroom}_j + \beta_3 + \epsilon_j$$

where ϵ_j is an error term. If we assume that each ϵ_j is independent and identically distributed as a normal random variable with mean zero and variance σ^2, we have what is often called normal linear regression; and we can fit the model by ML. As derived in section 15.3, the log likelihood for the jth observation assuming homoskedasticity (constant variance) is

$$\ln \ell_j = \ln \phi \left(\frac{\text{turn}_j - \beta_1 \text{length}_j - \beta_2 \text{headroom}_j - \beta_3}{\sigma} \right) - \ln \sigma$$

There are four parameters in our model: β_1, β_2, β_3, and σ, so we will specify our `ml model` statement so that our likelihood evaluator receives a vector of coefficients with four columns. As a matter of convention, we will use the four elements of that vector in the order we just listed so that, for example, β_2 is the second element and σ is the fourth element. Our likelihood-evaluator program is

```
───────────────────────────────────────── begin mynormal1_gf0.ado ─────────
    program mynormal1_gf0
            args todo b lnfj
            tempvar xb
            quietly generate double `xb' = `b'[1,1]*length +          ///
                              `b'[1,2]*headroom + `b'[1,3]
            quietly replace `lnfj' = ln(normalden((turn - `xb')/`b'[1,4])) - ///
                              ln(`b'[1,4])
    end
──────────────────────────────────────────── end mynormal1_gf0.ado ─────────
```

In our previous example, when we typed

```
. ml model gf0 myprobit_gf0 (foreign = weight price)
```

`ml` knew to create a coefficient vector with three elements because we specified two right-hand-side variables, and by default `ml` includes a constant term unless we specify the `noconstant` option, which we discuss in the next chapter. How do we get `ml` to include a fourth parameter for σ? The solution is to type

```
. ml model gf0 mynormal1_gf0 (turn = length headroom) /sigma
```

The notation `/sigma` tells `ml` to include a fourth element in our coefficient vector and to label it `sigma` in the output. Having identified our model, we can now maximize the log-likelihood function:

```
. ml maximize
initial:        log likelihood =       -<inf>   (could not be evaluated)
feasible:       log likelihood =  -8418.567
rescale:        log likelihood = -327.16314
rescale eq:     log likelihood = -215.53986
Iteration 0:    log likelihood = -215.53986   (not concave)
Iteration 1:    log likelihood = -213.33272   (not concave)
Iteration 2:    log likelihood = -211.10519   (not concave)
Iteration 3:    log likelihood =  -209.6059   (not concave)
Iteration 4:    log likelihood = -207.93809   (not concave)
Iteration 5:    log likelihood = -206.43891   (not concave)
Iteration 6:    log likelihood =  -205.1962   (not concave)
Iteration 7:    log likelihood = -204.11317   (not concave)
Iteration 8:    log likelihood = -203.00323   (not concave)
Iteration 9:    log likelihood =  -202.1813   (not concave)
Iteration 10:   log likelihood = -201.42353   (not concave)
Iteration 11:   log likelihood = -200.64586   (not concave)
Iteration 12:   log likelihood =  -199.9028   (not concave)
Iteration 13:   log likelihood = -199.19009   (not concave)
Iteration 14:   log likelihood = -198.48271   (not concave)
Iteration 15:   log likelihood = -197.78686   (not concave)
Iteration 16:   log likelihood = -197.10722   (not concave)
Iteration 17:   log likelihood = -196.43923   (not concave)
Iteration 18:   log likelihood = -195.78098   (not concave)
Iteration 19:   log likelihood = -195.13352   (not concave)
Iteration 20:   log likelihood = -194.49664   (not concave)
Iteration 21:   log likelihood = -193.86938   (not concave)
Iteration 22:   log likelihood = -193.25148   (not concave)
Iteration 23:   log likelihood = -192.64285   (not concave)
Iteration 24:   log likelihood = -192.04319   (not concave)
Iteration 25:   log likelihood = -191.45242   (not concave)
Iteration 26:   log likelihood = -190.87034   (not concave)
Iteration 27:   log likelihood = -190.29685   (not concave)
Iteration 28:   log likelihood = -189.73203   (not concave)
Iteration 29:   log likelihood = -189.17561   (not concave)
Iteration 30:   log likelihood = -188.62745
Iteration 31:   log likelihood = -177.20678   (backed up)
Iteration 32:   log likelihood = -163.35109
Iteration 33:   log likelihood = -163.18766
Iteration 34:   log likelihood = -163.18765
```

```
                                        Number of obs   =        74
                                        Wald chi2(2)    =    219.18
Log likelihood = -163.18765             Prob > chi2     =    0.0000
```

turn	Coef.	Std. Err.	z	P>|z|	[95% Conf. Interval]	
eq1						
length	.1737845	.0134739	12.90	0.000	.1473762	.2001929
headroom	-.1542077	.3546293	-0.43	0.664	-.8492684	.540853
_cons	7.450477	2.197352	3.39	0.001	3.143747	11.75721
sigma						
_cons	2.195259	.1804491	12.17	0.000	1.841585	2.548932

The point estimates match those we obtain from typing

```
. regress turn length headroom
```

The standard errors differ by a factor of sqrt(71/74) because `regress` makes a small-sample adjustment in estimating the error variance. In the special case of linear regression, the need for a small-sample adjustment is not difficult to prove. However, in general ML estimators are only justified asymptotically, so small-sample adjustments have dubious value.

❏ **Technical note**

The log-likelihood function is only defined for $\sigma > 0$—standard deviations must be nonnegative and $\ln(0)$ is not defined. `ml` assumes that all coefficients can take on any value, but it is designed to gracefully handle situations where this is not the case. For example, the output indicates that at the initial values (all four coefficients set to zero), `ml` could not evaluate the log-likelihood function; but `ml` was able to find alternative values with which it could begin the optimization process. In other models, you may have coefficients that are restricted to be in the range $(0, 1)$ or $(-1, 1)$, and in those cases, `ml` is often unsuccessful in finding feasible initial values. The best solution is to reparameterize the likelihood function so that all the parameters appearing therein are unrestricted; subsequent chapters contain examples where we do just that.

❏

2.3   Robust standard errors

Robust standard errors are commonly reported nowadays along with linear regression results because they allow for correct statistical inference even when the tenuous assumption of homoskedasticity is not met. Cluster–robust standard errors can be used when related observations' errors are correlated. Obtaining standard errors with most estimation commands is trivial: you just specify the option `vce(robust)` or `vce(cluster id)`, where *id* is the name of a variable identifying groups. Using our previous regression example, you might type

```
. regress turn length headroom, vce(robust)
```

For the evaluator functions we have written so far, both of which have been method `gf0`, obtaining robust or cluster–robust standard errors is no more difficult than with other estimation commands. To refit our linear regression model, obtaining robust standard errors, we type

```
. ml model gf0 mynormal1_gf0 (turn = length headroom) /sigma, vce(robust)

. ml maximize, nolog
initial:       log pseudolikelihood =     -<inf>  (could not be evaluated)
feasible:      log pseudolikelihood = -8418.567
rescale:       log pseudolikelihood = -327.16314
rescale eq:    log pseudolikelihood = -215.53986
                                            Number of obs   =         74
                                            Wald chi2(2)    =     298.85
Log pseudolikelihood = -163.18765           Prob > chi2     =     0.0000
```

		Robust					
turn	Coef.	Std. Err.	z	P>|z|	[95% Conf. Interval]	
eq1							
length	.1737845	.0107714	16.13	0.000	.152673	.1948961
headroom	-.1542077	.2955882	-0.52	0.602	-.73355	.4251345
_cons	7.450477	1.858007	4.01	0.000	3.80885	11.0921
sigma							
_cons	2.195259	.2886183	7.61	0.000	1.629577	2.76094

ml model accepts vce(cluster *id*) with method-gf0 evaluators just as readily as it accepts vce(robust).

Being able to obtain robust standard errors just by specifying an option to ml model should titillate you. When we discuss other types of evaluator programs, we will see that in fact there is a lot of work happening behind the scenes to produce robust standard errors. With method-gf0 evaluators (and other linear-form evaluators), ml does all the work for you.

2.4 Weighted estimation

Stata provides four types of weights that the end-user can apply to estimation problems. Frequency weights, known as fweights in the Stata vernacular, represent duplicated observations; instead of having five observations that record identical information, fweights allow you to record that observation once in your dataset along with a frequency weight of 5, indicating that observation is to be repeated a total of five times. Analytic weights, called aweights, are inversely proportional to the variance of an observation and are used with group-mean data. Sampling weights, called pweights, denote the inverse of the probability that an observation is sampled and are used with survey data where some people are more likely to be sampled than others. Importance weights, called iweights, indicate the relative "importance" of the observation and are intended for use by programmers who want to produce a certain computation.

Obtaining weighted estimates with method-gf0 likelihood evaluators is the same as with most other estimation commands. Suppose that in auto.dta, rep78 is actually a frequency weight variable. To obtain frequency-weighted estimates of our probit model, we type

```
. ml model gf0 myprobit_gf0 (foreign = weight price) [fw = rep78]

. ml maximize
initial:       log likelihood = -162.88959
alternative:   log likelihood = -159.32929
rescale:       log likelihood = -156.55825
Iteration 0:   log likelihood = -156.55825
Iteration 1:   log likelihood = -72.414357
Iteration 2:   log likelihood =  -66.82292
Iteration 3:   log likelihood = -66.426129
Iteration 4:   log likelihood = -66.424675
Iteration 5:   log likelihood = -66.424675
```

```
                                          Number of obs  =        235
                                          Wald chi2(2)   =      58.94
Log likelihood = -66.424675               Prob > chi2    =     0.0000
```

foreign	Coef.	Std. Err.	z	P>|z|	[95% Conf. Interval]	
weight	-.0027387	.0003576	-7.66	0.000	-.0034396	-.0020379
price	.0004361	.0000718	6.07	0.000	.0002953	.0005768
_cons	4.386445	.5810932	7.55	0.000	3.247523	5.525367

Just like with obtaining robust standard errors, we did not have to do anything to our likelihood-evaluator program. We just added a weight specification, and `ml` did all the heavy lifting to make that work. You should be impressed. Other evaluator types require you to account for weights yourself, which is not always a trivial task.

2.5   Other features of method-gf0 evaluators

In addition to easily obtaining robust standard errors and weighted estimates, method-`gf0` likelihood evaluators provide several other features. By specifying the `svy` option to `ml model`, you can obtain results that take into account the complex survey design of your data. Before using the `svy` option, you must first `svyset` your data; see [U] **26.19 Survey data**.

You can restrict the estimation sample by using `if` and `in` conditions in your `ml` `model` statement. Again, method-`gf0` evaluators require you to do nothing special to make them work. See [U] **11 Language syntax** to learn about `if` and `in` qualifiers.

2.6   Limitations

We have introduced `ml` using method-`gf0` evaluators because they align most closely with the way you would write the likelihood function for a specific model. However, writing your likelihood evaluator in terms of a particular model with prespecified variables severely limits your flexibility.

For example, say that we had a binary variable `good` that we wanted to use instead of `foreign` as the dependent variable in our probit model. If we simply change our `ml model` statement to read

    `. ml model gf0 myprobit_gf0 (good = weight price)`

the output from `ml maximize` will label the dependent variable as `good`, but the output will otherwise be unchanged! When we wrote our likelihood-evaluator program, we hardcoded in the name of the dependent variable. As far as our likelihood-evaluator program is concerned, changing the dependent variable in our `ml model` statement did nothing.

When you specify the dependent variable in your `ml model` statement, `ml` stores the variable name in the global macro `$ML_y1`. Thus a better version of our `myprobit_gf0` program would be

```
─────────────────────────────── begin myprobit_gf0_good.ado ───────────────
program myprobit_gf0_good
        args todo b lnfj
        tempvar xb
        quietly generate double 'xb' = 'b'[1,1]*weight + 'b'[1,2]*price + ///
                                'b'[1,3]
        quietly replace 'lnfj' = ln(normal('xb')) if $ML_y1 == 1
        quietly replace 'lnfj' = ln(normal(-1*'xb')) if $ML_y1 == 0
end
────────────────────────────────── end myprobit_gf0_good.ado ──────────────
```

With this change, we can specify dependent variables at will.

Adapting our program to accept an arbitrary dependent variable was straightforward. Unfortunately, making it accept an arbitrary set of independent variables is much more difficult. We wrote our likelihood evaluator assuming that the coefficient vector `b` had three elements, and we hardcoded the names of our independent variables in the likelihood-evaluator program. If we were hell-bent on making our method-`gf0` evaluator work with an arbitrary number of independent variables, we could examine the column names of `b` and deduce the number of variables, their names, and even the number of equations. In the next chapter, we will learn a better way to approach problems using `ml` that affords us the ability to change regressors without having to modify our evaluator program in any way.

3   Overview of ml

In the previous chapter, we gave an introduction to `ml`, programming our likelihood evaluator much like you would write the likelihood function for a particular model and dataset on paper. In this chapter, we show how `ml`'s syntax allows us to write likelihood evaluators in more general terms, without having to specify variable names or refer to individual coefficients within our evaluator program. By using the methods in this chapter, we will overcome the shortcomings we encountered in the last chapter.

3.1   The terminology of ml

Before we can really discuss using `ml` for model fitting, we must first settle on some notation and terminology.

Any likelihood function can be written as a function of parameters and data variables (both dependent and independent). Thus in the most general case, we can write

$$\ln L = \ln L\{(\theta_{1j}, \theta_{2j}, \ldots, \theta_{Ej}; y_{1j}, y_{2j}, \ldots, y_{Dj}) : j = 1, 2, \ldots, N\}$$
$$= \ln L(\boldsymbol{\theta}_1, \boldsymbol{\theta}_2, \ldots, \boldsymbol{\theta}_E; \mathbf{y}_1, \mathbf{y}_2, \ldots, \mathbf{y}_D)$$

$$\theta_{ij} = \mathbf{x}_{ij}\boldsymbol{\beta}_i = \beta_{i0} + x_{ij1}\beta_{i1} + \cdots + x_{ijk}\beta_{ik}$$
$$\boldsymbol{\theta}_i = \mathbf{X}_i\boldsymbol{\beta}_i$$

where

- j is the observation index and N is the number of observations;
- D is the number of dependent variables ($D \geq 0$);
- $\boldsymbol{\theta}'_i = (\theta_{i1}, \theta_{i2}, \ldots, \theta_{iN})$ is the ith model parameter, which we will refer to as the ith equation because it may vary from one observation to another as a linear combination of independent variables;
- i is the equation index and E is the number of equations ($E \geq 1$);
- $\boldsymbol{\beta}_i$ is the vector of coefficients for the ith equation; and
- $\mathbf{X}_i$ is the data matrix of independent variables for the ith equation, with each row representing an observation so that $\mathbf{x}_{ij}$ is the row vector of values from the jth observation of the independent variables in the ith equation.

We use θ to denote probability model parameters and β to denote coefficients when discussing log-likelihood functions in general. By probability model parameters, we mean parameters that appear directly in the density function we use to derive the likelihood function. For example, the normal density function depends on the mean μ and variance σ^2; here we have $\theta_1 = \mu$ and $\theta_2 = \sigma^2$.

The β coefficients are the weights of a linear combination of variables that determines the value of a model parameter. For example, if θ_1 represents the mean of a distribution and the mean depends on two variables that you specify, then $\theta_{1j} = \beta_1 x_{1j} + \beta_2 x_{2j} + \beta_0$. [ml includes a constant term (β_0 here) in linear combinations unless you tell it not to.]

We will use β to denote coefficients throughout this book; however, we will try to use the conventional parameter symbols when discussing a specific density function, such as μ and σ for the normal density instead of θ_1 and θ_2.

In the special case when

$$\ln \ell_j = \ln \ell(\theta_{1j}, \theta_{2j}, \ldots, \theta_{Ej}; y_{1j}, y_{2j}, \ldots, y_{Dj})$$

$$\ln L = \sum_{j=1}^{N} \ln \ell_j$$

that is, when 1) the log-likelihood contribution can be calculated separately for each observation and 2) the sum of the individual contributions equals the overall log likelihood, the likelihood function is said to meet the linear-form restrictions.

As an example, the probit likelihood function meets the linear-form restrictions because we can express the log likelihood for a single observation; and the log likelihood for the entire sample is just the sum of the observations' log-likelihood values. Conditional logistic regression, on the other hand, violates the linear-form requirements because log-likelihood values can be calculated only for groups of observations.

3.2   Equations in ml

The key to understanding ml is getting comfortable with ml's equation notation. To do this, we will continue to focus on the normal linear regression model.

The log likelihood for the normal linear regression model is

$$\ln L = \sum_{j=1}^{N} [\ln \phi\{(y_j - \mathbf{x}_j \boldsymbol{\beta})/\sigma\} - \ln \sigma]$$

where the point is to estimate $\boldsymbol{\beta}$, the vector of coefficients, and σ, the standard deviation of the residuals. Identifying the regression equation

$$\mu_j = E(y_j) = \mathbf{x}_j \boldsymbol{\beta}$$

we can rewrite the above log likelihood as

$$\ln L = \sum_{j=1}^{N} [\ln \phi \{ (y_j - \theta_{1j})/\theta_{2j} \} - \ln \theta_{2j}]$$

$$\theta_{1j} = \mu_j = \mathbf{x}_{1j} \boldsymbol{\beta}_{1j}$$

$$\theta_{2j} = \sigma_j = \mathbf{x}_{2j} \boldsymbol{\beta}_{2j}$$

So we see that our log likelihood is a function of y_j, μ_j, and σ_j. Notice that we identify the σ_j parameter as an `ml` equation. We emphasize this so that you will think about and write your likelihood evaluator in terms of the inputs supplied by `ml`. Later, when it comes time to fit a model, you will specify the independent variables that go in each linear equation, and `ml` will produce the corresponding coefficient estimates.

After deriving the log-likelihood function from the probability model, we write a program to evaluate the log-likelihood function

```
program mynormal1_lf
        This program receives data variables for μ_j and σ_j as arguments and y_j in $ML_y1.
        It returns ln L as output.
end
```

We can then specify the particular model to be fit by typing

```
. ml model lf mynormal1_lf (y = equation for μ) (equation for σ)
```

and finally fit the model by typing

```
. ml maximize
```

Using `auto.dta`, we would fit a linear regression of `mpg` on `weight` and `displacement` by typing

```
. ml model lf mynormal1_lf (mpg = weight displacement) ()
. ml maximize
```

The part of the `ml model` statement that reads (`mpg = weight displacement`) specifies that

$$y_j = \texttt{mpg}$$

$$\mu_j = \beta_1 \texttt{weight}_j + \beta_2 \texttt{displacement}_j + \beta_3$$

The odd-looking () at the end of the `ml model` statement specifies that $\sigma_j = \beta_4$.

Once the `ml model` statement is specified, `ml` now has everything it needs to begin optimizing the log likelihood. It has a program, `mynormal1_lf`, that defines the log likelihood $\ln L(\mu_j, \sigma_j; y_j)$, and it has definitions for y_j, μ_j, and σ_j.

Our likelihood evaluator is

```
─────────────────────────────── begin mynormal1_lf.ado ───────────
program mynormal1_lf
        version 11
        args lnfj mu sigma
        quietly replace `lnfj' = ln(normalden($ML_y1,`mu',`sigma'))
end
                                            ─────── end mynormal1_lf.ado ───────────
```

In our evaluator, we took advantage of the fact that the `normalden()` function allows us to specify the mean and standard deviation of the normal distribution rather than restricting us to use the "standard normal" variant. Indeed, we could have written the penultimate line of our evaluator program as

```
quietly replace `lnf' = ln(normalden(($ML_y1 - `mu')/`sigma')) - ln(`sigma')
```

though arguably the version we used in our evaluator program is easier. The result from typing the above `ml` commands is

```
. sysuse auto
(1978 Automobile Data)

. ml model lf mynormal1_lf (mpg = weight displacement) ()

. ml maximize
initial:       log likelihood =      -<inf>  (could not be evaluated)
feasible:      log likelihood = -10383.274
rescale:       log likelihood = -292.89564
rescale eq:    log likelihood = -238.45986
Iteration 0:   log likelihood = -238.45986  (not concave)
Iteration 1:   log likelihood = -225.48871  (not concave)
Iteration 2:   log likelihood = -215.66781
Iteration 3:   log likelihood = -199.18065
Iteration 4:   log likelihood = -195.29859
Iteration 5:   log likelihood =   -195.24
Iteration 6:   log likelihood = -195.2398
Iteration 7:   log likelihood = -195.2398

                                        Number of obs   =         74
                                        Wald chi2(2)    =     139.21
Log likelihood = -195.2398              Prob > chi2     =     0.0000
```

mpg	Coef.	Std. Err.	z	P>|z|	[95% Conf. Interval]	
eq1						
weight	-.0065671	.0011424	-5.75	0.000	-.0088061	-.0043281
displacement	.0052808	.0096674	0.55	0.585	-.0136671	.0242286
_cons	40.08452	1.978738	20.26	0.000	36.20627	43.96278
eq2						
_cons	3.385282	.2782684	12.17	0.000	2.839886	3.930678

In the output above, `eq1` stands for equation 1, corresponding to μ, and `eq2` stands for equation 2, corresponding to σ. We just fit the model defined by

$$\ln L = \sum_{j=1}^{N} [\ln \phi\{(y_j - \theta_{1j})/\theta_{2j}\} - \ln \theta_{2j}]$$

$$y_j = \texttt{mpg}_j$$
$$\theta_{1j} = \mu_j = \beta_{1,1}\texttt{weight}_j + \beta_{1,2}\texttt{displacement}_j + \beta_{1,3}$$
$$\theta_{2j} = \sigma_j = \beta_{2,1}$$

because we wrote a program to calculate

$$\ln L = \sum_{j=1}^{N} [\ln \phi\{(y_j - \theta_{1j})/\theta_{2j}\} - \ln \theta_{2j}]$$

and we defined y_j, θ_{1j}, and θ_{2j} by typing

```
. ml model lf mynormal1_lf (mpg=weight displacement) ()
```

In general, any likelihood problem can be written as

$$\ln L = \ln L\{(\theta_{1j}, \theta_{2j}, \ldots, \theta_{Ej}; y_{1j}, y_{2j}, \ldots, y_{Dj}) : \ j = 1, 2, \ldots, N\}$$
$$y_{1j} = \ldots$$
$$y_{2j} = \ldots$$
$$\vdots$$
$$y_{Dj} = \ldots$$
$$\theta_{1j} = \ldots$$
$$\theta_{2j} = \ldots$$
$$\vdots$$
$$\theta_{Ej} = \ldots$$

The program you write fills in the definition of $\ln L$. The `ml model` statement you issue fills in the dots.

Equations are specified in the `ml model` statement:

```
. ml model ... (equation for θ₁) (equation for θ₂) ...
```

The syntax for specifying an equation is

$$([\,equation\_name{:}\,]\,[\,varlist_y\, =\,]\,[\,varlist_x\,]\,[\,,\ \underline{\text{noconst}}\text{ant}\,])$$

equation_name is optional. If you do not specify equation names, the first equation will
be named `eq1`, the second `eq2`, etc. Specifying an equation name usually makes
the output more readable. Had we, in our linear regression example, specified the
equation for θ_{2j} as (`sigma:`) rather than (), the output would have shown the name
`sigma` where it now has `eq2`.

varlist$_y$ specifies the dependent variables and must always be followed by `=`.

varlist$_x$ specifies the independent variables that make up the linear combination for the
equation. If no independent variables are specified, the equation is constant across
all observations.

`noconstant` determines whether the linear equation includes a constant term. It is a
syntax error to specify this option in an equation where no independent variables
are specified.

Let's forget the linear regression example for a minute. Without the context of a particular likelihood function, here are some `ml model` specifications with their translations
into dependent variables and linear equations:

```
. ml model ... (foreign = mpg weight)
```

$$y_{1j} = \texttt{foreign}_j$$
$$\theta_{1j} = \beta_1\texttt{mpg}_j + \beta_2\texttt{weight}_j + \beta_3$$

```
. ml model ... (mpg = weight displacement) ()
```

$$y_{1j} = \texttt{mpg}_j$$
$$\theta_{1j} = \beta_{1,1}\texttt{weight}_j + \beta_{1,2}\texttt{displacement}_j + \beta_{1,3}$$
$$\theta_{2j} = \beta_{2,1}$$

```
. ml model ... (mpg = weight displacement) (price)
```

$$y_{1j} = \texttt{mpg}_j$$
$$\theta_{1j} = \beta_{1,1}\texttt{weight}_j + \beta_{1,2}\texttt{displacement}_j + \beta_{1,3}$$
$$\theta_{2j} = \beta_{2,1}\texttt{price}_j + \beta_{2,2}$$

```
. ml model ... (mpg = weight displacement, nocons) (price)
```

$$y_{1j} = \mathtt{mpg}_j$$
$$\theta_{1j} = \beta_{1,1}\mathtt{weight}_j + \beta_{1,2}\mathtt{displacement}_j$$
$$\theta_{2j} = \beta_{2,1}\mathtt{price}_j + \beta_{2,2}$$

```
. ml model ... (mpg = weight displacement) (price, nocons)
```

$$y_{1j} = \mathtt{mpg}_j$$
$$\theta_{1j} = \beta_{1,1}\mathtt{weight}_j + \beta_{1,2}\mathtt{displacement}_j + \beta_{1,3}$$
$$\theta_{2j} = \beta_{2,1}\mathtt{price}_j$$

```
. ml model ... (mpg = weight displacement) (price = weight) ()
```

$$y_{1j} = \mathtt{mpg}_j$$
$$y_{2j} = \mathtt{price}_j$$
$$\theta_{1j} = \beta_{1,1}\mathtt{weight}_j + \beta_{1,2}\mathtt{displacement}_j + \beta_{1,3}$$
$$\theta_{2j} = \beta_{2,1}\mathtt{weight}_j + \beta_{2,2}$$
$$\theta_{3j} = \beta_{3,1}$$

```
. ml model ... (mpg price = weight displacement) (weight) ()
```

$$y_{1j} = \mathtt{mpg}_j$$
$$y_{2j} = \mathtt{price}_j$$
$$\theta_{1j} = \beta_{1,1}\mathtt{weight}_j + \beta_{1,2}\mathtt{displacement}_j + \beta_{1,3}$$
$$\theta_{2j} = \beta_{2,1}\mathtt{weight}_j + \beta_{2,2}$$
$$\theta_{3j} = \beta_{3,1}$$

This is how `ml` equations work. Note especially the last two examples:

```
. ml model ... (mpg = weight displacement) (price = weight) ()
. ml model ... (mpg price = weight displacement) (weight) ()
```

Both resulted in the same definitions of y_{1j}, y_{2j}, θ_{1j}, θ_{2j}, and θ_{3j}. When you specify the dependent variables, it really does not matter in which equation they appear. The first y variable is y_1, the second is y_2, and so on.

In our example above, we never bothered to specify equation names. Let's go back and show the result of specifying them.

```
. sysuse auto
(1978 Automobile Data)

. ml model lf mynormal1_lf (mu: mpg = weight displacement) (sigma:)

. ml maximize
initial:       log likelihood =     -<inf>  (could not be evaluated)
feasible:      log likelihood = -10383.274
rescale:       log likelihood = -292.89564
rescale eq:    log likelihood = -238.45986
Iteration 0:   log likelihood = -238.45986  (not concave)
Iteration 1:   log likelihood = -225.48871  (not concave)
Iteration 2:   log likelihood = -215.66781
Iteration 3:   log likelihood = -199.18065
Iteration 4:   log likelihood = -195.29859
Iteration 5:   log likelihood =   -195.24
Iteration 6:   log likelihood = -195.2398
Iteration 7:   log likelihood = -195.2398
```

			Number of obs	=	74
			Wald chi2(2)	=	139.21
Log likelihood = -195.2398			Prob > chi2	=	0.0000

mpg	Coef.	Std. Err.	z	P>|z|	[95% Conf. Interval]
mu					
weight	-.0065671	.0011424	-5.75	0.000	-.0088061 -.0043281
displacement	.0052808	.0096674	0.55	0.585	-.0136671 .0242286
_cons	40.08452	1.978738	20.26	0.000	36.20627 43.96278
sigma					
_cons	3.385282	.2782684	12.17	0.000	2.839886 3.930678

Now, instead of `eq1` and `eq2`, our equation names are labeled `mu` and `sigma`.

Actually, specifying an equation name does more than just change how the output is labeled. If—after we fit a model—we wish to refer to the coefficient on `weight` in an expression, we would refer to it as [*equation_name*]_b[`weight`]. For example, after the model we just fit, we would type [mu]_b[`weight`]. In fact, we would probably simply type _b[`weight`] because Stata assumes we are referring to the first equation when an equation is not specified. Thus we only need to specify the equation name to refer to a coefficient from an equation other than the first equation.

There is another way `ml model` will let us specify equations. Instead of typing (*equation_name*:), which is one way of specifying a named constant equation, we can type /*equation_name*. We tend to specify equations this way instead; thus in our linear regression example, we would probably specify the model as

```
. ml model lf mynormal1_lf (mu: mpg = weight displacement) /sigma
```

rather than

```
. ml model lf mynormal1_lf (mu: mpg = weight displacement) (sigma:)
```

or

```
. ml model lf mynormal1_lf (mpg = weight displacement) ()
```

However, it really makes no substantive difference.

Now that we have explained `ml` equations, let's show their real power. In the last chapter, we lamented that because our method-`gf0` linear regression likelihood evaluator worked directly with the β coefficients and variables of our model, changing the specification of our regression required us to modify our likelihood evaluator. With a method-`lf` evaluator, we instead express our likelihood function in terms of the θ parameters. As the previous `ml model` specifications we considered indicate, all we need to do to change the definition of a θ parameter is change the variables specified in the `ml model` command.

We have shown that

―――――――――――――――――――――――― begin mynormal1_lf.ado ―――――――

```
program mynormal1_lf
        version 11
        args lnfj mu sigma
        quietly replace 'lnfj' = ln(normalden($ML_y1,'mu','sigma'))
end
```

――――――――――――――――――――――――― end mynormal1_lf.ado ―――――――

```
. ml model lf mynormal1_lf (mu: mpg = weight displacement) /sigma
. ml maximize
```

will produce the maximum likelihood estimates of

$$\mathtt{mpg}_j = \beta_1 \mathtt{weight}_j + \beta_2 \mathtt{displacement}_j + \beta_3 + \epsilon_j, \quad \epsilon_j \sim N(0, \sigma^2)$$

Without changing our likelihood evaluator, we can just as easily fit a model in which σ varies with `price`; that is, we can fit a linear regression model with conditional heteroskedasticity. To illustrate, we will fit $\sigma_j = \beta_{2,1} \mathtt{price}_j + \beta_{2,2}$:

(Continued on next page)

```
. sysuse auto
(1978 Automobile Data)

. ml model lf mynormal1_lf (mu: mpg = weight displacement) (sigma: price)

. ml maximize, nolog
initial:       log likelihood =      -<inf>  (could not be evaluated)
feasible:      log likelihood = -10383.274
rescale:       log likelihood = -292.89564
rescale eq:    log likelihood = -238.45986
                                              Number of obs   =         74
                                              Wald chi2(2)    =     126.71
Log likelihood = -195.14254                   Prob > chi2     =     0.0000

         mpg |     Coef.   Std. Err.      z    P>|z|     [95% Conf. Interval]
-------------+----------------------------------------------------------------
mu           |
      weight | -.0065258   .0011369    -5.74   0.000    -.008754   -.0042975
displacement |  .0041644   .0099923     0.42   0.677     -.01542    .0237489
       _cons |   40.1942   1.976217    20.34   0.000    36.32089    44.06752
-------------+----------------------------------------------------------------
sigma        |
       price |  .0000438   .0001061     0.41   0.680    -.0001641   .0002517
       _cons |   3.11326   .6915815     4.50   0.000    1.757785    4.468735
```

3.3   Likelihood-evaluator methods

`ml` provides a total of 16 types of likelihood-evaluator functions that can be grouped
into four families. Each evaluator has both a short name as well as a longer, more
descriptive name. We tend to use the short names exclusively, though you may specify
either the short or the long name for each evaluator. Table 3.1 classifies the evaluators:

Table 3.1. Likelihood-evaluator family names

Family	Short name	Long name
Linear form	lf	linearform
	lf0	linearform0
	lf1	linearform1
	lf2	linearform2
	lf1debug	linearform1debug
	lf2debug	linearform2debug
Derivative	d0	derivative0
	d1	derivative1
	d2	derivative2
	d1debug	derivative1debug
	d2debug	derivative2debug
General form	gf0	generalform0
	gf1	generalform1
	gf2	generalform2
	gf1debug	generalform1debug
	gf2debug	generalform2debug

Evaluators whose names end in a number indicate the analytic derivatives you are to provide: 0 indicates no derivatives are required; 1 indicates first derivatives are required; and 2 indicates both first and second derivatives are required. The suffix debug indicates the evaluator is used for debugging purposes. These evaluators compute derivatives numerically and provide a report comparing those derivatives to the ones computed using the analytic formulas you provided; see chapter 7 for details and examples.

Among the four families,

1. Use method lf when the likelihood function meets the linear-form restrictions and you do not wish to program first or second derivatives. This is the most popular method among users and is often the only method one ever needs.

2. Also use methods lf0, lf1, and lf2 when the likelihood function meets the linear-form restrictions. In exchange for a slightly more complicated syntax, methods lf1 and lf2 allow you to provide first and second derivatives, respectively. lf0 does not require you to provide any derivatives and is used as a stepping stone toward writing lf1 and lf2. If you have no intention of providing derivatives, use method lf instead because it has a simpler syntax but is otherwise comparable to method lf0.

3. Use d-family evaluators with any likelihood function—not just those meeting the linear-form restrictions. Method d0 does not require you to program any derivatives. Method d1 requires you to program first derivatives of the overall likelihood

function with respect to model parameters. Method `d2` requires you to program first and second derivatives of the overall likelihood function with respect to the model parameters. Robust standard errors and survey data support are not available with `d`-family evaluators.

4. Use `gf`-family evaluators similarly to methods `lf0`, `lf1`, and `lf2`, except that the log-likelihood function for `gf`-family evaluators may exist only at the group level rather the observation level. (This also impacts how the gradient and Hessian are computed.) The overall log likelihood is equal to the sum of the group-level log-likelihood functions. Virtually every model you are likely to encounter can be fit using one of the other evaluator families. The only exceptions we can think of are panel-data estimators and the Cox proportional-hazards model, and unless robust standard errors are required, these models can be implemented using `d`-family evaluators, as we illustrate in chapter 6. Moreover, no Stata commands are currently implemented using a `gf`-family evaluator, so we do not discuss this family further in this book.

There is a clear tradeoff between speed of programming and speed of execution. Evaluators that do not require you to specify first or second derivatives are easier to write, but `ml` will take longer to execute because it must compute derivatives numerically. Evaluators that require both first and second derivatives execute most quickly, though obtaining the second derivatives of complicated likelihood functions can be difficult. First derivatives are generally easier to obtain that second derivatives, and providing just first derivatives will typically still have a noticeable impact on speed.

Although the numerical derivative engine in `ml` is quite good, analytic derivatives are still more accurate. Therefore, evaluators that require second derivatives are most accurate, followed by those that require first derivatives; evaluators that compute all derivatives numerically tend to be least accurate. An exception to this rule is the method-`lf` evaluator, which we have carefully programmed to produce accurate results even though it does not require you to specify any derivatives.

In general, we heartily recommend method `lf` because it is easy to program, quick to execute, and accurate. Use one of the other methods only when method `lf` does not provide the capabilities you need.

You specify the method and the name of the program you write to evaluate the likelihood function in the `ml model` statement:

$$\text{ml model } \textit{method progname } \ldots$$

In the next several chapters, we discuss each of the four families of likelihood evaluators in detail. However, in the previous chapter, we introduced maximum likelihood estimation using method-`gf0` evaluators, and in this chapter, we have used a method-`lf` evaluator for linear regression. We want to emphasize right now is that this method-`lf` evaluator is short but entirely functional.

```
————————————————————————————— begin mynormal1_lf.ado ——————
program mynormal1_lf
        version 11
        args lnfj mu sigma
        quietly replace 'lnfj' = ln(normalden($ML_y1,'mu','sigma'))
end
————————————————————————————————— end mynormal1_lf.ado ——————
```

In the rest of this chapter, we will continue using this example, and we want to direct your attention to the many things ml will do for you once you have written your evaluator and without your having to write more code.

3.4   Tools for the ml programmer

As you write and improve your likelihood evaluator, the following utilities may be helpful:

- ml check helps verify that your program has no syntax errors.
- ml init allows you to specify initial values.
- ml search allows you to hunt for initial values.
- ml plot graphs the log-likelihood function and helps you find initial values.

What you do about initial values is up to you. In most cases, ml maximize will succeed even if the log-likelihood function cannot be evaluated at the default initial values, which are a vector of zeros. Providing better initial values usually speeds maximization.

The following utilities can be used after ml maximize finishes fitting the model:

- ml graph graphs the log-likelihood iteration values.
- ml display redisplays the estimation results.

3.5   Common ml options

The syntax and options for ml are described in appendix A. The following sections give an overview of the most commonly used options of the ml model statement.

3.5.1   Subsamples

If you wish to fit your model on a subsample of the data, specify the restriction in the ml model statement using if or in. Continuing with our linear regression example, suppose that we want to fit the model for data in which trunk>8. Using if, we type

```
. sysuse auto
(1978 Automobile Data)
. ml model lf mynormal1_lf (mpg = weight displacement) /sigma if trunk>8
. ml maximize
initial:       log likelihood =     -<inf>  (could not be evaluated)
feasible:      log likelihood = -8455.0133
rescale:       log likelihood = -252.88251
rescale eq:    log likelihood = -203.43609
Iteration 0:   log likelihood = -203.43609  (not concave)
Iteration 1:   log likelihood = -194.40235  (not concave)
Iteration 2:   log likelihood = -189.62396
Iteration 3:   log likelihood = -173.44871
Iteration 4:   log likelihood = -169.91692
Iteration 5:   log likelihood = -169.85719
Iteration 6:   log likelihood = -169.85702
Iteration 7:   log likelihood = -169.85702
```

		Number of obs	=	64
		Wald chi2(2)	=	109.55
Log likelihood = -169.85702		Prob > chi2	=	0.0000

mpg	Coef.	Std. Err.	z	P>|z|	[95% Conf. Interval]	
eq1						
weight	-.0064416	.0011849	-5.44	0.000	-.008764	-.0041191
displacement	.0062196	.0099341	0.63	0.531	-.0132509	.0256901
_cons	39.38041	2.134564	18.45	0.000	35.19674	43.56408
sigma						
_cons	3.438645	.3039361	11.31	0.000	2.842941	4.034349

You can specify one or both of `if` and `in` in the `ml model` statement.

3.5.2 Weights

If you wish to perform weighted estimation, specify the weights in the `ml model` statement. For example,

```
. sysuse auto
(1978 Automobile Data)

. ml model lf mynormal1_lf (mpg = weight displacement) /sigma [aweight=trunk]

. ml maximize, nolog
initial:       log likelihood =      -<inf>  (could not be evaluated)
feasible:      log likelihood =   -9382.79
rescale:       log likelihood = -288.88285
rescale eq:    log likelihood = -233.20956
```

	Number of obs	=	74
	Wald chi2(2)	=	138.35
Log likelihood = -193.1554	Prob > chi2	=	0.0000

mpg	Coef.	Std. Err.	z	P>|z|	[95% Conf. Interval]	
eq1						
weight	-.0064548	.0011702	-5.52	0.000	-.0087484	-.0041612
displacement	.0060624	.0096706	0.63	0.531	-.0128917	.0250165
_cons	39.48953	2.072494	19.05	0.000	35.42751	43.55154
sigma						
_cons	3.291257	.2705397	12.17	0.000	2.761009	3.821505

ml allows `fweights`, `pweights`, `aweights`, and `iweights`; however, you must decide if a given type of weight is meaningful for your probability model. Specifying `pweights` implies robust estimates of variance. See [U] **11.1.6 weight** for more information on weights.

3.5.3   OPG estimates of variance

The theory and formulas for computing variance estimates using the outer product of gradients (OPG) were explained in section 1.2.3. To obtain OPG variance estimates for the estimated coefficients, specify the `vce(opg)` option on the `ml model` statement. For example,

(Continued on next page)

```
. sysuse auto
(1978 Automobile Data)

. ml model lf mynormal1_lf (mpg = weight displacement) /sigma, vce(opg)

. ml maximize, nolog
initial:       log likelihood =      -<inf>  (could not be evaluated)
feasible:      log likelihood = -10383.274
rescale:       log likelihood = -292.89564
rescale eq:    log likelihood = -238.45986
                                                 Number of obs    =         74
                                                 Wald chi2(2)     =     128.47
Log likelihood =  -195.2398                      Prob > chi2      =     0.0000
```

mpg	Coef.	OPG Std. Err.	z	P>|z|	[95% Conf. Interval]
eq1					
weight	-.0065671	.0014932	-4.40	0.000	-.0094938 -.0036404
displacement	.0052808	.0132686	0.40	0.691	-.0207252 .0312868
_cons	40.08452	2.292743	17.48	0.000	35.59083 44.57822
sigma					
_cons	3.385282	.2181536	15.52	0.000	2.957709 3.812855

These are virtually the same results as were obtained on page 42, where we did not specify vce(opg). The differences are that the standard errors are different, albeit similar, and that the $\chi^2(2)$ test statistic reported in the header changes from 139.21 to 128.47. This test statistic is for the Wald test that all coefficients in the first equation (except the intercept) are zero, and it is based on the variance estimate chosen.

Method lf allows obtaining OPG variance estimates without your having to make any change to your likelihood-evaluation program—you need only specify the vce(opg) option on the ml model statement. As we will discuss later, you can also specify vce(opg) when you use lf-family evaluators. The OPG variance estimate is not available with d-family evaluators.

3.5.4 Robust estimates of variance

The theory and formulas for computing robust estimates of variance were explained in section 1.2.4. To obtain robust variance estimates for the estimated coefficients, specify the vce(robust) option on the ml model statement. For example,

```
. sysuse auto
(1978 Automobile Data)

. ml model lf mynormal1_lf (mpg = weight displacement) /sigma, vce(robust)

. ml maximize, nolog
initial:       log pseudolikelihood =      -<inf>  (could not be evaluated)
feasible:      log pseudolikelihood = -10383.274
rescale:       log pseudolikelihood = -292.89564
rescale eq:    log pseudolikelihood = -238.45986
                                           Number of obs    =         74
                                           Wald chi2(2)     =     111.39
Log pseudolikelihood =  -195.2398          Prob > chi2      =     0.0000
```

| | | Robust | | | | |
mpg	Coef.	Std. Err.	z	P>|z|	[95% Conf.	Interval]
eq1						
weight	-.0065671	.0009195	-7.14	0.000	-.0083692	-.004765
displacement	.0052808	.0073381	0.72	0.472	-.0091017	.0196633
_cons	40.08452	2.045516	19.60	0.000	36.07538	44.09366
sigma						
_cons	3.385282	.4544255	7.45	0.000	2.494624	4.275939

These are virtually the same results as reported on pages 42 and 54. The differences are in the standard errors and the $\chi^2(2)$ test—the test that the coefficients on weight and displacement are simultaneously zero—as shown in tables 3.2 and 3.3, respectively.

Table 3.2. Comparison of estimated standard errors

	Conventional	OPG	Robust
weight	0.0011	0.0015	0.0009
displacement	0.0097	0.0133	0.0073
_cons	1.9787	2.2927	2.0455
sigma: _cons	0.2783	0.2182	0.4544

Table 3.3. Comparison of the Wald tests

	Conventional	OPG	Robust
Wald $\chi^2(2)$	139.21	128.47	111.39

As with OPG variance estimates, robust estimates are available for all likelihood evaluators except those in the d family. To obtain cluster–robust standard errors instead, just specify vce(cluster *varname*), where *varname* identifies the clusters or groups. Cluster–robust variance estimates relax the assumption that each observation is independently distributed and only assumes that observations in different clusters are independent. Obtaining cluster–robust variance estimates requires no additional programming beyond what is required to obtain robust variance estimates.

If you specify pweights, ml uses vce(robust) automatically.

3.5.5   Survey data

The svy option of ml model causes ml to use the survey design characteristics set by svyset.

```
. use http://www.stata-press.com/data/r11/nhanes2a
. svyset [pweight=leadwt], strata(stratid) psu(psuid)
      pweight: leadwt
          VCE: linearized
  Single unit: missing
     Strata 1: stratid
         SU 1: psuid
        FPC 1: <zero>
. ml model lf mynormal1_lf (loglead = age female) /sigma, svy
. ml maximize
initial:       log pseudolikelihood =      -<inf>  (could not be evaluated)
feasible:      log pseudolikelihood = -1.040e+09
rescale:       log pseudolikelihood = -1.892e+08
rescale eq:    log pseudolikelihood = -1.324e+08
Iteration 0:   log pseudolikelihood = -1.324e+08  (not concave)
Iteration 1:   log pseudolikelihood =  -84998916  (not concave)
Iteration 2:   log pseudolikelihood =  -61420175
Iteration 3:   log pseudolikelihood =  -50138048
Iteration 4:   log pseudolikelihood =  -48525755
Iteration 5:   log pseudolikelihood =  -48522256
Iteration 6:   log pseudolikelihood =  -48522255
```

Number of strata	=	31	
Number of PSUs	=	62	

Number of obs	=	4948
Population size	=	112915016
Design df	=	31
F(2, 30)	=	646.47
Prob > F	=	0.0000

loglead	Coef.	Linearized Std. Err.	t	P>|t|	[95% Conf. Interval]
eq1					
age	.0024985	.0004326	5.78	0.000	.0016162 .0033807
female	-.3648611	.0101531	-35.94	0.000	-.3855684 -.3441537
_cons	2.663709	.0295542	90.13	0.000	2.603432 2.723985
sigma					
_cons	.3718685	.0109367	34.00	0.000	.349563 .3941741

In addition to the `svy` option, `ml model` also accepts the `subpop()` option, which is not a setting of the `svyset` command. For more information on `svyset`, see [SVY] **svyset**, and for information on survey data in general, see the *Survey Data Reference Manual*.

See chapter 14 for details on writing estimation commands with support for the `svy` prefix.

3.5.6   Constraints

To fit your model with constraints on the coefficients, use the `constraints()` option on the `ml model` statement. For example,

```
. sysuse auto
(1978 Automobile Data)
. constraint 1 weight = -displacement
. ml model lf mynormal1_lf (mpg = weight displacement) /sigma, constraints(1)
. ml maximize, nolog
initial:       log likelihood =      -<inf>  (could not be evaluated)
feasible:      log likelihood = -10383.274
rescale:       log likelihood = -292.89564
rescale eq:    log likelihood = -238.45986
                                            Number of obs   =         74
                                            Wald chi2(1)    =     139.15
Log likelihood = -195.25083                 Prob > chi2     =     0.0000

 ( 1)   [eq1]weight + [eq1]displacement = 0
```

mpg	Coef.	Std. Err.	z	P>|z|	[95% Conf. Interval]	
eq1						
weight	-.0067142	.0005692	-11.80	0.000	-.0078298	-.0055986
displacement	.0067142	.0005692	11.80	0.000	.0055986	.0078298
_cons	40.24594	1.653851	24.33	0.000	37.00445	43.48743
sigma						
_cons	3.385786	.2783099	12.17	0.000	2.840309	3.931264

Constraints are defined using the `constraint` command; see [R] **constraint**. For more examples of how to define constraints, see [R] **cnsreg**, [R] **mlogit**, and [R] **reg3**.

3.5.7   Choosing among the optimization algorithms

To fit your model using one of the three optimization algorithms described in section 1.3.2, specify the algorithm in the `technique()` option on the `ml model` statement.

- `technique(nr)` specifies the Newton–Raphson algorithm (the default).
- `technique(bhhh)` specifies the Berndt–Hall–Hall–Hausman (BHHH) algorithm.
- `technique(dfp)` specifies the Davidon–Fletcher–Powell (DFP) algorithm.

- `technique(bfgs)` specifies the Broydon–Fletcher–Goldbarb–Shanno (BFGS) algorithm.

Because the BHHH algorithm uses the (observation-level) OPG, it cannot be used with d-family evaluators. All of these algorithms can be used with the other three evaluator families.

Here is an example using the BHHH algorithm:

```
. sysuse auto
(1978 Automobile Data)
. ml model lf mynormal1_lf (mpg = weight displacement) /sigma, technique(bhhh)
. ml maximize
initial:       log likelihood =      -<inf>  (could not be evaluated)
feasible:      log likelihood = -10383.274
rescale:       log likelihood = -292.89564
rescale eq:    log likelihood = -238.45986
Iteration 0:   log likelihood = -238.45986
(output omitted )
Iteration 21:  log likelihood = -195.23982
Iteration 22:  log likelihood =  -195.2398
```

			Number of obs	=	74
			Wald chi2(2)	=	128.55
Log likelihood =	-195.2398		Prob > chi2	=	0.0000

mpg	Coef.	OPG Std. Err.	z	P>|z|	[95% Conf. Interval]
eq1					
weight	-.0065662	.0014926	-4.40	0.000	-.0094917 -.0036407
displacement	.0052777	.0132635	0.40	0.691	-.0207182 .0312736
_cons	40.08149	2.291861	17.49	0.000	35.58952 44.57345
sigma					
_cons	3.38453	.2180152	15.52	0.000	2.957228 3.811832

The BHHH algorithm produces the OPG variance estimate by default; the default variance estimate for the other algorithms is the inverse of the negative Hessian. If you want Hessian-based variance estimates when using the BHHH algorithm to fit your model, specify `technique(bhhh)` and `vce(oim)` together.

```
. sysuse auto
(1978 Automobile Data)
. ml model lf mynormal1_lf (mpg = weight displacement) /sigma, tech(bhhh)
> vce(oim)
. ml maximize
initial:       log likelihood =      -<inf>  (could not be evaluated)
feasible:      log likelihood = -10383.274
rescale:       log likelihood = -292.89564
rescale eq:    log likelihood = -238.45986
Iteration 0:   log likelihood = -238.45986
(output omitted )
Iteration 21:  log likelihood = -195.23982
Iteration 22:  log likelihood =  -195.2398
```

	Number of obs	=	74
	Wald chi2(2)	=	139.25
Log likelihood = -195.2398	Prob > chi2	=	0.0000

mpg	Coef.	Std. Err.	z	P>|z|	[95% Conf.	Interval]
eq1						
weight	-.0065662	.0011421	-5.75	0.000	-.0088047	-.0043277
displacement	.0052777	.0096653	0.55	0.585	-.013666	.0242213
_cons	40.08149	1.978299	20.26	0.000	36.20409	43.95888
sigma						
_cons	3.38453	.278114	12.17	0.000	2.839437	3.929623

The `technique()` option allows you to specify multiple algorithms, and in that case, `ml` will cycle between the specified algorithms. For example, specifying `technique(bhhh nr)` will cause `ml` to use BHHH for five iterations, Newton–Raphson for the next five, BHHH for the five after that, and so on:

```
. sysuse auto
(1978 Automobile Data)
. ml model lf mynormal1_lf (mpg = weight displacement) /sigma, tech(bhhh nr)
. ml maximize
initial:       log likelihood =      -<inf>  (could not be evaluated)
feasible:      log likelihood = -10383.274
rescale:       log likelihood = -292.89564
rescale eq:    log likelihood = -238.45986
(setting technique to bhhh)
Iteration 0:   log likelihood = -238.45986
Iteration 1:   log likelihood = -198.28587
Iteration 2:   log likelihood = -195.80276
Iteration 3:   log likelihood = -195.63467
Iteration 4:   log likelihood = -195.54062
(switching technique to nr)
Iteration 5:   log likelihood =  -195.3671
Iteration 6:   log likelihood = -195.23992
Iteration 7:   log likelihood =  -195.2398
Iteration 8:   log likelihood =  -195.2398
```

```
                                                    Number of obs   =        74
                                                    Wald chi2(2)    =    139.21
       Log likelihood =  -195.2398                  Prob > chi2     =    0.0000
```

mpg	Coef.	Std. Err.	z	P>|z|	[95% Conf. Interval]	
eq1						
weight	-.0065671	.0011424	-5.75	0.000	-.0088061	-.0043281
displacement	.0052808	.0096674	0.55	0.585	-.0136671	.0242286
_cons	40.08452	1.978739	20.26	0.000	36.20627	43.96278
sigma						
_cons	3.385282	.2782685	12.17	0.000	2.839886	3.930678

Switching between techniques is an excellent way of finding the maximum of a difficult-to-maximize function. By default, `ml` performs five iterations of each algorithm before switching to the next. You can specify a different number of iterations in the `technique()` option by placing your own count after each named algorithm. For example,

```
. sysuse auto
(1978 Automobile Data)
. ml model lf mynormal1_lf (mpg = weight displace) /sigma, tech(bhhh 2 nr 5)
. ml maximize
initial:       log likelihood =      -<inf>  (could not be evaluated)
feasible:      log likelihood = -10383.274
rescale:       log likelihood = -292.89564
rescale eq:    log likelihood = -238.45986
(setting technique to bhhh)
Iteration 0:   log likelihood = -238.45986
Iteration 1:   log likelihood = -198.28587
(switching technique to nr)
Iteration 2:   log likelihood = -195.80276
Iteration 3:   log likelihood = -195.24154
Iteration 4:   log likelihood =  -195.2398
Iteration 5:   log likelihood =  -195.2398
```

```
                                                    Number of obs   =        74
                                                    Wald chi2(2)    =    139.21
       Log likelihood =  -195.2398                  Prob > chi2     =    0.0000
```

mpg	Coef.	Std. Err.	z	P>|z|	[95% Conf. Interval]	
eq1						
weight	-.0065671	.0011424	-5.75	0.000	-.0088061	-.0043281
displacement	.0052808	.0096674	0.55	0.585	-.0136671	.0242286
_cons	40.08452	1.978738	20.26	0.000	36.20627	43.96278
sigma						
_cons	3.385282	.2782684	12.17	0.000	2.839886	3.930678

3.6 Maximizing your own likelihood functions

Here is an outline of the steps you will take when writing and debugging your likelihood evaluator.

1. Derive the log-likelihood function from your probability model.

2. Write a likelihood evaluator, the program that calculates the log-likelihood values. Depending on the method, you may have to include derivative calculations.

    See chapter 4 for the method-`lf` evaluator, chapter 5 for `lf`-family evaluators, and chapter 6 for `d`-family evaluators.

3. Specify the method (such as `lf`, `d0`, or `lf2`), the name of your program (*myprog*), and the model to be fit in the `ml model` statement.

    > . `ml model` *method myprog* (*equation for* θ_1) (*equation for* θ_2) ...

    At this point, you also specify options that affect how the model is fit. See section 3.5 for commonly specified options.

4. Verify that the program you have written is free of syntax errors and other common problems with

    > . `ml check`

    See section 7.1 for details, but the output produced by `ml check` is easy to interpret.

5. Optionally, specify starting values using `ml init`; see section 8.3. We do not recommend this unless you have an analytic approximation that gives good starting values. Otherwise, use `ml search`.

6. Optionally, look for starting values using `ml plot`; see section 8.2. This is not as useful as `ml search`.

7. Optionally—and we recommend this—tell `ml` to search for starting values with

    > . `ml search`

    See section 8.1.

8. Obtain maximum likelihood estimates with

    > . `ml maximize`

    See chapter 9.

9. To graph the convergence, type

    > . `ml graph`

    See section 10.1.

10. If you later wish to redisplay the results from the most recently fit model, type

    `. ml display`

    See section 10.2.

11. If you wish to add a new command to Stata to perform this kind of estimation, see chapter 13.

4　Method lf

Method `lf` is for use when the likelihood function meets the linear-form restrictions and you do not wish to provide first and second derivatives. Use this method whenever possible. Method `lf` is fast, accurate, and easy to use.

The likelihood function is assumed to be of the form

$$\ln \ell_j = \ln \ell_j(\theta_{1j}, \theta_{2j}, \ldots, \theta_{Ej}; y_{1j}, y_{2j}, \ldots, y_{Dj})$$

$$\ln L = \sum_{j=1}^{N} \ln \ell_j$$

$$\theta_{1j} = \mathbf{x}_{1j}\boldsymbol{\beta}_1$$

$$\theta_{2j} = \mathbf{x}_{2j}\boldsymbol{\beta}_2$$

$$\vdots$$

$$\theta_{Ej} = \mathbf{x}_{Ej}\boldsymbol{\beta}_E$$

where j indexes observations. The method-`lf` evaluator is required to calculate $\ln \ell_j$, the log likelihood, observation by observation. The outline for a method-`lf` likelihood evaluator is

```
program myprog
        version 11
        args lnfj theta1 theta2 ...
        // if you need to create any intermediate results
        tempvar tmp1 tmp2 ...
        quietly gen double 'tmp1' = ...
        ...
        quietly replace 'lnfj' = ...
end
```

You specify that you are using method `lf` with program *myprog* on the `ml model` statement,

```
. ml model lf myprog ...
```

which you issue after you have defined the likelihood-evaluation program.

1. `ml` places the names of any dependent variables in global macros; the first dependent variable is `$ML_y1`, the second is `$ML_y2`, ... Use `$ML_y1`, `$ML_y2`, ... just as you would any existing variable in the data.

2. `ml` supplies a variable for each equation specified in the `ml model` statement. Each variable contains the values of the linear combination of independent variables and current regression coefficients for the respective equation. Thus '*theta1*' contains the linear combination for the first equation, '*theta2*' contains the linear combination for the second equation, and so on.

   Do not change the contents of variables '*theta1*', '*theta2*', ... That is, never write

   ```
   . replace 'theta1' = ...
   ```

3. Final results are saved in '`lnfj`', which is just a double-precision variable that `ml` creates for you. When `ml` calls your program, '`lnfj`' contains missing values.

4. Any temporary variables you create for intermediate results must be `doubles`. Type

   ```
   . gen double name = ...
   ```

   Do not omit the word `double`.

Using method `lf` is simply a matter of writing the likelihood function for an observation, identifying the parameters that will make up `ml`'s equations, and coding it in a program.

4.1   The linear-form restrictions

Method `lf` requires that the physical observations in the dataset correspond to independent pieces of the likelihood function:

$$\ln L = \ln \ell_1 + \ln \ell_2 + \cdots + \ln \ell_N$$

This restriction is commonly (but not always) met in statistical analyses. It corresponds to

$$\Pr(\text{dataset}) = \Pr(\text{obs. 1}) \times \Pr(\text{obs. 2}) \times \cdots \times \Pr(\text{obs. } N)$$

with discrete data. Anytime you have a likelihood function and can write it as

"The log likelihood for the jth observation is $\ln \ell_j = \dots$."

and that formula uses data for observation j only, and it is obvious that

"The overall log likelihood function is $\ln L = \sum_{j=1}^{N} \ln \ell_j$."

you have a likelihood that meets this restriction. In other words, the observations are independent.

The following examples of likelihoods *do not* meet this criterion: the conditional (fixed-effects) logit model, Cox regression, and likelihoods for panel data. Basically, any likelihood that explicitly models independent groups rather than independent observations does not meet the `lf` criterion.

4.2   Examples

The following models meet the `lf` criterion that the physical observations in the dataset correspond to independent pieces of the likelihood function.

4.2.1   The probit model

The probit log-likelihood function for the jth observation is

$$\ln \ell_j = \begin{cases} \ln \Phi(\mathbf{x}_j \boldsymbol{\beta}) & \text{if } y_j = 1 \\ \ln \Phi(-\mathbf{x}_j \boldsymbol{\beta}) & \text{if } y_j = 0 \end{cases} \tag{4.1}$$

where $\Phi()$ denotes the cumulative probability distribution function of the standard normal distribution. This is a one-equation model, so we skip using θ and identify the equation directly as $\mathbf{x}_j \boldsymbol{\beta}$. Our method-`lf` likelihood evaluator is

---
———————————————————————— begin `myprobit_lf.ado` ————————

```
program myprobit_lf
        version 11
        args lnfj xb
        quietly replace `lnfj' = ln(normal( `xb')) if $ML_y1 == 1
        quietly replace `lnfj' = ln(normal(-`xb')) if $ML_y1 == 0
end
```

———————————————————————————— end `myprobit_lf.ado` ————————

`myprobit_lf` requires two arguments, as indicated by the line

```
        args lnfj xb
```

The first argument, `lnfj`, is the variable where we fill in the log-likelihood values, and the second, `xb`, is the variable that contains the values of $\mathbf{x}_j \boldsymbol{\beta}$ (the first and only equation). The name of the dependent variable is placed in the global macro `$ML_y1` by `ml`.

If we wanted to fit the probit regression of `died` on `i.drug` (a factor variable with three levels) and `age`, we would type

```
. ml model lf myprobit_lf (died = i.drug age)
. ml maximize
```

`died =` specifies that the dependent variable is `died`. `i.drug age` specifies that

$$\mathbf{x}_j \boldsymbol{\beta} = \beta_1 1.\text{drug}_j + \beta_2 2.\text{drug}_j + \beta_3 3.\text{drug}_j + \beta_4 \text{age} + \beta_5 \tag{4.2}$$

Our program never has to deal with the fact that $\mathbf{x}_j \boldsymbol{\beta}$ is given by (4.2), nor is it ever even aware of it. Let's pretend that at some point in the maximization process, $\beta_1 = -0.1$, $\beta_2 = -0.002$, $\beta_3 = 0.2$, and $\beta_4 = 8$. We never need to know any of that because all `lf` will ask our program is questions such as

"What is the log likelihood if the outcome is 1 and $\mathbf{x}_j \boldsymbol{\beta} = 0.5$?"

and all we have to do is reply "-0.36894642", which is $\ln\Phi(0.5)$, or

"What is the log likelihood if the outcome is 0 and $\mathbf{x}_j\boldsymbol{\beta} = 0.1$?"

to which we reply, "-0.77615459", which is $\ln\Phi(-0.1)$. In fact, lf asks both these questions at once, along with many others, when it asks us to fill in the variable 'lnfj' with the log likelihood for each of the observations, based on the values of the variables in $ML_y1 and 'xb'. Here we never need to know that the model being fit contains i.drug and age or whether the model has an intercept.

Here is the result of using this likelihood evaluator and some cancer data, modeling the incidence of death as a function of age and drug type (the example from chapter 1).

```
. sysuse cancer
(Patient Survival in Drug Trial)

. ml model lf myprobit_lf (died = i.drug age)

. ml maximize
initial:        log likelihood = -33.271065
alternative:    log likelihood = -31.427839
rescale:        log likelihood = -31.424556
Iteration 0:    log likelihood = -31.424556
Iteration 1:    log likelihood = -21.883453
Iteration 2:    log likelihood = -21.710899
Iteration 3:    log likelihood = -21.710799
Iteration 4:    log likelihood = -21.710799
```

		Number of obs	=	48
		Wald chi2(3)	=	13.58
Log likelihood = -21.710799		Prob > chi2	=	0.0035

died	Coef.	Std. Err.	z	P>|z|	[95% Conf. Interval]	
drug						
2	-1.941275	.597057	-3.25	0.001	-3.111485	-.7710644
3	-1.792924	.5845829	-3.07	0.002	-2.938686	-.6471626
age	.0666733	.0410666	1.62	0.104	-.0138159	.1471624
_cons	-2.044876	2.284283	-0.90	0.371	-6.521988	2.432235

Here we decided to use factor-variable notation for the categories of the drug variable. ml fully supports factor variables.

4.2.2 Normal linear regression

The jth observation's log-likelihood function for linear regression assuming normally distributed errors is

$$\ln\ell_j = \ln\phi\{(y_j - \theta_{1j})/\theta_{2j}\} - \ln\theta_{2j}$$
$$\theta_{1j} = \mu_j = \mathbf{x}_j\boldsymbol{\beta}$$
$$\theta_{2j} = \sigma$$

Most people think of linear regression as a single-equation model, $\mu_j = \mathbf{x}_j\boldsymbol{\beta}$, with an ancillary parameter σ that also must be estimated. This is just a matter of jargon, and in `ml` jargon, the ancillary parameter σ counts as an equation. Thus this is a two-equation model; our first equation represents the conditional mean, and the second equation is for the standard deviation. Our method-`lf` likelihood evaluator is

```
                                            ─── begin mynormal1_lf.ado ─────
   program mynormal1_lf
           version 11
           args lnfj mu sigma
           quietly replace 'lnfj' = ln(normalden($ML_y1,'mu','sigma'))
   end
                                            ─── end mynormal1_lf.ado ─────
```

`mynormal1_lf` requires three arguments, as indicated by the line

        args lnfj mu sigma

The first argument, '`lnfj`', is the variable where we fill in the log-likelihood values. The second argument, '`mu`', is the variable that contains the values of the regression equation. And the third argument, '`sigma`', is the variable that contains the values of the standard deviation. `ml` places the name of the dependent variable in the global macro `$ML_y1`.

If we wanted to regress `mpg` on `weight` and `displ` with constant σ, we would type

        . ml model lf mynormal1_lf (mu: mpg=weight displ) /sigma
        . ml maximize

Here `mpg=` specifies that the dependent variable is `mpg`, and `weight displ` specifies that

$$\mu_j = \beta_1\texttt{weight}_j + \beta_2\texttt{displ}_j + \beta_3$$

The second equation was specified as `/sigma`, so the standard deviation is modeled as a constant:

$$\sigma = \beta_4$$

`mynormal1_lf` can just as easily fit heteroskedastic models; we just specify an equation for σ, such as

$$\sigma_j = \beta_4\texttt{price}_j + \beta_5$$

which we fit using `auto.dta`:

(Continued on next page)

```
──────────────────────────────────────────────── begin ex_normal3 ────────
. sysuse auto
(1978 Automobile Data)

. ml model lf mynormal1_lf (mu: mpg = weight displacement) (sigma: price)

. ml maximize, nolog
initial:       log likelihood =      -<inf>  (could not be evaluated)
feasible:      log likelihood = -10383.274
rescale:       log likelihood = -292.89564
rescale eq:    log likelihood = -238.45986
                                              Number of obs   =        74
                                              Wald chi2(2)    =    126.71
          Log likelihood = -195.14254         Prob > chi2     =    0.0000
```

mpg	Coef.	Std. Err.	z	P>|z|	[95% Conf. Interval]
mu						
weight	-.0065258	.0011369	-5.74	0.000	-.008754	-.0042975
displacement	.0041644	.0099923	0.42	0.677	-.01542	.0237489
_cons	40.1942	1.976217	20.34	0.000	36.32089	44.06752
sigma						
price	.0000438	.0001061	0.41	0.680	-.0001641	.0002517
_cons	3.11326	.6915815	4.50	0.000	1.757785	4.468735

```
──────────────────────────────────────────────────── end ex_normal3 ────────
```

Had we not wanted the constant term in the second equation, such as

$$\sigma_j = \beta_4 \texttt{price}_j$$

we could have typed

```
. ml model lf mynormal1_lf (mu: mpg=weight displ) (sigma: price, nocons)
. ml maximize
```

❏ Technical note

With the way we have parameterized this model, the log likelihood is defined only for $\theta_{2j} = \sigma > 0$. `ml` is designed to deal with restricted ranges—there is nothing special you need to do—but you should be impressed and realize that you are asking a lot of `ml`. There could be convergence problems, and even if not, `ml` is going to perform more work, so the maximization process may not occur as quickly as it otherwise could. Note that `ml` may not be able to find feasible initial values for restricted ranges such as $(-1, 1)$ or $(0, 1)$.

It would be better if the model were parameterized so that $(-\infty, \infty)$ is a valid range for each element of $\boldsymbol{\theta}_i$. An alternative parameterization of the linear regression likelihood function would be

$$\ln \ell_j = \ln \phi\{(y_j - \theta_{1j})/\exp(\theta_{2j})\} - \theta_{2j}$$
$$\theta_{1j} = \mu_j = \mathbf{x}_j \boldsymbol{\beta}$$
$$\theta_{2j} = \ln \sigma$$

and that parameterization would lead to the following likelihood evaluator:

————————————————————————— begin `mynormal2_lf.ado` —————————

```
program mynormal2_lf
        version 11
        args lnfj mu lnsigma
        quietly replace 'lnfj' = ln(normalden($ML_y1,'mu',exp('lnsigma')))
end
```
————————————————————————— end `mynormal2_lf.ado` —————————

❑

4.2.3   The Weibull model

The Weibull log-likelihood function for the jth observation, where t_j is the time of failure or censoring and $d_j = 1$ if failure and 0 if censored, is

$$\ln \ell_j = -(t_j/e^{\theta_{1j}})^{e^{\theta_{2j}}} + d_j\{\theta_{2j} - \theta_{1j} + (e^{\theta_{2j}} - 1)(\ln t_j - \theta_{1j})\}$$
$$\theta_{1j} = \ln \eta_j = \mathbf{x}_{1j}\boldsymbol{\beta}_1$$
$$\theta_{2j} = \ln \gamma_j = \mathbf{x}_{2j}\boldsymbol{\beta}_2$$

For details on how this log likelihood is derived, see section 15.4. For now, it suffices to say that $\ln \eta_j$ is a scale parameter and $\ln \gamma_j$ is a shape parameter, and we wish to model the logs of both as linear combinations of covariates.

This is a two-equation model: the first equation models $\ln \eta_j$ and the second equation models $\ln \gamma_j$. The corresponding method-`lf` evaluator for this likelihood function is

---
————————————————————————— begin `myweibull_lf.ado` —————————

```
program myweibull_lf
        version 11
        args lnfj leta lgam
        tempvar p M R
        quietly {
                gen double 'p' = exp('lgam')
                gen double 'M' = ($ML_y1*exp(-'leta'))^'p'
                gen double 'R' = ln($ML_y1)-'leta'
                replace 'lnfj' = -'M' + $ML_y2*('lgam'-'leta'+('p'-1)*'R')
        }
end
```
————————————————————————— end `myweibull_lf.ado` —————————

`myweibull_lf` requires three arguments, as indicated by the line

```
args lnfj leta lgam
```

The first argument, '`lnfj`', is the variable where we fill in the log-likelihood values. The second argument, '`leta`', is the name of the variable that contains the values of $\ln \eta_j$. And the third argument, '`lgam`', is the name of the variable that contains the values of $\ln \gamma_j$.

We use temporary variables to evaluate some of the terms in the log likelihood. This is not necessary, but it makes the program more readable and keeps us from having to evaluate `exp('lgam')` more than once. It is important that these temporary variables were generated as `doubles`.

This model has two dependent variables: t_j and d_j. In `myweibull_lf`, we assume that t_j is the first dependent variable specified (so `$ML_y1` is coded for t_j) and d_j is the second (so `$ML_d2` is coded for d_j). We will arrange this by specifying the dependent variables in this order on the `ml model` statement.

Here is an example of our likelihood evaluator in action using `cancer.dta`, modeling time to death (with $t = $ `studytime` and $d = $ `died`) as a function of age and drug type:

```
. sysuse cancer
(Patient Survival in Drug Trial)
. ml model lf myweibull_lf (lneta: studytime died = i.drug age) /lngamma
. ml maximize
initial:       log likelihood =       -744
alternative:   log likelihood = -356.14276
rescale:       log likelihood = -200.80201
rescale eq:    log likelihood = -136.69232
Iteration 0:   log likelihood = -136.69232  (not concave)
Iteration 1:   log likelihood = -124.11726
Iteration 2:   log likelihood = -113.90508
Iteration 3:   log likelihood = -110.30519
Iteration 4:   log likelihood = -110.26747
Iteration 5:   log likelihood = -110.26736
Iteration 6:   log likelihood = -110.26736
```

```
                                              Number of obs   =        48
                                              Wald chi2(3)    =     35.25
Log likelihood = -110.26736                   Prob > chi2     =    0.0000
```

	Coef.	Std. Err.	z	P>|z|	[95% Conf. Interval]	
lneta						
drug						
2	1.012966	.2903917	3.49	0.000	.4438086	1.582123
3	1.45917	.2821195	5.17	0.000	.9062261	2.012114
age	-.0671728	.0205688	-3.27	0.001	-.1074868	-.0268587
_cons	6.060723	1.152845	5.26	0.000	3.801188	8.320259
lngamma						
_cons	.5573333	.1402154	3.97	0.000	.2825162	.8321504

4.3   The importance of generating temporary variables as doubles

In `myweibull_lf`, we used three temporary variables.

```
gen double 'p' = exp('lgam')
gen double 'M' = ($ML_y1*exp(-'leta'))^'p'
gen double 'R' = ln($ML_y1)-'leta'
```

We did not have to approach this problem in this way, but the temporary variables made the equation for the log likelihood,

```
replace 'lnfj' = -'M' + $ML_y2*('lgam'-'leta'+('p'-1)*'R')
```

easier and more readable and so reduced the chances we made an error. When we created these temporary variables, we created them as `doubles`. This is of great importance.

The program works as written. Just to show you what can happen if you do not declare temporary variables to be `doubles`, we made a modified copy of `myweibull_lf` in which the temporary variables are not declared as `double`:

```
─────────────────────────────── begin myweibull_bad_lf.ado ───────────
program myweibull_bad_lf
        version 11
        args lnfj leta lgam
        tempvar p M R
        quietly {
                gen 'p' = exp('lgam')
                gen 'M' = ($ML_y1*exp(-'leta'))^'p'
                gen 'R' = ln($ML_y1)-'leta'
                replace 'lnfj' = -'M' + $ML_y2*('lgam'-'leta'+('p'-1)*'R')
        }
end
──────────────────────────────── end myweibull_bad_lf.ado ───────────
```

Here is what happened when we used `myweibull_bad_lf`:

```
. sysuse cancer
(Patient Survival in Drug Trial)
. ml model lf myweibull_bad_lf (lneta: studyt died = i.drug age) /lngamma

. ml maximize, iter(100)

initial:       log likelihood =        -744
alternative:   log likelihood = -356.14277
rescale:       log likelihood = -200.80201
rescale eq:    log likelihood = -188.26758
Iteration 0:   log likelihood = -188.26758
Iteration 1:   log likelihood = -183.36968   (not concave)
Iteration 2:   log likelihood = -183.36817   (not concave)
   (output omitted, every line of which said "not concave")
Iteration 99:  log likelihood = -111.22265   (not concave)
Iteration 100: log likelihood = -111.21612   (not concave)
convergence not achieved
```

```
                                                Number of obs   =          48
                                                Wald chi2(1)    =           .
        Log likelihood = -111.21612            Prob > chi2     =           .
```

	Coef.	Std. Err.	z	P>|z|	[95% Conf.	Interval]
lneta						
drug						
2	.9506897	.	.	.	.	.
3	1.438377	.2580533	5.57	0.000	.9326021	1.944152
age	-.039302	.	.	.	.	.
_cons	4.509641	.	.	.	.	.
lngamma						
_cons	.5571306	.0627839	8.87	0.000	.4340764	.6801848

```
Warning: convergence not achieved
```

In the last iteration, the function was still "not concave". As we warned in chapter 1, when the iterations end in that way, you are in trouble. It is difficult to miss because when the last iteration is still "not concave", one or more of the standard errors will most likely be missing.

If you forget to specify `double`, one of five things will happen:

1. The maximization will have many "not concave" iterations, and importantly, the last iteration will be "not concave". (Intermediate "not concave" iterations should not concern you.)

2. `ml maximize` will get to iteration 1 or 2 and just sit. (In that case, press the *Break* key.)

3. `ml maximize` will iterate forever, never declaring convergence.

4. There is a small chance that `ml maximize` (and `ml check`) will report that it cannot calculate the derivatives. (In most cases when you see this error, however, it is due to a different problem: the likelihood evaluates to the same constant regardless of the value of the parameter vector because of some silly error in the evaluation program; often that constant is a missing value.)

5. There is a small chance that no mechanical problem will arise, but in that case, the reported standard errors are accurate to only a few digits.

If you include intermediate calculations, review the `generate` statements carefully to ensure that you have included the word `double` following `generate`.

❏ **Technical note**

Suppose that beforehand, we typed

```
. set type double
```

The result would be as if we had specified all the temporary variables as `double`. There are many reasons why Stata will not allow the default data type to be permanently set to `double`; suffice it to say, you cannot assume that the default data type will always be `double`, so make sure that you code `double` following every `generate`.

❏

4.4 Problems you can safely ignore

Method `lf` handles a variety of issues for you:

- You are not responsible for restricting your calculations to the estimation subsample, although your program will run slightly faster if you do. The variable `$ML_samp` contains 1 if the observation is to be used and 0 if it is to be ignored. It is necessary for you to fill in '`lnfj`' only in observations for which `$ML_samp==1`. Thus in our first example, the probit likelihood-evaluation program could read

```
————————————————————————— begin myprobit2_lf.ado ——————————
program myprobit2_lf
        version 11
        args lnfj xb
        quietly replace `lnfj' = ln(normal( `xb')) if $ML_y1==1 & $ML_samp==1
        quietly replace `lnfj' = ln(normal(-`xb')) if $ML_y1==0 & $ML_samp==1
end
——————————————————————————— end myprobit2_lf.ado ——————————
```

- This program would execute a little more quickly in some cases, but that is all. Even if you do not do this, results reported by `ml` method `lf` will be correct.

- Whether you specify `ml model`'s `nopreserve` option (discussed in appendix A) makes no difference. This is a technical issue; do not bother to see the appendix over this. As we just said, restricting calculations to the estimation subsample is handled by `ml` method `lf` itself.

- You can perform weighted estimation without doing anything special in your program. When you read about `$ML_w` in other parts of this text, ignore it if you are using `ml` method `lf`. Weights are automatically handled by `ml` method `lf`.

- You can specify the `vce(opg)`, `vce(robust)`, `vce(cluster` *varname*`)`, and `svy` options on the `ml model` statement without doing anything special in your program to handle this.

4.5   Nonlinear specifications

The term *linear-form restrictions* suggests that the parameterization of the likelihood must be linear. That is not true. The linear-form restrictions require only that the likelihood function be calculable observation by observation and that the individual equations each be linear.

For instance, suppose you wished to fit

$$y_j = \beta_1 x_{1j} + \beta_2 x_{2j} + \beta_3 x_{3j}^{\beta_4} + \beta_5 + \epsilon_j$$

where ϵ_j is distributed $N(0, \sigma^2)$. Note that the term involving x_3 is nonlinear. Here is how you could write this model:

$$\ln \ell_j = \ln \phi \{ (y_j - \theta_{1j} - \theta_{2j} x_{3j}^{\theta_{3j}}) / \theta_{4j} \} - \ln \theta_{4j}$$
$$\theta_{1j} = \beta_1 x_{1j} + \beta_2 x_{2j} + \beta_5$$
$$\theta_{2j} = \beta_3$$
$$\theta_{3j} = \beta_4$$
$$\theta_{4j} = \sigma$$

In `ml` jargon, this is a four-equation model even though, conceptually, there is only one equation. Nonlinear terms add equations. Our method-`lf` evaluator for this likelihood function is

```
                                                    begin mynonlin_lf.ado
program mynonlin_lf
        version 11
        args lnfj theta1 theta2 theta3 sigma
        quietly replace 'lnfj' = ///
                ln(normalden($ML_y1,'theta1'+'theta2'*X3^'theta3','sigma'))
end
                                                      end mynonlin_lf.ado
```

The variable name corresponding to x_{3j} is hardcoded in the evaluation program. We used the name X3, but you would substitute the actual variable name. For instance, if you wished to fit $\mathsf{bp} = \beta_1 \mathsf{age} + \beta_2 \mathsf{sex} + \beta_3 \mathsf{bp0}^{\beta_4} + \epsilon$, you could change the program to read

```
program mynonlin_lf
        version 11
        args lnfj theta1 theta2 theta3 sigma
        quietly replace 'lnfj' = ///
                ln(normalden($ML_y1,'theta1'+'theta2'*bp0^'theta3','sigma'))
end
```

and then fit the model by typing

```
. ml model lf mynonlin_lf (bp=age sex) /beta3 /beta4 /sigma
. ml maximize
```

Alternatively, you could create a global macro holding the name of the variable corresponding to x_{3j} and refer to that global macro in your evaluator. Thus the linear-form restriction is not really a linear restriction in the usual sense because any nonlinear parameter can be split into an equation that contains a constant.

4.6   The advantages of lf in terms of execution speed

Because method `lf` works with the model parameters (the θ's) instead of the coefficients (the β's), it provides a considerable speed advantage over d-family evaluators and approaches the performance of method `d2`, which requires you to specify first- and second-order derivatives. For similar reasons, the `lf`-family evaluators we discuss in the next chapter will also execute more quickly than d-family evaluators.

For the purpose of this discussion, and without loss of generality, assume that our model contains only one equation,

$$\theta_j = x_{1j}\beta_1 + x_{2j}\beta_2 + \cdots + x_{kj}\beta_k$$

In addition to the log-likelihood function itself, the `ml` optimizer needs the first and second derivatives with respect to the coefficients:

$$\mathbf{g} = \begin{pmatrix} \frac{\partial \ln L}{\partial \beta_1} & \frac{\partial \ln L}{\partial \beta_2} & \cdots & \frac{\partial \ln L}{\partial \beta_k} \end{pmatrix}$$

$$\mathbf{H} = \begin{pmatrix} \frac{\partial^2 \ln L}{\partial \beta_1^2} & \frac{\partial^2 \ln L}{\partial \beta_1 \partial \beta_2} & \cdots & \frac{\partial^2 \ln L}{\partial \beta_1 \partial \beta_k} \\ \frac{\partial^2 \ln L}{\partial \beta_2 \partial \beta_1} & \frac{\partial^2 \ln L}{\partial \beta_2^2} & \cdots & \frac{\partial^2 \ln L}{\partial \beta_2 \partial \beta_k} \\ \vdots & \vdots & \ddots & \vdots \\ \frac{\partial^2 \ln L}{\partial \beta_k \partial \beta_1} & \frac{\partial^2 \ln L}{\partial \beta_k \partial \beta_2} & \cdots & \frac{\partial^2 \ln L}{\partial \beta_k^2} \end{pmatrix}$$

We write $\mathbf{g}$ as a row vector for this discussion because that is what `ml` expects of likelihood evaluators that are required to return the gradient.

If we do not have analytic formulas for $\mathbf{g}$ and $\mathbf{H}$, `ml` must calculate each of these $k + k(k+1)/2$ derivatives numerically. For instance, `ml` could calculate the first element of $\mathbf{g}$ by

$$\frac{\ln L(\beta_1 + h/2, \beta_2, \ldots, \beta_k; \mathbf{Z}) - \ln L(\beta_1 - h/2, \beta_2, \ldots, \beta_k; \mathbf{Z})}{h}$$

This involves evaluating the log-likelihood function $\ln L()$ twice. The second element of $\mathbf{g}$ could be obtained by

$$\frac{\ln L(\beta_1, \beta_2 + h/2, \ldots, \beta_k; \mathbf{Z}) - \ln L(\beta_1, \beta_2 - h/2, \ldots, \beta_k; \mathbf{Z})}{h}$$

which are two more evaluations, and so on. Using centered derivatives, we would need to evaluate the log-likelihood function $2k$ times, assuming we know the optimum value of h. If we have to search for an optimum h in each calculation, we will need even more evaluations of $\ln L()$.

Next we will have to evaluate the $k(k+1)/2$ second derivatives, and as we showed in section 1.3.5, that takes at least three function evaluations each, although one of the evaluations is $\ln L(\beta_1, \beta_2, \ldots, \beta_k; \mathbf{Z})$ and is common to all the terms. Furthermore, $2k$ of the required function evaluations are in common with the function evaluations for the first derivative. Thus we need "only" $k(k-1)$ more function evaluations. The grand total for first and second derivatives is thus $2k + k(k-1)$.

In the case of `lf`, however, $\ln \ell_j = \ln \ell(\theta_j)$ where $\theta_j = \mathbf{x}_j \boldsymbol{\beta}$, and this provides a shortcut for the calculation of the derivatives. To simplify notation, let's write $f(\theta_j)$ for $\ln \ell(\theta_j)$, $f'(\theta_j)$ for its first derivative with respect to θ_j, and $f''(\theta_j)$ for its second. Using the chain rule,

$$\mathbf{g}_j = \frac{\partial f(\theta_j)}{\partial \boldsymbol{\beta}} = \frac{\partial f(\theta_j)}{\partial \theta_j} \frac{\partial \theta_j}{\partial \boldsymbol{\beta}} = f'(\theta_j)\mathbf{x}_j$$

Thus

$$\mathbf{g} = \sum_{j=1}^{N} \mathbf{g}_j = \sum_{j=1}^{N} f'(\theta_j)\mathbf{x}_j$$

The k derivatives of the gradient vector can be obtained by calculating just one derivative numerically: $f'(\theta_j)$. In fact, we must calculate N numeric derivatives because we need $f'(\theta_j)$ separately evaluated for each observation, but that is not as bad as it sounds. The major cost of evaluating numeric derivatives is the time it takes to evaluate the likelihood function. We can obtain all $f'(\theta_j)$, $j = 1, \ldots, N$, with just two calls to the user-written likelihood-evaluation program, at least if we are willing to use the same h for each observation, and if we assume we already know the optimal h to use.

Similarly, the matrix of second derivatives can be written as

$$\mathbf{H} = \sum_{j=1}^{N} f''(\theta_j)\mathbf{x}_j'\mathbf{x}_j$$

and these $k(k+1)/2$ derivatives can similarly be obtained by evaluating one second derivative numerically (requiring two more calls).

This approach results in substantial computer time savings, especially for large values of k.

5 Methods lf0, lf1, and lf2

In the previous chapter, we used method `lf`, which for many applications is the only method you will need. One shortcoming of method `lf` is that there is no facility through which you can specify analytic first and second derivatives. In this chapter, we introduce methods `lf0`, `lf1`, and `lf2`, which we refer to collectively as the `lf`-family evaluators. In exchange for a more complicated syntax for evaluator programs, this family of evaluators allows you to specify derivatives. Method `lf0` is a stepping stone used while developing method-`lf1` and method-`lf2` evaluators. If you do not intend to specify analytic derivatives, you should use the easier method `lf` instead. Method `lf1` requires you to program analytic first derivatives, and method `lf2` requires you to program analytic first and second derivatives.

5.1 Comparing these methods

Methods `lf0`, `lf1`, and `lf2` differ from method `lf` in the arguments that `ml` passes to the likelihood evaluator and what the likelihood evaluator returns:

- In method `lf`, you are passed $\theta_{ij} = \mathbf{x}_{ij}\boldsymbol{\beta}_i$, the linear combinations that have already been evaluated.

  Methods `lf0`, `lf1`, and `lf2` pass your likelihood evaluator the coefficients themselves combined into a single vector $\boldsymbol{\beta} = (\boldsymbol{\beta}_1, \boldsymbol{\beta}_2, \ldots)$. It is your responsibility to obtain the θ_{ij} values from $\boldsymbol{\beta}$. The utility command `mleval`, described below, helps do this.

- Methods `lf0`, `lf1`, and `lf2` pass your program an additional argument 'todo' indicating whether your program is to evaluate the likelihood, the likelihood and first derivatives, or the likelihood, first, and second derivatives.

- Methods `lf1` and `lf2` pass your program additional arguments into which you place the observation-level first derivatives (scores) and the Hessian matrix when requested.

To fit a model using an evaluator of method `lf0`, `lf1`, or `lf2`, you type the command

```
. ml model lf0 myprog ...
```

or

```
. ml model lf1 myprog ...
```

or

    . ml model lf2 *myprog* ...

Write *myprog*—your likelihood evaluator—according to the `lf0`, `lf1`, or `lf2` specifications.

  As long as you do not specify the `nopreserve` option to `ml model`, `ml` automatically restricts computations to the relevant estimation sample when used with `lf`-family evaluators. Therefore, you do not need to handle `if` or `in` conditions or exclude observations with missing values. Weights are also handled automatically. If you do specify the `nopreserve` option, you will need to restrict your calculations to the estimation sample. Also, if you calculate the Hessian matrix for an `lf2` evaluator and do not use the `mlmatsum` command (discussed later), you will need to account for weights and restrict your computations for the Hessian to the relevant subsample.

5.2   Outline of evaluators of methods lf0, lf1, and lf2

Methods `lf0`, `lf1`, and `lf2` can be used with likelihood functions that meet the linear-form restrictions

$$\ln \ell_j = \ln \ell(\theta_{1j}, \theta_{2j}, \ldots, \theta_{Ej}; y_{1j}, y_{2j}, \ldots, y_{Dj})$$

$$\ln L = \sum_{j=1}^{N} \ln \ell_j$$

$$\theta_{ij} = \mathbf{x}_{ij}\boldsymbol{\beta}_i, \qquad i = 1, \ldots, E \text{ and } j = 1, \ldots, N$$

where i is the equation index and j is the observation index.

The outline of an evaluator of method `lf0`, `lf1`, or `lf2` is

```
program myprog
      version 11
      args todo b lnfj g1 g2 ... H
      // if you need to create any intermediate results:
      tempvar tmp1 tmp2
      quietly gen double 'tmp1' = ...
      ...
      quietly replace 'lnfj' = ...
      if ('todo'==0 | 'lnfj' >= .) exit      // method-lf0 evaluators end here
      // fill in score variables
      quietly replace 'g1' = ∂ ln ℓ_j/∂θ_1j
      quietly replace 'g2' = ∂ ln ℓ_j/∂θ_2j
      ...
      if ('todo'==1) exit                     // method-lf1 evaluators end here
```

```
// compute Hessian matrix
tempname d11 d12 d22 ...
mlmatsum 'lnfj' 'd11' = ∂² ln L/∂θ₁²
mlmatsum 'lnfj' 'd12' = ∂² ln L/∂θ₁∂θ₂
mlmatsum 'lnfj' 'd22' = ∂² ln L/∂θ₂²
matrix 'H' = ('d11', 'd12', ... 'd12'', 'd22', ...)
```
end

5.2.1   The todo argument

The first argument an evaluator of method `lf0`, `lf1`, or `lf2` receives is 'todo'.

- If 'todo' = 0, the evaluator must fill in $\ln \ell_j$, the observation-level contributions to the log likelihood.
- If 'todo' = 1, the evaluator must fill in $\ln \ell_j$ and the observation-level scores, stored as a set of E variables.
- If 'todo' = 2, the evaluator must fill in $\ln \ell_j$, the observation-level scores, and the Hessian matrix.

Method-`lf1` evaluators receive 'todo' = 0 or 'todo' = 1. When called with 'todo' = 0, the evaluator may calculate the gradient, but it will be ignored; doing so needlessly increases execution time.

Method-`lf2` evaluators receive 'todo' = 0, 'todo' = 1, or 'todo' = 2. Regardless of the value received, the evaluator may calculate more than is requested, but the extra will be ignored; doing so needlessly increases execution time.

5.2.2   The b argument

The second argument an evaluator of method `lf0`, `lf1`, or `lf2` receives is 'b', the current guess for $\boldsymbol{\beta}$ (the vector of coefficients) as a row vector. If you type

    . ml model lf1 *myprog* (foreign=mpg weight)

when specifying your model, then the vector 'b' that *myprog* receives would contain three elements: the coefficient on `mpg`, the coefficient on `weight`, and the intercept. The interpretation would be

$$\theta_j = \beta_1 \mathtt{mpg}_j + \beta_2 \mathtt{weight}_j + \beta_3$$

If you type

    . ml model lf1 *myprog* (foreign=mpg weight) (displ=weight)

then 'b' would contain five elements representing two equations:

$$\theta_{1j} = \beta_{1,1} \mathtt{mpg}_j + \beta_{1,2} \mathtt{weight}_j + \beta_{1,3}$$
$$\theta_{2j} = \beta_{2,1} \mathtt{weight}_j + \beta_{2,2}$$

If you type

```
. ml model lf1 myprog (foreign=mpg weight) (displ=weight) /sigma
```

'b' would contain six elements representing three equations:

$$\theta_{1j} = \beta_{1,1}\mathtt{mpg}_j + \beta_{1,2}\mathtt{weight}_j + \beta_{1,3}$$
$$\theta_{2j} = \beta_{2,1}\mathtt{weight}_j + \beta_{2,2}$$
$$\theta_{3j} = \beta_{3,1}$$

The equation order and the order of coefficients within the equation are exactly as you specify them in the `ml model` statement. The intercept coefficient is always the last coefficient of a given equation. Also, `ml` sets the column names of 'b' with the respective equation and variable names.

If you type

```
. ml model lf1 myprog (foreign=mpg weight)
```

and *myprog* is called with 'b' = (1, 2, 3), then your program will receive the following vector:

```
. matrix list 'b', noheader
        eq1:    eq1:    eq1:
        mpg   weight   _cons
   r1     1        2       3
```

If you type

```
. ml model lf1 myprog (foreign=mpg weight) (displ=weight) /sigma
```

and 'b' = (1, 2, 3, 4, 5, 6), then your program would receive

```
. matrix list 'b', noheader
        eq1:    eq1:    eq1:    eq2:    eq2:  sigma:
        mpg   weight   _cons  weight   _cons   _cons
   r1     1        2       3       4       5       6
```

Using mleval to obtain values from each equation

The `mleval` utility calculates linear combinations using a coefficient vector. The syntax of `mleval` is

mleval *newvarname* = *vecname* $\big[$, eq($\#$) $\big]$

newvarname, stored as a `double`, is created containing $\theta_{ij} = \mathbf{x}_{ij}\boldsymbol{\beta}_i$, the linear combination of independent variables corresponding to the *i*th equation.

vecname refers to the coefficient vector passed by `ml` (that is, 'b').

eq() specifies *i*, the equation to be evaluated; it defaults to eq(1).

`mleval` is typically used as follows:

```
program myprog
        version 11
        args todo b lnfj g1 g2 ... H
        tempvar theta1 theta2 ...
        mleval 'theta1' = 'b', eq(1)
        mleval 'theta2' = 'b', eq(2)
        ...
end
```

The rest of the program can then be written in terms of 'theta1', 'theta2', ...

If we type

```
. ml model lf2 myprog (foreign=mpg weight)
```

then our evaluator can be written as

```
program myprog
        version 11
        args todo b lnfj g1 H
        tempvar xb
        mleval 'xb' = 'b'
        ...
end
```

because `mleval` defaults to the first equation.

Consider another model:

```
. ml model lf2 myprog (foreign=mpg weight) (displ=weight) /sigma
```

We might write

```
program myprog
        version 11
        args todo b lnfj g1 g2 g3 H
        tempvar theta1 theta2 theta3
        mleval 'theta1' = 'b', eq(1)
        mleval 'theta2' = 'b', eq(2)
        mleval 'theta3' = 'b', eq(3)
        ...
end
```

In this example, $\theta_3 = \sigma$ is a constant. Thus there would be an efficiency gain to making 'theta3' a scalar rather than a variable in the dataset. `mleval` has a second syntax for such instances:

`mleval` *scalarname* = *vecname*, `scalar` $\big[$ `eq(`$\#$`)` $\big]$

With the `scalar` option, `mleval` generates a scalar instead of a variable.

```
program myprog3
        version 11
        args todo b lnfj g1 g2 g3 H
        tempvar theta1 theta2
        tempname theta3
        mleval 'theta1' = 'b', eq(1)
        mleval 'theta2' = 'b', eq(2)
        mleval 'theta3' = 'b', eq(3) scalar
        ...
end
```

If you tried to use the `scalar` option on an inappropriate equation (such as `eq(1)` in the example above), `mleval` would issue the following error:

```
. mleval 'theta1' = 'b', eq(1) scalar
mleval, scalar: eq(1) not constant
r(198);
```

When evaluating equations, use `mleval` without the `scalar` option for equations that produce values that vary across observations or that might vary across observations. To produce more efficient code, specify the `scalar` option for equations that produce values that are constant.

5.2.3   The lnfj argument

The third argument an evaluator of method `lf0`, `lf1`, or `lf2` receives is '`lnfj`', the name of a variable whose values you are to replace with the observation-level log-likelihood values evaluated at '`b`'. If the log likelihood cannot be calculated for an observation, that observation's value of '`lnfj`' is to contain a missing value, ".". If `ml` encounters any missing values of '`lnfj`' in the estimation sample, it considers the current value of '`b`' to be inadmissible and searches for a new value.

To illustrate how to use the '`b`' and '`lnfj`' arguments, here is how we would write a method-`lf0` evaluator for the probit model:

```
───────────────────────────────────────────── begin myprobit_lf0.ado ─────────
program myprobit_lf0
        version 11
        args todo b lnfj
        tempvar xb
        mleval 'xb' = 'b'
        quietly replace 'lnfj' = ln(normal( 'xb')) if $ML_y1 == 1
        quietly replace 'lnfj' = ln(normal(-'xb')) if $ML_y1 == 0
end
─────────────────────────────────────────────── end myprobit_lf0.ado ─────────
```

Because the probit model includes just one equation, we did not bother to specify the `eq(1)` option with `mleval` because `eq(1)` is the default.

❏ **Technical note**

The probability of observing a positive outcome in the probit model is $\Phi(\mathbf{x}_{ij}\boldsymbol{\beta}_i)$, so the probability of observing a negative outcome is $1 - \Phi(\mathbf{x}_{ij}\boldsymbol{\beta}_i)$. You might therefore be tempted to write the penultimate line of `myprobit_lf0.ado` as

```
quietly replace 'lnfj' = ln(1-normal('xb')) if $ML_y1 == 0
```

However, calculating the log likelihood that way has poor numerical properties when the outcome is zero. `ln(1-normal(9))` evaluates to missing, whereas `ln(normal(-9))` evaluates to the correct answer of -43.63. The difficulty arises because `normal(9)` is close to one, and evaluating the natural logarithm of one minus a number close to one is inherently unstable.

❏

5.2.4   Arguments for scores

A method-`lf1` evaluator receives $3 + E$ arguments, and a method-`lf2` evaluator receives $4 + E$ arguments, where E is the number of equations in the model. We have already discussed the first three arguments: `todo`, `b`, and `lnfj`. The next E arguments these two evaluators receive are 'g1', 'g2', ..., variables into which you are to place the observation-level scores for the E equations.

The general outline for a method-`lf1` evaluator is

```
program myprog
        version 11
        args todo b lnfj g1 g2 ...
        // if you need to create any intermediate results:
        tempvar tmp1 tmp2
        quietly gen double 'tmp1' = ...
        ...
        quietly replace 'lnfj' = ...
        if ('todo'==0) exit                    // method-lf0 evaluators end here
        // fill in score variables
        quietly replace 'g1' = ∂ ln ℓⱼ/∂θ₁ⱼ
        quietly replace 'g2' = ∂ ln ℓⱼ/∂θ₂ⱼ
        ...
    end
```

The argument `g1` is the name of a temporary variable whose values we are to replace with the scores for the first equation, $\partial \ln \ell_j / \partial \theta_{1j}$, evaluated at the current values of the coefficient vector `b`. Similarly, if our model had two or more equations, `g2`'s values would need to be replaced with $\partial \ln \ell_j / \partial \theta_{2j}$.

In some cases, you might be able to evaluate the likelihood function at the current value of '`b`' but cannot evaluate one or more of the scores, perhaps because of a division-by-zero error. In these cases, you can place a missing value in any observation of any of the score variables, or else exit the evaluator program as soon as you realize a problem because the score variables contain missing values when passed to your program. If `ml` encounters any missing values in any of the score variables in the estimation sample, it considers the current value of 'b' to be inadmissible and searches for a new value.

▷ **Example**

Using the probit likelihood function in (4.1), the score equation is

$$\frac{d\ln\ell_j}{d\theta_j} = \begin{cases} \phi(\theta_j)/\Phi(\theta_j) & \text{if } y_j = 1 \\ -\phi(\theta_j)/\Phi(-\theta_j) & \text{if } y_j = 0 \end{cases} \tag{5.1}$$

In our likelihood-evaluator program, we would therefore write

```
replace 'g1' =  normalden('xb')/normal( 'xb') if $ML_y1 == 1
replace 'g1' = -normalden('xb')/normal(-'xb') if $ML_y1 == 0
```

◁

Later in the chapter, when we demonstrate complete examples, we will show how to fill in the score variables for multiple-equation models.

5.2.5   The H argument

The final argument a method-lf2 evaluator receives is 'H', the name of a matrix you are to create containing the Hessian (the matrix of second derivatives). 'H' is to be filled in when 'todo' is 2.

The second derivatives are to be stored as a matrix that is conformable with the 'b' vector you were passed. This means that if 'b' is $1 \times k$, 'H' is to be $k \times k$, and it is to have the derivatives in the same order as the coefficients in 'b'.

'H' is to contain the second partial derivatives of $\ln L$ with respect to each of the coefficients:

$$\mathbf{H} = \begin{pmatrix} \frac{\partial^2 \ln L}{\partial \beta_1^2} & \frac{\partial^2 \ln L}{\partial \beta_1 \partial \beta_2} & \cdots & \frac{\partial^2 \ln L}{\partial \beta_1 \partial \beta_k} \\ \frac{\partial^2 \ln L}{\partial \beta_2 \partial \beta_1} & \frac{\partial^2 \ln L}{\partial \beta_2^2} & \cdots & \frac{\partial^2 \ln L}{\partial \beta_2 \partial \beta_k} \\ \vdots & \vdots & \ddots & \vdots \\ \frac{\partial^2 \ln L}{\partial \beta_k \partial \beta_1} & \frac{\partial^2 \ln L}{\partial \beta_k \partial \beta_2} & \cdots & \frac{\partial^2 \ln L}{\partial \beta_k^2} \end{pmatrix}$$

The row and column names placed on 'H' do not matter because ml will set them.

Unlike scores, which are observation-specific, 'H' represents the second derivative of the overall log-likelihood function. However, likelihoods that meet the linear-form restrictions, as both lf1 and lf2 evaluators must, afford a major simplification in computing the first and second derivatives of the overall log likelihood with respect to β. We have

$$\ln L = \sum_{j=1}^{N} \ln \ell(\theta_{1j}, \theta_{2j}, \ldots, \theta_{Ej})$$

$$\theta_{ij} = \mathbf{x}_{ij}\boldsymbol{\beta}_i$$

and $\boldsymbol{\beta} = (\boldsymbol{\beta}_1, \boldsymbol{\beta}_2, \ldots, \boldsymbol{\beta}_E)$. Thus the gradient vector can be written as a sum

$$\frac{\partial \ln L}{\partial \boldsymbol{\beta}} = \sum_{j=1}^{N} \frac{\partial \ln \ell_j}{\partial \boldsymbol{\beta}}$$

where

$$\frac{\partial \ln \ell_j}{\partial \boldsymbol{\beta}} = \left(\frac{\partial \ln \ell_j}{\partial \theta_{1j}} \frac{\partial \theta_{1j}}{\partial \boldsymbol{\beta}_1}, \quad \frac{\partial \ln \ell_j}{\partial \theta_{2j}} \frac{\partial \theta_{2j}}{\partial \boldsymbol{\beta}_2}, \quad \ldots, \quad \frac{\partial \ln \ell_j}{\partial \theta_{Ej}} \frac{\partial \theta_{Ej}}{\partial \boldsymbol{\beta}_E} \right)$$

$$= \left(\frac{\partial \ln \ell_j}{\partial \theta_{1j}} \mathbf{x}_{1j}, \quad \frac{\partial \ln \ell_j}{\partial \theta_{2j}} \mathbf{x}_{2j}, \quad \ldots, \quad \frac{\partial \ln \ell_j}{\partial \theta_{Ej}} \mathbf{x}_{Ej} \right)$$

The last equation shows why we can define our scores with respect to $\theta_{1j}, \theta_{2j}, \ldots, \theta_{Ej}$ and have our likelihood evaluator work regardless of the number of regressors we specify for each equation. For notational simplicity, we focus on the second derivative of $\ln L$ first with respect to β_i and then with respect to β_k. We have

$$\frac{\partial^2 \ln L}{\partial \beta_k \partial \beta_i} = \frac{\partial}{\partial \beta_k} \left(\frac{\partial \ln L}{\partial \beta_i} \right) = \sum_{j=1}^{N} \frac{\partial^2 \ln \ell_j}{\partial \beta_k \partial \beta_i}$$

where

$$\frac{\partial^2 \ln \ell_j}{\partial \beta_k \partial \beta_i} = \frac{\partial}{\partial \beta_k} \left(\frac{\partial \ln \ell_j}{\partial \theta_{ij}} \mathbf{x}_{ij} \right)$$

$$= \frac{\partial^2 \ln \ell_j}{\partial \theta_{kj} \partial \theta_{ij}} \mathbf{x}'_{ij} \mathbf{x}_{kj}$$

This result provides us with a considerable reduction in complexity, both mathematically and with respect to our likelihood-evaluator program. As with scores, we need only obtain second derivatives with respect to the θs instead of the individual βs; and because the chain rule applies with linear-form models, we can change the independent variables in our equations at will without having to touch our likelihood-evaluator program.

Like with scores, in some cases you might be able to evaluate the log likelihood at the given value of 'b' but not the second derivative for one or more observations. In those cases, you can either set one or more observations on 'lnfj' to missing, or you can set one or more elements of 'H' to missing. ml then considers the current value of 'b' to be infeasible and searches for a new value.

The second derivative of the overall log-likelihood function is a sum over the observations in the dataset. Were we to compute this sum ourselves, we would need to limit our sum to only those observations within the estimation sample, we would need to take into account any weights if specified, and we would need to know the variables represented by the $\mathbf{x}_i$'s. Fortunately, `ml` provides the `mlmatsum` command, which makes computing $\mathbf{H}$ easy once we obtain the second derivatives with respect to the θs. `mlmatsum` obviates the three details we would need to consider if we were to compute the sum ourselves.

Using mlmatsum to define H

`mlmatsum` assists in calculating the Hessian matrix of the overall log-likelihood function when the linear-form restrictions are met. As we just saw, in this case we can write the Hessian as

$$\mathbf{H} = \sum_{j=1}^{N} \frac{\partial^2 \ln \ell_j}{\partial \boldsymbol{\beta} \partial \boldsymbol{\beta}'}$$

$$\frac{\partial^2 \ln \ell_j}{\partial \boldsymbol{\beta} \partial \boldsymbol{\beta}'} = \begin{pmatrix} \frac{\partial^2 \ln \ell_j}{\partial \theta_{1j}^2} \mathbf{x}'_{1j} \mathbf{x}_{1j} & \frac{\partial^2 \ln \ell_j}{\partial \theta_{1j} \partial \theta_{2j}} \mathbf{x}'_{1j} \mathbf{x}_{2j} & \cdots & \frac{\partial^2 \ln \ell_j}{\partial \theta_{1j} \partial \theta_{kj}} \mathbf{x}'_{1j} \mathbf{x}_{kj} \\ \frac{\partial^2 \ln \ell_j}{\partial \theta_{2j} \partial \theta_{1j}} \mathbf{x}'_{2j} \mathbf{x}_{1j} & \frac{\partial^2 \ln \ell_j}{\partial \theta_{2j}^2} \mathbf{x}'_{2j} \mathbf{x}_{2j} & \cdots & \frac{\partial^2 \ln \ell_j}{\partial \theta_{2j} \partial \theta_{kj}} \mathbf{x}'_{2j} \mathbf{x}_{kj} \\ \vdots & \vdots & \ddots & \vdots \\ \frac{\partial^2 \ln \ell_j}{\partial \theta_{kj} \partial \theta_{1j}} \mathbf{x}'_{kj} \mathbf{x}_{1j} & \frac{\partial^2 \ln \ell_j}{\partial \theta_{kj} \partial \theta_{2j}} \mathbf{x}'_{kj} \mathbf{x}_{2j} & \cdots & \frac{\partial^2 \ln \ell_j}{\partial \theta_{kj}^2} \mathbf{x}'_{kj} \mathbf{x}_{kj} \end{pmatrix}$$

Now $\partial^2 \ln \ell_j / \partial \theta_{ij} \partial \theta_{kj}$ is just a scalar value for each observation, so we can store the values for all the observations in a variable. Moreover, in general, log-likelihood functions satisfy Young's Theorem so that $\partial^2 \ln \ell_j / \partial \theta_{ij} \partial \theta_{kj} = \partial^2 \ln \ell_j / \partial \theta_{kj} \partial \theta_{ij}$. We have E equations, so we have a total of $E(E+1)/2$ distinct submatrices in $\mathbf{H}$.

The syntax for `mlmatsum` is

`mlmatsum` $varname_{\text{lnfj}}$ $matrixname$ = exp $\big[\,if\,\big]$ $\big[\,$`,` `eq(`$\#\big[\,,\#\big]$`)`$\big]$

Obtaining $\partial^2 \ln L / \partial \boldsymbol{\beta} \partial \boldsymbol{\beta}'$ is simply a matter of issuing one `mlmatsum` per equation pair and then joining the results into a single matrix.

In a single-equation system, we have

   `mlmatsum 'lnfj' 'H'` = *formula for* $\partial^2 \ln \ell_j / \partial \theta_{1j}^2$

In a two-equation system, we have

```
tempname d11 d12 d22
mlmatsum 'lnfj' 'd11' = formula for ∂² ln ℓ_j/∂θ²_1j    , eq(1)
mlmatsum 'lnfj' 'd12' = formula for ∂² ln ℓ_j/∂θ_1j∂θ_2j, eq(1,2)
mlmatsum 'lnfj' 'd22' = formula for ∂² ln ℓ_j/∂θ²_2j    , eq(2)
matrix 'H' = ('d11','d12' \ ('d12')','d22')
```

In the last line, we wrote ('d12')' instead of 'd12'' to emphasize that we are taking the transpose of 'd12'.

Note that `mlmatsum` has two left-hand-side arguments:

```
. mlmatsum `lnfj' `whatever' = exp
```

`mlmatsum` checks that every observation in the estimation subsample has a nonmissing value for *exp* before computing the sum. If there are any missing values for *exp*, `mlmatsum` replaces the first observation on `lnfj' to missing to indicate to `ml` that the Hessian cannot be computed at the current value of `b'; `ml` will then search for a different value of `b'. Further advantages of using `mlmatsum` are that it restricts its calculations to the estimation sample and automatically applies the user's weights if specified.

▷ **Example**

Using the log likelihood from the probit model (4.1) and its first derivative (5.1), the Hessian is

$$\frac{d^2 \ln \ell_j}{d\theta_j^2} = \begin{cases} -R(\theta_j)\{R(\theta_j) + \theta_j\} & \text{if } y_j = 1 \\ -S(\theta_j)\{S(\theta_j) - \theta_j\} & \text{if } y_j = 0 \end{cases}$$

where

$$R(\theta_j) = \phi(\theta_j)/\Phi(\theta_j)$$
$$S(\theta_j) = \phi(\theta_j)/\Phi(-\theta_j)$$

Thus `mlmatsum` can be used to obtain the Hessian by programming

```
tempvar R S h
gen double `R' = normalden(`xb')/normal(`xb')
gen double `S' = normalden(`xb')/normal(-`xb')
gen double `h' = -`R'*(`R'+`xb') if $ML_y1 == 1
replace     `h' = -`S'*(`S'-`xb') if $ML_y1 == 0
mlmatsum `lnfj' `H' = `h'
```

Because the right-hand side of `mlmatsum` can be an expression, the above could be compressed into fewer lines using the `cond()` function, if preferred.

◁

5.2.6   Aside: Stata's scalars

To use methods `lf1` and `lf2` effectively—and other `ml` methods—you must use Stata's scalars if you make intermediate scalar calculations. Instead of programming

```
local eta = ...
generate double ... = ... `eta' ...
```

you should program

```
tempname eta
scalar `eta' = ...
generate double ... = ... `eta' ...
```

You do this for the same reason that you specify that the intermediate variables you `generate` are `doubles`: scalars are more accurate than macros. Macros provide about 12 digits of accuracy, sometimes more but never fewer. Scalars are full double-precision binary numbers, meaning that they provide about 16.5 digits of accuracy in all cases.

A Stata scalar is just that—it contains one number (which could be a missing value). The `scalar` command defines scalars, and you use scalars in expressions just as you would any variable:

```
. scalar x = 3
. display x
3
. scalar y = 1/sqrt(2)
. display y
.70710678
. scalar z = 1/ln(x-3)
. display z
.
```

We have been sloppy in the above example, although there is no reason you would know that. The example would have better read

```
. scalar x = 3
. display scalar(x)            // we type scalar(x), not x
3
. scalar y = 1/sqrt(2)
. display scalar(y)            // we type scalar(y), not y
.70710678
. scalar z = 1/ln(scalar(x)-3) // we type scalar(x), not x
. display scalar(z)            // we type scalar(z), not z
.
```

The `scalar()` function says "use the value of the named scalar here", and as you have seen, we can omit it. We should not do this, however, unless we are certain there is no variable in the dataset by the same name. Pretend that we had a variable named `x` and a scalar named `x`:

```
. clear all
. input x                      // create the variable name x

           x
  1. 1
  2. 2
  3. 3
  4. end
. scalar x = 57               // create the scalar name x
. gen y = x+1                 // what is in y?
```

Notice that `gen y = x+1` did not result in an error, nor did defining a scalar named `x` when a variable named `x` already existed; neither would it have been an error to do things the other way around and create a variable named `x` when a scalar named x already existed. This, however, leads to an interesting question: to which are we referring when we type x in `gen y = x+1`? Is it the data variable, in which case y would contain 2, 3, and 4? Or is it the scalar, in which case y would contain 58, 58, and 58?

When Stata has a variable and a scalar of the same name, it chooses the variable over the scalar:

```
. list

     +-------+
     | x   y |
     |-------|
  1. | 1   2 |
  2. | 2   3 |
  3. | 3   4 |
     +-------+
```

We previously typed `gen y = x+1`. There is a scalar named x that contains 57; nevertheless, we obtained y = variable x + 1. If we really want Stata to use the scalar called x, we must tell it:

```
. gen z = scalar(x) + 1
. list

     +--------------+
     | x   y    z   |
     |--------------|
  1. | 1   2    58  |
  2. | 2   3    58  |
  3. | 3   4    58  |
     +--------------+
```

The same logic applies to all Stata commands, including `scalar` itself:

```
. scalar new = x
. display new
1
```

Why 1 and not 57? Because `scalar new = x` was interpreted as

```
. scalar new = /*VARIABLE*/ x
```

and that in turn was interpreted as

```
. scalar new = /*VARIABLE*/ x /*IN THE FIRST OBSERVATION*/
```

Had there instead been no variable called x, `scalar new = x` would have been interpreted as

```
. scalar new = /*SCALAR*/ x
```

An error would result if there were no variable and no scalar named x:

```
. scalar new = x
x not found
r(111);
```

Stata's preference for variables over scalars is even stronger than we have indicated. Here is another example:

```
. clear all
. input mpg weight
          mpg      weight
  1. 22 2100
  2. 23 1900
  3. 12 3800
  4. end
. scalar m = 9.5
. scalar z = m
. display z
22
```

Scalar z contains 22 and not 9.5 because m was taken as an abbreviation for variable mpg! Thus it appears that about the only safe way we can use scalars is to reference them inside the scalar() function:

```
. scalar z = scalar(m)
. display scalar(z)
9.5
```

In fact, we seldom do that. Instead, we write our programs using the tempname statement:

```
tempname x y
scalar 'x' = (some complicated function)
scalar 'y' = ... 'x' ...
```

The last line could have read

```
scalar 'y' = ... scalar('x') ...
```

but the scalar() is not necessary. tempname will create a unique name for the scalar, and we can be certain that there is no variable by the same name, abbreviated or not.

5.3   Summary of methods lf0, lf1, and lf2

5.3.1   Method lf0

The method-lf0 evaluator calculates the observation-level log likelihood but does not compute any derivatives. Method lf0 is a stepping stone toward writing an lf1 or lf2 evaluator. If you have no intention of every programming the derivatives, you should use method lf instead, because it is easier to use.

The outline of a method-lf0 evaluator is

```
program myprog
        version 11
        args todo b lnfj
        // if you need to create any intermediate results:
        tempvar tmp1 tmp2
        quietly gen double 'tmp1' = ...
        ...
        quietly replace 'lnfj' = ...
end
```

In section 5.2, we showed that for methods lf0, lf1, and lf2, the `args` statement has the form

```
args todo b lnfj g1 g2 ... H
```

Method-lf0 evaluators are always called with 'todo' = 0 only and are never requested to compute scores or the Hessian matrix, so the shorter version

```
args todo b lnfj
```

will work just as well, though you must remember to modify that line when you convert your lf0 evaluator to an lf1 or lf2 evaluator.

You specify that you are using method lf0 with program *myprog* on the `ml model` statement:

```
. ml model lf0 myprog ...
```

Unlike method lf, which passes to your evaluator the parameters $\boldsymbol{\theta}_i$, method-lf0 evaluators receive the coefficient vector 'b'. Because the likelihood function is generally written as a function of the parameters $\boldsymbol{\theta}_i$ from the probability model, the first step is to obtain their values from the coefficients vector 'b':

```
program myprog
        version 11
        args todo b lnf g
        tempvar theta1 theta2 ...
        mleval 'theta1' = 'b'
        mleval 'theta2' = 'b', eq(2)          // if there is a θ2
        ...
end
```

The following points enumerate what `ml` supplies and expects of a method-lf0 likelihood evaluator.

- You are passed the vector of coefficients. Use `mleval` to obtain the values of each of the linear predictors of each equation from it; see section 5.2.2.

- The names of the dependent variables (if any) are placed in global macros by `ml`; the first dependent variable is $ML_y1, the second is $ML_y2, ... Use $ML_y1, $ML_y2, ... just as you would any existing variable in the data.

- You must create any temporary variables for intermediate results as `doubles`. Type

      gen double *name* = ...

  Do not omit the word `double`.

- The estimation subsample is identified by `$ML_samp==1`. You may safely ignore this if you do not specify `ml model`'s `nopreserve` option. When your program is called, only relevant data will be in memory. Even if you do specify `nopreserve`, `mleval` automatically restricts itself to the estimation subsample; it is not necessary to include `if $ML_samp==1` with this command. However, you will need to restrict other calculation commands to the `$ML_samp==1` subsample if you specify `nopreserve`.

- The observation-level log-likelihood value is to be saved in the variable '`lnfj`'.

5.3.2   Method lf1

The method-`lf1` evaluator calculates the observation-level log likelihood and scores. The outline for a method-`lf1` likelihood evaluator is

```
program myprog
        version 11
        args todo b lnfj g1 g2 ...
        // if you need to create any intermediate results:
        tempvar tmp1 tmp2
        quietly gen double 'tmp1' = ...
        ...
        quietly replace 'lnfj' = ...
        if ('todo'==0 | 'lnfj' >= .) exit        // method-lf0 evaluators end here
        // fill in score variables
        quietly replace 'g1' = ∂ln ℓ_j/∂θ_{1j}
        quietly replace 'g2' = ∂ln ℓ_j/∂θ_{2j}
end
```

The final argument that `lf`-family evaluators receive is H, used to store the Hessian matrix. Method-`lf1` evaluators are never required to compute the Hessian, so we can omit H from the `args` statement. You specify that you are using method `lf1` with program *myprog* on the `ml model` statement, which you issue after you have defined the likelihood-evaluation program.

```
    . ml model lf1 myprog ...
```

As with method `lf0`, the likelihood evaluator receives the coefficient vector '`b`', so typically, the first part of your program will need to extract the θ parameters using `ml eval`.

Method `lf1` also requires you to calculate the scores $\partial \ln \ell_j/\partial \theta_{1j}$, $\partial \ln \ell_j/\partial \theta_{2j}$, ... In many cases, the log-likelihood function and scores contain the same terms [such as

$\phi(\mathbf{x}\boldsymbol{\beta})$], so to increase readability of your program, you should first create temporary variables (as type `double`) to hold those terms.

```
program myprog
        version 11
        args todo b lnfj g1 g2 ...
        tempvar theta1 theta2 ...
        mleval 'theta1' = 'b'
        mleval 'theta2' = 'b', eq(2)                    // if there is a θ₂
        ...
        tempvar tmp1 tmp2
        quietly gen double 'tmp1' = ...
        quietly gen double 'tmp2' = ...
        ...
        quietly replace 'lnfj' = expression in terms of 'tmp1' and 'tmp2'
        if ('todo'==0 | 'lnfj'>=.) exit

        replace 'g1' = formula for ∂ln ℓⱼ/∂θ₁ⱼ in terms of 'tmp1' and 'tmp2'
        replace 'g2' = formula for ∂ln ℓⱼ/∂θ₂ⱼ in terms of 'tmp1' and 'tmp2'
        ...
end
```

The following points enumerate what `ml` supplies and expects of a method-`lf1` likelihood evaluator.

- You are passed the vector of coefficients. Use `mleval` to obtain the values of each of the linear predictors of each equation from it; see section 5.2.2.

- The names of the dependent variables (if any) are placed in global macros by `ml`; the first dependent variable is `$ML_y1`, the second is `$ML_y2`, ... Use `$ML_y1`, `$ML_y2`, ... just as you would any existing variable in the data.

- You must create any temporary variables for intermediate results as `doubles`. Type

      gen double *name* = ...

  Do not omit the word `double`.

- The estimation subsample is identified by `$ML_samp==1`. You may safely ignore this if you do not specify `ml model`'s `nopreserve` option. When your program is called, only relevant data will be in memory. Even if you do specify `nopreserve`, `mleval` automatically restricts itself to the estimation subsample; it is not necessary to include `if $ML_samp==1` with this command. However, you will need to restrict other calculation commands to the `$ML_samp==1` subsample if you specify `nopreserve`.

- The observation-level log-likelihood value is to be saved in the variable `'lnfj'`.

- Observation-level scores for each equation are to be stored in variables `'g1'`, `'g2'`, and so on.

5.3.3   Method lf2

The method-`lf2` evaluator calculates the observation-level log likelihood and scores and the Hessian matrix of the overall log likelihood. The outline for the program is

```
program myprog
        version 11
        args todo b lnfj g1 g2 ... H
        // if you need to create any intermediate results:
        tempvar tmp1 tmp2
        quietly gen double 'tmp1' = ...
        ...
        quietly replace 'lnfj' = ...
        if ('todo'==0 | 'lnfj' >= .) exit          // method-lf0 evaluators end here
        // fill in score variables
        quietly replace 'g1' = ∂ ln ℓ_j/∂θ_{1j}
        quietly replace 'g2' = ∂ ln ℓ_j/∂θ_{2j}
        ...
        if ('todo'==1) exit                        // method-lf1 evaluators end here
        // compute Hessian matrix
        tempname d11 d12 d22 ...
        mlmatsum 'lnfj' 'd11' = ∂² ln L/∂θ²_1
        mlmatsum 'lnfj' 'd12' = ∂² ln L/∂θ_1∂θ_2
        mlmatsum 'lnfj' 'd22' = ∂² ln L/∂θ²_2
        matrix 'H' = ('d11', 'd12', ...  'd12'', 'd22', ...)
    end
```

You specify that you are using method `lf2` on the `ml model` statement, which you issue after you have defined the likelihood-evaluation program.

```
. ml model lf2 myprog ...
```

The method-`lf2` evaluator continues from where the method-`lf1` evaluator left off. Taking out the common code, here is how the evaluators differ:

```
program myprog
        version 11
        args todo b lnfj g1 g2 ... H
        ...
        if ('todo'==1 | 'lnfj'==.) exit            // from here down is new
        tempname d11 d12 d22 ...
        mlmatsum 'lnfj' 'd11' = formula for ∂² ln ℓ_j/∂θ²_{1j},  eq(1)
        mlmatsum 'lnfj' 'd12' = formula for ∂² ln ℓ_j/∂θ_{1j}∂θ_{2j},  eq(1,2)
        mlmatsum 'lnfj' 'd22' = formula for ∂² ln ℓ_j/∂θ²_{2j},  eq(2)
        ...
        matrix 'H' = ('d11','d12',... \ 'd12'','d22',...)
    end
```

The Hessian matrix you are to define is

$$
\mathbf{H} = \frac{\partial^2 \ln L}{\partial \boldsymbol{\beta} \partial \boldsymbol{\beta}'} = \begin{pmatrix}
\frac{\partial^2 \ln L}{\partial \beta_1^2} & \frac{\partial^2 \ln L}{\partial \beta_1 \partial \beta_2} & \cdots & \frac{\partial^2 \ln L}{\partial \beta_1 \partial \beta_k} \\
\frac{\partial^2 \ln L}{\partial \beta_2 \partial \beta_1} & \frac{\partial^2 \ln L}{\partial \beta_2^2} & \cdots & \frac{\partial^2 \ln L}{\partial \beta_2 \partial \beta_k} \\
\vdots & \vdots & \ddots & \vdots \\
\frac{\partial^2 \ln L}{\partial \beta_k \partial \beta_1} & \frac{\partial^2 \ln L}{\partial \beta_k \partial \beta_2} & \cdots & \frac{\partial^2 \ln L}{\partial \beta_k^2}
\end{pmatrix}
$$

ml provides the coefficient vector as a $1 \times k$ row vector and expects the Hessian matrix to be returned as a $k \times k$ matrix. Because method-lf2 evaluators satisfy the linear-form restrictions, you simply need to compute the second derivatives with respect to the equation-level $\boldsymbol{\theta}$s. Then you can use mlmatsum to create the required submatrices of $\mathbf{H}$ without having to worry about the dimensions of β or the variables in the model; see section 5.2.5 for more information.

The following points enumerate what ml supplies and expects of a method-lf2 likelihood evaluator.

- You are passed the vector of coefficients. Use mleval to obtain the values of each of the linear predictors of each equation from it; see section 5.2.2.

- The names of the dependent variables (if any) are placed in global macros by ml; the first dependent variable is \$ML_y1, the second is \$ML_y2, ... Use \$ML_y1, \$ML_y2, ... just as you would any existing variable in the data.

- You must create any temporary variables for intermediate results as doubles. Type

      gen double *name* = ...

  Do not omit the word double.

- The estimation subsample is identified by \$ML_samp==1. You may safely ignore this if you do not specify ml model's nopreserve option. When your program is called, only relevant data will be in memory. Even if you do specify nopreserve, mleval automatically restricts itself to the estimation subsample; it is not necessary to include if \$ML_samp==1 with this command. However, you will need to restrict other calculation commands to the \$ML_samp==1 subsample if you specify nopreserve.

- The observation-level log-likelihood value is to be saved in the variable 'lnfj'.

- Observation-level scores for each equation are to be stored in variables 'g1', 'g2', and so on.

- The Hessian of the overall log likelihood is to be saved in matrix 'H'. Use the mlmatsum command to produce the various submatrices; it automatically handles weights and properly restricts the summations to the estimation sample.

5.4   Examples

We will now illustrate the implementation of method-`1f1` and method-`1f2` likelihood evaluators for the examples of chapter 4: probit, linear regression, and Weibull models. All of these models meet the linear-form restrictions. Because there is no point in implementing these estimators with method `1f0` evaluator functions other than as a first step toward implementing `1f1` and `1f2` evaluators, we do not consider `1f0` evaluators except for debugging purposes.

5.4.1   The probit model

The probit log-likelihood function for the jth observation is

$$\ln \ell_j = \begin{cases} \ln \Phi(\mathbf{x}_j \boldsymbol{\beta}) & \text{if } y_j = 1 \\ \ln \Phi(-\mathbf{x}_j \boldsymbol{\beta}) & \text{if } y_j = 0 \end{cases}$$

Taking derivatives with respect to $\mathbf{x}_j \boldsymbol{\beta}$, the jth term for the gradient is

$$g_j = \frac{\partial \ln \ell_j}{\partial \mathbf{x}_j \boldsymbol{\beta}} = \begin{cases} \phi(\mathbf{x}_j \boldsymbol{\beta})/\Phi(\mathbf{x}_j \boldsymbol{\beta}) & \text{if } y_j = 1 \\ -\phi(\mathbf{x}_j \boldsymbol{\beta})/\Phi(-\mathbf{x}_j \boldsymbol{\beta}) & \text{if } y_j = 0 \end{cases}$$

where $\phi()$ is the probability density function for the standard normal distribution. Thus our method-`1f1` likelihood evaluator is

———————————————————————————————— begin `myprobit_lf1.ado` ————————

```
program myprobit_lf1
        version 11
        args todo b lnfj g1
        tempvar xb lj
        mleval 'xb' = 'b'
        quietly {
                // Create temporary variable used in both likelihood
                // and scores
                gen double 'lj'  = normal( 'xb')  if $ML_y1 == 1
                replace    'lj'  = normal(-'xb')  if $ML_y1 == 0
                replace    'lnfj' = log('lj')
                if ('todo'==0) exit
                replace 'g1' =  normalden('xb')/'lj'  if $ML_y1 == 1
                replace 'g1' = -normalden('xb')/'lj'  if $ML_y1 == 0
        }
end
```

———————————————————————————————————— end `myprobit_lf1.ado` ————————

Taking the second derivative with respect to $\mathbf{x}_j \boldsymbol{\beta}$, the jth term of the Hessian is

$$H_j = \frac{\partial^2 \ln \ell_j}{\partial^2 \mathbf{x}_j \boldsymbol{\beta}} = -g_j(g_j + \mathbf{x}_j \boldsymbol{\beta})$$

and our method-lf2 likelihood evaluator is

```
─────────────────────────────────── begin myprobit_lf2.ado ───────────
program myprobit_lf2
        version 11
        args todo b lnfj g1 H
        tempvar xb lj
        mleval 'xb' = 'b'
        quietly {
                // Create temporary variable used in both likelihood
                // and scores
                gen double 'lj'  = normal( 'xb')  if $ML_y1 == 1
                replace    'lj'  = normal(-'xb')  if $ML_y1 == 0
                replace    'lnfj' = log('lj')
                if ('todo'==0) exit

                replace 'g1' =  normalden('xb')/'lj'  if $ML_y1 == 1
                replace 'g1' = -normalden('xb')/'lj'  if $ML_y1 == 0
                if ('todo'==1) exit

                mlmatsum 'lnfj' 'H' = -'g1'*('g1'+'xb'), eq(1,1)
        }
end
─────────────────────────────────── end myprobit_lf2.ado ──────────────
```

▷ Example

Using `cancer.dta` and `myprobit_lf2`, we fit a probit model of the incidence of death on drug type and age:

```
. sysuse cancer, clear
(Patient Survival in Drug Trial)

. ml model lf2 myprobit_lf2 (died = i.drug age)

. ml maximize
initial:       log likelihood = -33.271065
alternative:   log likelihood = -31.427839
rescale:       log likelihood = -31.424556
Iteration 0:   log likelihood = -31.424556
Iteration 1:   log likelihood = -21.884411
Iteration 2:   log likelihood = -21.710897
Iteration 3:   log likelihood = -21.710799
Iteration 4:   log likelihood = -21.710799

                                          Number of obs   =          48
                                          Wald chi2(3)    =       13.58
Log likelihood = -21.710799               Prob > chi2     =      0.0035
```

died	Coef.	Std. Err.	z	P>|z|	[95% Conf. Interval]	
drug						
2	-1.941275	.597057	-3.25	0.001	-3.111485	-.7710644
3	-1.792924	.5845829	-3.07	0.002	-2.938686	-.6471626
age	.0666733	.0410666	1.62	0.104	-.0138159	.1471624
_cons	-2.044876	2.284283	-0.90	0.371	-6.521988	2.432235

◁

The first time we tried fitting this model, here is what we obtained:

```
. ml model lf2 myprobit_lf2 (died = i.drug age)

. ml maximize
initial:        log likelihood = -33.271065
alternative:    log likelihood = -31.427839
rescale:        log likelihood = -31.424556
varlist required
r(100);
```

How did we go about finding our mistake? In chapter 7, we will learn about tools available to check our derivative calculations, but in this case, we did not need them. `myprobit_lf2.ado` is an `lf2` evaluator, but it also satisfies the requirements of `lf0` and `lf1` evaluators. If we type

```
. ml model lf1 myprobit_lf2 (died = i.died age)
. ml maximize
```

and do not receive any errors, we can deduce that the problem with our likelihood evaluator must lie in the portion of the program that computes the Hessian. If we do receive an error, our next step would be to try

```
. ml model lf0 myprobit_lf2 (died = i.died age)
. ml maximize
```

If that works, we at least know we are computing the log likelihood correctly.

5.4.2   Normal linear regression

The linear regression log-likelihood function for the jth observation is

$$\ln \ell_j = \ln \left\{ \frac{1}{\sigma_j} \phi \left(\frac{y_j - \mu_j}{\sigma_j} \right) \right\}$$

For this example, we will model the standard deviation in the log space. Using the notation from section 3.2, our second equation is

$$\theta_{2j} = \ln \sigma_j = \mathbf{x}_{2j} \boldsymbol{\beta}_j$$

We can use `normalden(y,`μ`,`σ`)` to quickly compute

$$\frac{1}{\sigma} \phi \left(\frac{y - \mu}{\sigma} \right)$$

The derivative of $\phi()$ is

$$\phi'(z) = -z\phi(z)$$

Taking derivatives with respect to μ_j and $\ln \sigma_j$ yields the scores for the jth observation:

$$g_{1j} = \frac{\partial \ln \ell_j}{\partial \mu_j} = \frac{y_j - \mu_j}{\sigma_j^2}$$

$$g_{2j} = \frac{\partial \ln \ell_j}{\partial \ln \sigma_j} = \left(\frac{y_j - \mu_j}{\sigma_j} \right)^2 - 1$$

Our method-`lf1` likelihood evaluator is

```
——————————————————————————————— begin mynormal_lf1.ado ——————————
    program mynormal_lf1
          version 11
          args todo b lnfj g1 g2
          tempvar mu lnsigma sigma
          mleval 'mu' = 'b', eq(1)
          mleval 'lnsigma' = 'b', eq(2)
          quietly {
                  gen double 'sigma' = exp('lnsigma')
                  replace 'lnfj' = ln( normalden($ML_y1,'mu','sigma') )
                  if ('todo'==0) exit

                  tempvar z
                  tempname dmu dlnsigma
                  gen double 'z' = ($ML_y1-'mu')/'sigma'
                  replace 'g1' = 'z'/'sigma'
                  replace 'g2' = 'z'*'z'-1
          }
    end
  ——————————————————————————————— end mynormal_lf1.ado ——————————
```

Taking second derivatives yields the jth element's observations to the Hessian matrix:

$$H_{11j} = \frac{\partial^2 \ell_j}{\partial^2 \mu_j} = -\frac{1}{\sigma_j^2}$$

$$H_{12j} = \frac{\partial^2 \ell_j}{\partial \mu_j \partial \ln \sigma_j} = -2 \frac{y_j - \mu_j}{\sigma_j^2}$$

$$H_{22j} = \frac{\partial^2 \ell_j}{\partial^2 \ln \sigma_j} = -2 \left(\frac{y_j - \mu_j}{\sigma_j} \right)^2$$

(Continued on next page)

Our method-lf2 likelihood evaluator is

```
——————————————————————————————————— begin mynormal_lf2.ado ———————
program mynormal_lf2
        version 11
        args todo b lnfj g1 g2 H
        tempvar mu lnsigma sigma
        mleval 'mu' = 'b', eq(1)
        mleval 'lnsigma' = 'b', eq(2)
        quietly {
                gen double 'sigma' = exp('lnsigma')
                replace 'lnfj' = ln( normalden($ML_y1,'mu','sigma') )
                if ('todo'==0) exit

                tempvar z
                tempname dmu dlnsigma
                gen double 'z' = ($ML_y1-'mu')/'sigma'
                replace 'g1' = 'z'/'sigma'
                replace 'g2' = 'z'*'z'-1
                if ('todo'==1) exit

                tempname d11 d12 d22
                mlmatsum 'lnfj' 'd11' = -1/'sigma'^2      , eq(1)
                mlmatsum 'lnfj' 'd12' = -2*'z'/'sigma'    , eq(1,2)
                mlmatsum 'lnfj' 'd22' = -2*'z'*'z'        , eq(2)
                matrix 'H' = ('d11', 'd12' \ 'd12'', 'd22')
        }
end
——————————————————————————————————— end mynormal_lf2.ado ———————
```

▷ Example

Using `auto.dta` and `mynormal_lf1`, we fit a linear regression with the `vce(robust)` option.

```
                                                      begin ex_normal_lf1 ─────────
. sysuse auto
(1978 Automobile Data)

. ml model lf1 mynormal_lf1 (mu: mpg = weight displacement) /lnsigma,
> vce(robust)

. ml maximize
initial:       log pseudolikelihood =      -<inf>  (could not be evaluated)
feasible:      log pseudolikelihood = -6441.8168
rescale:       log pseudolikelihood = -368.12496
rescale eq:    log pseudolikelihood =  -257.3948
Iteration 0:   log pseudolikelihood =  -257.3948  (not concave)
Iteration 1:   log pseudolikelihood = -241.78314  (not concave)
Iteration 2:   log pseudolikelihood =  -231.8097  (not concave)
Iteration 3:   log pseudolikelihood = -222.78552  (not concave)
Iteration 4:   log pseudolikelihood = -212.75192
Iteration 5:   log pseudolikelihood = -197.06449
Iteration 6:   log pseudolikelihood = -195.24283
Iteration 7:   log pseudolikelihood =  -195.2398
Iteration 8:   log pseudolikelihood =  -195.2398
                                         Number of obs   =          74
                                         Wald chi2(2)    =      111.39
Log pseudolikelihood =  -195.2398        Prob > chi2     =      0.0000
```

mpg	Coef.	Robust Std. Err.	z	P>|z|	[95% Conf. Interval]
mu						
weight	-.0065671	.0009195	-7.14	0.000	-.0083692	-.004765
displacement	.0052808	.0073381	0.72	0.472	-.0091017	.0196633
_cons	40.08452	2.045516	19.60	0.000	36.07538	44.09366
lnsigma						
_cons	1.219437	.1342357	9.08	0.000	.9563401	1.482534

```
                                                      ─── end ex_normal_lf1 ─────────
```

◁

▷ **Example**

In our development, we wrote about the parameter $\ln \sigma$ as if it were a scalar, but note our parameterization of the second equation,

$$\theta_{2j} = \ln \sigma_j = \mathbf{x}_{2j}\boldsymbol{\beta}_2$$

By examining evaluators `mynormal_lf1` and `mynormal_lf2`, you will notice that we have followed all the ordinary prescriptions, meaning that a linear equation can be estimated for $\ln \sigma_j$.

In the example below, we fit a model in which we assume that the variance of the disturbance is given by

$$\ln \sigma_j = \text{price}\beta_{2,1} + \beta_{2,2}$$

```
. sysuse auto
(1978 Automobile Data)

. ml model lf1 mynormal_lf1 (mu: mpg = weight displacement) (lnsigma: price),
> vce(robust)

. ml maximize
initial:       log pseudolikelihood =    -<inf>  (could not be evaluated)
feasible:      log pseudolikelihood = -6441.8168
rescale:       log pseudolikelihood = -368.12496
rescale eq:    log pseudolikelihood =  -257.3948
Iteration 0:   log pseudolikelihood =  -257.3948  (not concave)
Iteration 1:   log pseudolikelihood =  -241.3443  (not concave)
Iteration 2:   log pseudolikelihood = -230.49029  (not concave)
Iteration 3:   log pseudolikelihood = -222.66544  (not concave)
Iteration 4:   log pseudolikelihood = -218.79323
Iteration 5:   log pseudolikelihood = -202.46255
Iteration 6:   log pseudolikelihood = -195.15622
Iteration 7:   log pseudolikelihood = -195.13889
Iteration 8:   log pseudolikelihood = -195.13887
Iteration 9:   log pseudolikelihood = -195.13887
                                               Number of obs   =        74
                                               Wald chi2(2)    =    115.98
Log pseudolikelihood = -195.13887              Prob > chi2     =    0.0000
```

mpg	Coef.	Robust Std. Err.	z	P>|z|	[95% Conf. Interval]
mu						
weight	-.0065244	.0009002	-7.25	0.000	-.0082887	-.0047601
displacement	.0041361	.0071429	0.58	0.563	-.0098638	.018136
_cons	40.19516	2.032263	19.78	0.000	36.212	44.17832
lnsigma						
price	.0000129	.0000267	0.48	0.630	-.0000396	.0000653
_cons	1.138677	.2280395	4.99	0.000	.6917278	1.585626

This is an exciting example: we have just fit a model with heteroskedasticity without changing our code one bit. And in fact, when we did the development, we did not even think about issues of generalizing the parameterization of $\ln \sigma$! The way `ml` is designed, you have to think twice about how to impose limitations.

In the case of `mynormal_lf2`, the sixth line reads

```
mleval 'lnsigma' = 'b', eq(2)
```

If we wanted to force the limitation that our estimate of $\ln \sigma$ must be a simple constant, we could have coded

```
mleval 'lnsigma' = 'b', eq(2) scalar
```

That is the only change we would need to make. The advantage would be that our program would run a little faster. The disadvantage would be that we would lose the ability to estimate heteroskedastic models. We like our program as it is.

◁

❏ **Technical note**

In the example above, we said that we need to make only one change to our program if we want to force $\ln \sigma$ to be estimated as a scalar:

```
mleval 'lnsigma' = 'b', eq(2)
```

to

```
mleval 'lnsigma' = 'b', eq(2) scalar
```

There is a second change we should make. Earlier in the program, where it says

```
tempvar mu lnsigma sigma
```

we should change it to read

```
tempvar mu sigma
tempname lnsigma
```

We should make that change because 'lnsigma' will no longer be a temporary variable in our dataset but instead will be a temporary scalar. That was, after all, the point of adding the `scalar` option to the `mleval` command. It does not matter whether you make that extra change. The way Stata works, you may use `tempvars` for `tempnames` and vice versa. What you must not do is use the same name twice for two different things, and we are not doing that.

So because `tempvar` and `tempname` are interchangeable, why does Stata have both? The answer concerns programming style: your program will be more readable if it clearly indicates how the name 'lnsigma' will be used. On that basis alone, you ought to make the change.

There is yet another change you ought to make, too. This one has to do with efficiency. Later in the code, after

```
mleval 'lnsigma' = 'b', eq(2) scalar
```

is a line that reads

```
gen double 'sigma' = exp('lnsigma')
```

That line ought to be change to read

```
scalar 'sigma' = exp('lnsigma')
```

Making that change will make your program run faster. The idea here is that if we are going to force the parameter 'lnsigma' to be a scalar, then we should look throughout the rest of our code and exploit the speed advantages associated with scalars wherever we can.

So here is a better version of `mynormal_lf2`, done right:

─────────────────────────────────── begin `mynormal_nohet_lf2.ado` ───────

```
program mynormal_nohet_lf2
        version 11
        args todo b lnfj g1 g2 H
        tempvar mu
        tempname lnsigma sigma
        mleval 'mu' = 'b', eq(1)
        mleval 'lnsigma' = 'b', eq(2) scalar
        scalar 'sigma' = exp('lnsigma')
        quietly {
                replace 'lnfj' = ln( normalden($ML_y1,'mu','sigma') )
                if ('todo'==0) exit

                tempvar z
                tempname dmu dlnsigma
                gen double 'z' = ($ML_y1-'mu')/'sigma'
                replace 'g1' = 'z'/'sigma'
                replace 'g2' = 'z'*'z'-1
                if ('todo'==1) exit

                tempname d11 d12 d22
                mlmatsum 'lnfj' 'd11' = -1/'sigma'^2    , eq(1)
                mlmatsum 'lnfj' 'd12' = -2*'z'/'sigma'  , eq(1,2)
                mlmatsum 'lnfj' 'd22' = -2*'z'*'z'      , eq(2)
                matrix 'H' = ('d11', 'd12' \ 'd12'', 'd22')
        }
end
```

─────────────────────────────────── end `mynormal_nohet_lf2.ado` ───────

The program `mynormal_nohet_lf2` is not strictly better than the original `mynormal_lf2`. Indeed, one could argue that it is not as good because it does not let us estimate heteroskedastic models. `mynormal_nohet_lf2` is better only in that it is faster, and in some situations, speed may be more important than flexibility.

❏

5.4.3   The Weibull model

A Weibull log-likelihood function for the jth observation, where t_j is the time of failure or censoring and $d_j = 1$ if failure and 0 if censored, is

$$\ln \ell_j = -(t_j/e^{\theta_{1j}})^{e^{\theta_{2j}}} + d_j\{\theta_{2j} - \theta_{1j} + (e^{\theta_{2j}} - 1)(\ln t_j - \theta_{1j})\}$$
$$\theta_{1j} = \ln \eta_j = \mathbf{x}_{1j}\boldsymbol{\beta}_2$$
$$\theta_{2j} = \ln \gamma_j = \mathbf{x}_{2j}\boldsymbol{\beta}_1$$

Before we take derivatives of the log likelihood, let's rewrite this function using slightly different notation. Define the variables

$$p_j = e^{\theta_{2j}}$$
$$M_j = (t_j e^{-\theta_{1j}})^{e^{\theta_{2j}}}$$
$$R_j = \ln t_j - \theta_{1j}$$

so that we can rewrite our log likelihood as

$$\ln \ell_j = -M_j + d_j\{\theta_{2j} - \theta_{1j} + (p_j - 1)R_j\}$$

Taking derivatives of p_j, R_j, and M_j with respect to θ_{1j} and θ_{2j} yields

$$\frac{\partial p_j}{\partial \theta_{1j}} = 0 \qquad\qquad\qquad \frac{\partial p_j}{\partial \theta_{2j}} = p_j$$

$$\frac{\partial M_j}{\partial \theta_{1j}} = -p_j M_j \qquad\qquad\qquad \frac{\partial M_j}{\partial \theta_{2j}} = R_j p_j M_j$$

$$\frac{\partial R_j}{\partial \theta_{1j}} = -1 \qquad\qquad\qquad \frac{\partial R_j}{\partial \theta_{2j}} = 0$$

Then by using the chain rule, the jth observation's scores are

$$g_{1j} = \frac{\partial \ln \ell_j}{\partial \theta_{1j}} = p_j(M_j - d_j)$$

$$g_{2j} = \frac{\partial \ln \ell_j}{\partial \theta_{2j}} = d_j - R_j p_j(M_j - d_j)$$

and our method-lf1 likelihood evaluator is

——————————————————————— begin myweibull_lf1.ado ———————

```
program myweibull_lf1
        version 11
        args todo b lnf g1 g2
        tempvar leta lgam p M R
        mleval `leta' = `b', eq(1)
        mleval `lgam' = `b', eq(2)
        local t "$ML_y1"
        local d "$ML_y2"
        quietly {
                gen double `p' = exp(`lgam')
                gen double `M' = (`t'*exp(-`leta'))^`p'
                gen double `R' = ln(`t')-`leta'
                replace `lnf' = -`M' + `d'*(`lgam'-`leta' + (`p'-1)*`R')
                if (`todo'==0) exit
                replace `g1' = `p'*(`M'-`d')
                replace `g2' = `d' - `R'*`p'*(`M'-`d')
        }
end
```

——————————————————————— end myweibull_lf1.ado ———————

Taking second derivatives yields the jth observation's contribution to the Hessian:

$$H_{11j} = \frac{\partial^2 \ln \ell_j}{\partial^2 \theta_{1j}} = -p_j^2 M_j$$

$$H_{22j} = \frac{\partial^2 \ln \ell_j}{\partial^2 \theta_{2j}} = -p_j R_j(R_j p_j M_j + M_j - d_j)$$

$$H_{12j} = \frac{\partial^2 \ln \ell_j}{\partial \theta_{1j}\partial \theta_{2j}} = p_j(M_j - d_j + R_j p_j M_j)$$

Our method-lf2 likelihood evaluator is

```
──────────────────────────────────────── begin myweibull_lf2.ado ────────
program myweibull_lf2
        version 11
        args todo b lnf g1 g2 H
        tempvar leta lgam p M R
        mleval 'leta' = 'b', eq(1)
        mleval 'lgam' = 'b', eq(2)
        local t "$ML_y1"
        local d "$ML_y2"
        quietly {
                gen double 'p' = exp('lgam')
                gen double 'M' = ('t'*exp(-'leta'))^'p'
                gen double 'R' = ln('t')-'leta'
                replace 'lnf' = -'M' + 'd'*('lgam'-'leta' + ('p'-1)*'R')
                if ('todo'==0) exit

                replace 'g1' = 'p'*('M'-'d')
                replace 'g2' = 'd' - 'R'*'p'*('M'-'d')
                if ('todo'==1) exit

                tempname d11 d12 d22
                mlmatsum 'lnf' 'd11' = -'p'^2 * 'M'                 , eq(1)
                mlmatsum 'lnf' 'd12' = 'p'*('M'-'d'+'R'*'p'*'M')    , eq(1,2)
                mlmatsum 'lnf' 'd22' = -'p'*'R'*('R'*'p'*'M'+'M'-'d') , eq(2)
                matrix 'H' = ('d11','d12' \ 'd12','d22')
        }
end
──────────────────────────────────────── end myweibull_lf2.ado ────────
```

▷ Example

Using `cancer.dta` and `myweibull_lf2`, we fit the Weibull model of time of death with censoring on drug type and age.

```
. sysuse cancer
(Patient Survival in Drug Trial)
. ml model lf2 myweibull_lf2 (lneta: studytime died = i.drug age) /lngamma
. ml maximize, nolog
```

			Number of obs	=	48
			Wald chi2(3)	=	35.25
Log likelihood = -110.26736			Prob > chi2	=	0.0000

	Coef.	Std. Err.	z	P>|z|	[95% Conf. Interval]
lneta					
drug					
2	1.012966	.2903917	3.49	0.000	.4438086 1.582123
3	1.45917	.2821195	5.17	0.000	.9062261 2.012114
age	-.0671728	.0205688	-3.27	0.001	-.1074868 -.0268587
_cons	6.060723	1.152845	5.26	0.000	3.801188 8.320259
lngamma					
_cons	.5573333	.1402154	3.97	0.000	.2825163 .8321504

◁

▷ **Example**

Just as with the second example of section 5.4.2, we can parameterize the second equation of our model ($\ln \gamma_j$) with a linear equation simply by modifying what we type on the `ml model` statement:

```
. ml model lf2 myweibull_lf2
>        (lneta: studytime died = i.drug age)
>        (lngamma: female)
```

In the above command, we would be specifying

$$\ln \gamma_j = \texttt{female } \beta_{2,1} + \beta_{2,2}$$
$$= \begin{cases} \beta_{2,2} & \text{if male} \\ \beta_{2,1} + \beta_{2,2} & \text{if female} \end{cases}$$

Linear equations can be specified for any of the estimated equations. As in the linear regression example previously, if we wanted to limit our program to allow only simple constants for the second equation, we would change

```
mleval 'lgam' = 'b', eq(2)
```

to

```
mleval 'lgam' = 'b', eq(2) scalar
```

and we would declare `lgam` to be a `tempname` rather than a `tempvar`. After those changes, our program would run slightly faster.

◁

6   Methods d0, d1, and d2

Methods d0, d1, and d2 are used when the likelihood function does not meet the linear-form restrictions. Method d0 requires you to evaluate the overall log likelihood, and it calculates the gradient and Hessian numerically. Method d1 requires you to evaluate the overall log likelihood and analytic gradient, and it calculates the Hessian numerically based on your analytic gradient. Method d2 requires you to do all the work yourself, evaluating the overall log likelihood and the analytic gradient and Hessian.

These d-family evaluators can be used to fit models that satisfy the linear-form requirements, but you are better off using method lf or one of the lf-family evaluators. The d-family evaluators do not use observation-level scores, so you cannot obtain robust or cluster–robust standard errors, work with survey data and sampling weights, or use the BHHH maximization algorithm. Also, d-family evaluators are slightly more difficult to write. Use the d-family evaluators only when the other alternatives are insufficient.

We assume that you have already read chapter 5.

6.1   Comparing these methods

Methods d0, d1, and d2 differ from lf-family evaluators in several respects.

- In lf-family evaluators, you fill in the variable 'lnfj' with the log likelihoods $\ln \ell_j$ for each observation. Those methods sum the components to form the overall log likelihood $\ln L = \ln \ell_1 + \ln \ell_2 + \cdots + \ln \ell_N$.

  Methods d0, d1, and d2 require you to fill in the *scalar* 'lnf' with $\ln L$, the overall log likelihood. The utility command mlsum, described below, helps you do this.

- Methods lf1 and lf2 receive arguments g1, g2, etc., into which you store observation-level scores $\partial \ln \ell_j / \partial \theta_{ij}$. In contrast, methods d1 and d2 receive one argument, g, into which you place the $1 \times k$ row vector $(\partial \ln L / \partial \beta_1, \partial \ln L / \partial \beta_2, \ldots, \partial \ln L / \partial \beta_k)$ where k is the number of coefficients in the model. The utility command mlvecsum, described below, helps you do this.

- lf-family evaluators gracefully handle many problems for you, such as when the likelihood or scores cannot be calculated for a given set of parameter values. Utility routines mleval, mlsum, mlvecsum, mlmatsum, and mlmatbysum can assist you in handling these problems when you use d-family evaluators. These utilities are documented in this chapter.

Methods `d0`, `d1`, and `d2` differ from each other only in the way they calculate derivatives.

- Method `d0` requires that you fill in only the overall log likelihood. No analytical derivatives are required.
- Method `d1` requires that you fill in the overall log likelihood and the overall gradient vector (`mlvecsum` helps do this).
- Method `d2` requires that you fill in the overall log likelihood, the overall gradient vector, and the overall negative Hessian (`mlmatsum` and `mlmatbysum` help do this).

There are two reasons why you would use method `d0` instead of methods `lf`, `lf0`, `lf1`, or `lf2`:

1. Method `d0` does not assume that the overall log likelihood ln L is the sum of the log likelihoods from each observation. Some models—such as the Cox proportional hazards model, conditional logistic regression, and panel-data models—produce log likelihoods for groups of observations and so do not lend themselves to methods that assume the likelihood function satisfies the linear-form restrictions.

2. Method `d0` is the first step along the path to writing a method-`d1` or method-`d2` evaluator.

You would choose methods `d1` and `d2` for the same reasons. Method `d1` is faster and more accurate than method `d0`—but it is more work to program—and method `d2` is faster and more accurate than that—and even more work to program.

To specify method `d0`, `d1`, or `d2`, you type `d0`, `d1`, or `d2` on the `ml model` line:

```
. ml model d0 myprog ...
```

or

```
. ml model d1 myprog ...
```

or

```
. ml model d2 myprog ...
```

Write *myprog*—your likelihood evaluator—according to the `d0`, `d1`, or `d2` specifications.

6.2   Outline of method d0, d1, and d2 evaluators

Methods `d0`, `d1`, and `d2` can be used with any likelihood function,

$$\ln L = \ln L\{(\theta_{1j}, \theta_{2j}, \ldots, \theta_{Ej}; y_{1j}, y_{2j}, \ldots, y_{Dj}) : j = 1, \ldots, N\}$$
$$\theta_{ij} = \mathbf{x}_{ij}\boldsymbol{\beta}_i, \qquad i = 1, \ldots, E \text{ and } j = 1, \ldots, N$$

where i is the equation index and j is the observation index. We wrote the log likelihood in terms of θs because most models include regressors, but we emphasize that by letting $x_{ik} = 1$, a constant term, our notation allows θ_{ik} to represent a scalar parameter in the log likelihood function.

The outline of the evaluator of method d0, d1, or d2 is

```
program myprog
        version 11
        args todo b lnf g H
        Form scalar 'lnf' = ln L
        if ('todo'==0 | 'lnf' >= .) exit         // method-d0 evaluators end here
        Form row vector 'g' = ∂ ln L/∂b
        if ('todo'==0 | 'lnf' >= .) exit         // method-d1 evaluators end here
        Form matrix 'H' = ∂² ln L/∂b∂b'
end
```

6.2.1   The todo argument

The first argument an evaluator of method d0, d1, or d2 receives is 'todo'.

- If 'todo' = 0, the evaluator must fill in the log likelihood.

- If 'todo' = 1, the evaluator must fill in the log likelihood and the gradient vector (stored as a *row* vector).

- If 'todo' = 2, the evaluator must fill in the log likelihood, the gradient vector, and the Hessian matrix.

Method-d0 evaluators receive 'todo' = 0 only.

Method-d1 evaluators receive 'todo' = 0 or 'todo' = 1. When called with 'todo' = 0, the evaluator may calculate the gradient vector, but it will be ignored; calculating the gradient vector needlessly increases execution time.

Method-d2 evaluators receive 'todo' = 0, 'todo' = 1, or 'todo' = 2. Regardless of the value received, the evaluator may calculate more than is requested, but the extra will be ignored; calculating more than requested needlessly increases execution time.

6.2.2   The b argument

The second argument an evaluator of method d0, d1, or d2 receives is 'b', the current guess for $\boldsymbol{\beta}$ (the vector of coefficients) as a row vector. The 'b' argument for these evaluators works the same way it does for lf-family evaluators, so we do not dwell on it here. You will again typically use the mleval utility command to calculate linear combinations (the θ_{ij}'s) based on the coefficient vector 'b'. See section 5.2.2.

6.2.3   The lnf argument

The third argument an evaluator of method d0, d1, or d2 receives is 'lnf', the name of a scalar you will fill in with the overall log-likelihood value evaluated at 'b'. If the log likelihood cannot be calculated, 'lnf' is to contain a missing value, ".".

To aid understanding, here is how you might write the method-d0 likelihood evaluator for the probit model:

```
―――――――――――――――――――――――――――――――――――――――― begin myprobit_bad_d0.ado ――――――――――
program myprobit_bad_d0
        version 11
        args todo b lnf
        tempvar xb lnfj
        mleval 'xb' = 'b'
        quietly {
                gen double 'lnfj' = ln(  normal('xb'))  if $ML_y1 == 1
                replace    'lnfj' = ln(1-normal('xb'))  if $ML_y1 == 0
                replace    'lnfj' = sum('lnfj')
                scalar 'lnf' = 'lnfj'[_N]
        }
end
―――――――――――――――――――――――――――――――――――――――――― end myprobit_bad_d0.ado ――――――――――
```

Do not take this example too seriously; it has several problems:

- When you use methods d0, d1, or d2, it is your responsibility to restrict your calculations to the estimation subsample. Note how we sum over all the observations, including, possibly, observations that should be excluded. Perhaps the user typed

  ```
  . ml model d0 myprobit (foreign=mpg weight) if mpg>20
  ```

  Do not make too much of this problem because it is your responsibility to restrict calculations to the estimation subsample only when you specify ml model's nopreserve option. Otherwise, ml automatically saves your data and then drops the irrelevant observations, so you can be sloppy. (ml does not do this with method lf and lf-family methods because they carefully sum the observations.) Nevertheless, preserving and restoring the data takes time, and you might be tempted to specify ml model's nopreserve option. With the nopreserve option, the program above would then produce incorrect results.

- We did nothing to account for weights, should they be specified. That will not be a problem as long as no weights are specified on the ml model statement, but a better draft of this program would replace

  ```
  replace 'lnfj' = sum('lnfj')
  ```

  with

  ```
  replace 'lnfj' = sum($ML_w*'lnfj')
  ```

  If you use method d0, d1, or d2, weights are your responsibility.

- We merely assumed that the log likelihood can be calculated for every observation. Let's pretend, however, that the values we received for 'b' are odd and result in the calculation of the log likelihood being missing in the third observation. The sum() function used at the last step will treat the missing value as contributing 0 and so calculate an overall log-likelihood value that will be incorrect. Worse, treating $\ln \ell_3 = 0$ is tantamount to $\ell_3 = 1$. In this case, because likelihoods correspond to probabilities, the impossible value is treated as producing a probability 1 result.

  You might think this unlikely, but it is not. We direct your attention to line 8 of our program:

$$\text{replace 'lnfj'} = \ln(1-\text{normal('xb')}) \text{ if } \$\text{ML\_y1} == 0$$

  As we discussed in the technical note in section 5.2.3, $\ln(1-\text{normal('xb')})$ turns out to be a poor way of calculating $\ln \ell$ when the observed outcome is 0. In fact, $\ln(1-\text{normal}(9))$ evaluates to missing though the correct value is -43.63. Method lf and the lf-family methods protected you by watching for when things went badly; methods d0, d1, and d2 do not protect you because they cannot—they never see the results for each observation, and indeed, for applications that require d-family evaluators, there often is nothing to watch.

This third problem is potentially serious, and therefore we strongly urge you to avoid Stata's sum() function and instead use the mlsum command, described below, to fill in the log likelihood. Using mlsum, an adequate version of our probit evaluator would read

──────────────── begin myprobit_d0.ado ────────────────
```
program myprobit_d0
        version 11
        args todo b lnf
        tempvar xb lj
        mleval 'xb' = 'b'
        quietly {
                gen double 'lj'  = normal( 'xb')  if $ML_y1 == 1
                replace    'lj'  = normal(-'xb')  if $ML_y1 == 0
                mlsum 'lnf' = ln('lj')
        }
end
```
──────────────── end myprobit_d0.ado ────────────────

mlsum addresses all three of the shortcomings we have mentioned.

Using lnf to indicate that the likelihood cannot be calculated

Your evaluator might be called with values of 'b' for which the log likelihood cannot be evaluated for substantive as well as numerical reasons; for instance, perhaps one of the parameters to be estimated is a variance, and the maximizer attempts to evaluate the function with this parameter set to a negative value.

If the likelihood function cannot be evaluated at the coefficient values 'b' supplied by ml, your evaluator is to return a missing value in 'lnf'. At this point, it is best to exit and skip any remaining calculations.

This issue—not being able to calculate the log likelihood—never arose with methods designed for models that satisfy the linear-form restrictions because ml could detect the problem by examining what your evaluator returned. With those methods, you calculated log-likelihood values for each observation. If some of the returned values turned out to be missing, whether for numerical or substantive reasons, ml could spot the missing values and take the appropriate action.

Methods d0, d1, and d2 cannot do this because they never see the values from each observation. The evaluator you write returns the overall log likelihood, so it is your responsibility to watch for impossible coefficient values and let the maximizer know. If you encounter impossible values, set 'lnf' to contain a missing value and exit.

Using mlsum to define lnf

Most likelihood-evaluation programs generate a log likelihood for each observation or group of observations and then sum the contributions to obtain the overall log-likelihood value. mlsum will perform the summation. The syntax of mlsum is

mlsum *scalarname*$_{\text{lnf}}$ = *exp* $\left[\, if \,\right]$

There are other ways you could make this sum in Stata, but there are three reasons why you should use mlsum:

1. mlsum automatically restricts the summation to the estimation subsample (observations for which $ML_samp==1$) and thus makes it more likely that results are correct even if you specify ml model's nopreserve option.

2. mlsum automatically applies weights, if they were specified, in forming the sum.

3. mlsum verifies that what is being summed contains no missing values or sets *scalarname*$_{\text{lnf}}$ to missing if it does.

mlsum is typically used as follows:

```
program myprog
        version 11
        args todo b lnf g H
        ...
        mlsum 'lnf' = ...
        if ('todo'==0 | 'lnf'>=.) exit
        ...
end
```

The right side of mlsum is filled in with something that, when summed across the observations, yields the overall log-likelihood value. The above outline is relevant when log-likelihood values are calculated for each observation in the estimation subsample.

`mlsum` itself restricts the summation to the relevant observations (the observations for which `$ML_samp==1`).

When log-likelihood values instead exist for only groups of observations, as is typically true for models that require d-family evaluators, the outline is

```
program myprog
        version 11
        args todo b lnf g H
        ...
        mlsum 'lnf' = ... if ...
        if ('todo'==0 | 'lnf'>=.) exit
        ...
end
```

The `if` *exp* must be filled in to be true for observations where you expect log-likelihood values to exist. `mlsum` does *not* skip over missing values in the estimation subsample. It takes missing as an indication that the log-likelihood value could not be calculated, so it sets the `'lnf'` also to contain missing.

▷ Example

Using the θ notation, the log likelihood for the probit model is

$$\ln \ell_j = \begin{cases} \ln \Phi(\theta_j) & \text{if } y_j = 1 \\ \ln \Phi(-\theta_j) & \text{if } y_j = 0 \end{cases} \tag{6.1}$$

$$\theta_j = \mathbf{x}_j \boldsymbol{\beta}$$

The dots in `mlsum` must be filled in with the formula for $\ln \ell_j$:

```
tempvar xb lnfj
mleval 'xb' = 'b'
gen double 'lnfj' = ln(normal( 'xb')) if $ML_y1 == 1
replace    'lnfj' = ln(normal(-'xb')) if $ML_y1 == 0
mlsum 'lnf' = 'lnfj'
```

Note that we can specify an expression following `mlsum`, so the above lines could more compactly be written as

```
tempvar xb
mleval 'xb' = 'b'
mlsum 'lnf' = ln(normal(cond($ML_y1==1,'xb',-'xb')))
```

◁

▷ Example

As mentioned above, when log-likelihood values exist not observation by observation but instead for groups of observations, you must include the `if` restriction with the `mlsum` command:

```
mlsum 'lnf' = ... if ...
```

This step is necessary, for example, when coding likelihoods for panel-data models. We will show examples of this in sections 6.4 and 6.5.

◁

6.2.4   The g argument

The fourth argument an evaluator of method d0, d1, or d2 receives is 'g', the name of a row vector you are to fill in with the gradient of the overall log likelihood. You need only create 'g' if 'todo' is 1 or 2 (which would happen, of course, only if you are using methods d1 or d2).

Method-d0 evaluators never need to fill in 'g', and most people would write a method-d0 evaluator as

```
program myprog
        version 11
        args todo b lnf
        Form scalar 'lnf' = ln L
end
```

In fact, even method-d0 evaluators receive the fourth argument 'g' and fifth argument 'H'; however, these evaluators are called only with 'todo' = 0, so they can ignore the fourth and fifth arguments. This allows a method-d1 or method-d2 evaluator to be used with method d0. Like we saw with lf-family evaluators, this is a useful feature for debugging and checking results.

A method-d1 or method-d2 evaluator will be called upon to fill in 'g'. The first derivatives (gradient) are to be stored as a row vector that is conformable with the 'b' vector passed by ml. This means that if 'b' is $1 \times k$, 'g' will be $1 \times k$ and will have the derivatives in the same order as the coefficients in 'b'.

'g' will contain the partial derivatives of $\ln L$ with respect to each of the coefficients:

$$\mathbf{g} = \left(\frac{\partial \ln L}{\partial \beta_1}, \frac{\partial \ln L}{\partial \beta_2}, \dots, \frac{\partial \ln L}{\partial \beta_k} \right)$$

The row and column names placed on 'g' do not matter because they will be ignored by ml. If 'g' cannot be calculated at the current values of the coefficient vector 'b', then 'lnf' is to be set to contain a missing value, and the contents of 'g', or even whether 'g' is defined, will not matter.

Using mlvecsum to define g

mlvecsum assists in calculating the gradient vector when the linear-form restrictions are met. In that case,

$$\ln L = \sum_{j=1}^{N} \ln \ell(\theta_{1j}, \theta_{2j}, \ldots, \theta_{Ej})$$

$$\theta_{ij} = \mathbf{x}_{ij}\boldsymbol{\beta}_i$$

and $\boldsymbol{\beta} = (\boldsymbol{\beta}_1, \boldsymbol{\beta}_2, \ldots, \boldsymbol{\beta}_E)$. Thus the gradient vector can be written as a sum:

$$\frac{\partial \ln L}{\partial \boldsymbol{\beta}} = \sum_{i=1}^{N} \frac{\partial \ln \ell_j}{\partial \boldsymbol{\beta}}$$

$$\frac{\partial \ln \ell_j}{\partial \boldsymbol{\beta}} = \left(\frac{\partial \ln \ell_j}{\partial \theta_{1j}} \frac{\partial \theta_{1j}}{\partial \boldsymbol{\beta}_1}, \quad \frac{\partial \ln \ell_j}{\partial \theta_{2j}} \frac{\partial \theta_{2j}}{\partial \boldsymbol{\beta}_2}, \quad \ldots, \quad \frac{\partial \ln \ell_j}{\partial \theta_{Ej}} \frac{\partial \theta_{Ej}}{\partial \boldsymbol{\beta}_E} \right)$$

$$= \left(\frac{\partial \ln \ell_j}{\partial \theta_{1j}} \mathbf{x}_{1j}, \quad \frac{\partial \ln \ell_j}{\partial \theta_{2j}} \mathbf{x}_{2j}, \quad \ldots, \quad \frac{\partial \ln \ell_j}{\partial \theta_{Ej}} \mathbf{x}_{Ej} \right)$$

$\partial \ln \ell_j / \partial \theta_{ij}$ is simply one value per observation. You supply $\partial \ln \ell_j / \partial \theta_{ij}$, and `mlvecsum` returns

$$\partial \ln L / \partial \boldsymbol{\beta}_i = \sum_{j=1}^{N} \frac{\partial \ln \ell_j}{\partial \theta_{ij}} \mathbf{x}_{ij}$$

which is one component of the vector above.

`mlvecsum`'s syntax is

`mlvecsum` *scalarname*$_{\text{lnf}}$ *rowvecname* = *exp* $\big[\, if \,\big]$ $\big[\,$, eq(#)$\,\big]$

where *exp* evaluates to $\partial \ln \ell_j / \partial \theta_{ij}$ for equation $i = \#$. Thus obtaining $\partial \ln L / \partial \boldsymbol{\beta}$ is simply a matter of issuing one `mlvecsum` per equation and then joining the results into a single row vector.

In a single-equation system,

> `mlvecsum 'lnf' 'g'` = formula for $\partial \ln \ell_j / \partial \theta_{1j}$

In a two-equation system,

```
tempname d1 d2
mlvecsum 'lnf' 'd1' = formula for ∂ ln ℓj/∂θ1j, eq(1)
mlvecsum 'lnf' 'd2' = formula for ∂ ln ℓj/∂θ2j, eq(2)
matrix 'g' = ('d1','d2')
```

In a three-equation system,

```
tempname d1 d2 d3
mlvecsum 'lnf' 'd1' = formula for ∂ ln ℓj/∂θ1j, eq(1)
mlvecsum 'lnf' 'd2' = formula for ∂ ln ℓj/∂θ2j, eq(2)
mlvecsum 'lnf' 'd3' = formula for ∂ ln ℓj/∂θ3j, eq(3)
matrix 'g' = ('d1','d2','d3')
```

Note that `mlvecsum` has two left-hand-side arguments:

```
. mlvecsum 'lnf' 'whatever' = exp
```

That is so `mlvecsum` can reset 'lnf' to be a missing value if *exp* cannot be calculated for some relevant observation. Evaluators of methods d0, d1, and d2 are to set 'lnf' to missing when called with impossible values of the coefficient vector. It is possible that 'lnf' can be calculated but the gradient cannot. `mlvecsum` appropriately handles such cases, and it applies weights if they were specified.

▷ Example

Using the probit likelihood from (6.1), the derivative is

$$\frac{d \ln \ell_j}{d\theta_j} = \begin{cases} \phi(\theta_j)/\Phi(\theta_j) & \text{if } y_j = 1 \\ -\phi(\theta_j)/\Phi(-\theta_j) & \text{if } y_j = 0 \end{cases} \tag{6.2}$$

Thus `mlvecsum` could be used to obtain the entire gradient vector by coding

```
tempvar gj
gen double 'gj' =  normalden('xb')/normal( 'xb') if $ML_y1 == 1
replace      'gj' = -normalden('xb')/normal(-'xb') if $ML_y1 == 0
mlvecsum 'lnf' 'g' = 'gj'
```

Because the right-hand side of `mlvecsum` can be an expression, the above could be reduced to one line using the `cond()` function if preferred.

◁

Like with `mlsum`, the real power of `mlvecsum` lies in its acceptance of an `if` clause. As we will see by way of example later in the chapter, this acceptance that simplifies the computation of gradients of some models that do not satisfy the linear-form restrictions.

6.2.5   The H argument

The fifth argument an evaluator of method d0, d1, or d2 receives is 'H'. You fill in this argument just as you do with `lf`-family evaluators; see section 5.2.5. The Hessian is to be stored as a matrix that is conformable with the 'b' vector you were passed. If 'b' is $1 \times k$, then 'H' is to be $k \times k$, and it is to have the derivatives in the same order as the coefficients in 'b'. If 'H' cannot be calculated at the current value of coefficient vector 'b', then 'lnf' is to be set to contain a missing value, and you need not fill in 'H'.

'H' is to contain the second partial derivatives of $\ln L$ with respect to each of the coefficients:

$$\mathbf{H} = \begin{pmatrix} \frac{\partial^2 \ln L}{\partial \beta_1^2} & \frac{\partial^2 \ln L}{\partial \beta_1 \partial \beta_2} & \cdots & \frac{\partial^2 \ln L}{\partial \beta_1 \partial \beta_k} \\ \frac{\partial^2 \ln L}{\partial \beta_2 \partial \beta_1} & \frac{\partial^2 \ln L}{\partial \beta_2^2} & \cdots & \frac{\partial^2 \ln L}{\partial \beta_2 \partial \beta_k} \\ \vdots & \vdots & \ddots & \vdots \\ \frac{\partial^2 \ln L}{\partial \beta_k \partial \beta_1} & \frac{\partial^2 \ln L}{\partial \beta_k \partial \beta_2} & \cdots & \frac{\partial^2 \ln L}{\partial \beta_k^2} \end{pmatrix}$$

Unlike scores, which are observation-specific, 'H' represents the second derivative of the overall log-likelihood function. If the likelihood function satisfies the linear-form restrictions, then `mlmatsum` can be used to simplify the construction of 'H'. For likelihood functions that are defined at the group level, an additional command, `mlmatbysum`, exists to aide in computing the Hessian matrix. `mlmatbysum` is most easily understood in the context of a specific example, so we will wait until section 6.4.3 to discuss it.

▷ **Example**

Using the log likelihood from the probit model (6.1) and its first derivative (6.2), the second derivative is

$$\frac{d^2 \ln \ell_j}{d\theta_j^2} = \begin{cases} -R(\theta_j)\{R(\theta_j) + \theta_j\} & \text{if } y_j = 1 \\ -S(\theta_j)\{S(\theta_j) - \theta_j\} & \text{if } y_j = 0 \end{cases}$$

where

$$R(\theta_j) = \phi(\theta_j)/\Phi(\theta_j)$$
$$S(\theta_j) = \phi(\theta_j)/\Phi(-\theta_j)$$

Our likelihood function is written in terms of θ, so we can use `mlmatsum` to obtain the Hessian:

```
tempvar R S h
gen double 'R' = normalden('xb')/normal('xb')
gen double 'S' = normalden('xb')/normal(-'xb')
gen double 'h' = -'R'*('R'+'xb') if $ML_y1 == 1
replace     'h' = -'S'*('S'-'xb') if $ML_y1 == 0
mlmatsum 'lnf' 'H' = 'h'
```

Because the right-hand side of `mlmatsum` can be an expression, we could compress it into fewer lines by using the `cond()` function.

◁

6.3   Summary of methods d0, d1, and d2

6.3.1   Method d0

The method-`d0` evaluator calculates the overall log likelihood. The outline for a method-`d0` likelihood evaluator is

```
program myprog
        version 11
        args todo b lnf                          // 'todo' is subsequently ignored

        tempvar theta1 theta2 ...
        mleval 'theta1' = 'b'
        mleval 'theta2' = 'b', eq(2)             // if there is a θ₂
        ...
        // if you need to create any intermediate results
        tempvar tmp1 tmp2 ...
        gen double 'tmp1' = ...
        gen double 'tmp2' = ...
        ...
        mlsum 'lnf' = ...
end
```

You specify that you are using method d0 with program *myprog* on the `ml model` statement,

```
. ml model d0 myprog ...
```

which you issue after you have defined the likelihood-evaluation program.

Using method d0 is a matter of writing down the likelihood function and coding it. Unlike methods designed for models that satisfy the linear-form restrictions, however, you must sum the likelihood to produce the overall log-likelihood value.

Method-d0 evaluators will be called with 'todo' $= 0$ only, so method-d0 evaluators generally ignore 'todo' and also ignore the fact that they receive arguments 'g' and 'H'. That is, formally, a method-d0 evaluator could be written

```
program myprog
        version 11
        args todo b lnf g H
        ...
end
```

but it is perfectly acceptable to program

```
program myprog
        version 11
        args todo b lnf
        ...
end
```

Because the likelihood function is typically written as a function of the parameters $\boldsymbol{\theta}_i$ from the probability model, the first step is to obtain their values from the coefficients vector 'b':

```
program myprog
        version 11
        args todo b lnf

        tempvar theta1 theta2 ...
        mleval 'theta1' = 'b'
        mleval 'theta2' = 'b', eq(2)              // if there is a θ₂
        ...
end
```

The final step is to calculate the overall log-likelihood value and store it in the scalar 'lnf'. d-family evaluators are often used with models that produce group-level log-likelihood values, the leading examples being models for panel data. You can use mlsum in those cases to produce 'lnf':

```
program myprog
        version 11
        args todo b lnf

        tempvar theta1 theta2 ...
        mleval 'theta1' = 'b'
        mleval 'theta2' = 'b', eq(2)              // if there is a θ₂
        ...
        mlsum 'lnf' = ... if ...
end
```

where the first set of dots following mlsum is filled in with an expression producing the log-likelihood value for a group in terms of '*theta1*', '*theta2*', ..., and $ML_y1, $ML_y2, ... The dots following the keyword if restrict the sample so that mlsum uses just one observation from each group.

In more complicated cases, you will need to generate temporary variables containing partial calculations before invoking mlsum to sum the contributions:

```
program myprog
        version 11
        args todo b lnf

        tempvar theta1 theta2 ...
        mleval 'theta1' = 'b'
        mleval 'theta2' = 'b', eq(2)              // if there is a θ₂
        ...
        tempvar tmp1 tmp2
        quietly gen double 'tmp1' = ...
        quietly gen double 'tmp2' = ...
        mlsum 'lnf' = expression in terms of 'tmp1' and 'tmp2' if ...
end
```

The following points enumerate what ml supplies and expects of a method-d0 likelihood evaluator.

- You are passed the vector of coefficients. Use mleval to obtain the values of each of the linear predictors of each equation from it; see section 5.2.2.

- The names of the dependent variables (if any) are placed in global macros by ml; the first dependent variable is $ML_y1, the second is $ML_y2, ... Use $ML_y1, $ML_y2, ... just as you would any existing variable in the data.

- You must create any temporary variables for intermediate results as `doubles`. Type

    ```
    gen double name = ...
    ```

    Do not omit the word `double`.

- The estimation subsample is identified by `$ML_samp==1`. You may safely ignore this if you do not specify `ml model`'s `nopreserve` option. When your program is called, only relevant data will be in memory. If you do specify `nopreserve`, `mleval` and `mlsum` automatically restrict themselves to the estimation subsample; it is not necessary to include `if $ML_samp==1` with these commands. However, you will need to restrict other calculation commands to the `$ML_samp==1` subsample.

- `ml` stores the name of the weight variable in `$ML_w`, which contains 1 in every observation if no weights are specified. If you use `mlsum` to produce the log likelihood, you may ignore this because `mlsum` handles that itself.

- The resulting log-likelihood value is to be saved in scalar `'lnf'`. Use `mlsum` to produce this value if possible; see section 6.2.3.

- Though evaluators of methods d0, d1, and d2 can be used with `fweights`, `aweights`, and `iweights`, `pweights` cannot be specified. Moreover, you cannot specify `ml model`'s `vce(opg)`, `vce(robust)`, `vce(cluster` *clustvar*`)`, or `svy` options; nor can you use the BHHH optimization technique.

6.3.2 Method d1

The method-`d1` evaluator calculates the overall log likelihood and its first derivatives. The outline for the program is

```
program myprog
        version 11
        args todo b lnf g

        tempvar theta1 theta2 ...
        mleval 'theta1' = 'b'
        mleval 'theta2' = 'b', eq(2)              // if there is a θ₂
        ...

        // if you need to create any intermediate results:
        tempvar tmp1 tmp2 ...
        gen double 'tmp1' = ...
        gen double 'tmp2' = ...
        ...

        mlsum 'lnf' = ... if
        if ('todo'==0 | 'lnf'>=.) exit

        tempname d1 d2 ...
        mlvecsum 'lnf' 'd1' = formula for ∂ln ℓⱼ/∂θ₁ⱼ if ..., eq(1)
        mlvecsum 'lnf' 'd2' = formula for ∂ln ℓⱼ/∂θ₂ⱼ if ..., eq(2)
        ...
        matrix 'g' = ('d1','d2',...)
    end
```

You specify that you are using method **d1** with program *myprog* on the **ml model** statement,

```
. ml model d1 myprog ...
```

which you issue after you have defined the likelihood-evaluation program.

The method-**d1** evaluator simply picks up where the method-**d0** evaluator left off. Taking out the code in common, here is how they differ:

```
program myprog
        version 11
        args todo b lnf g                          // g is added
        ...
        if ('todo'==0 | 'lnf'>=.) exit             // from here down is new
        tempname d1 d2 ...
        mlvecsum 'lnf' 'd1' = formula for ∂ ln ℓⱼ/∂θ₁ⱼ if ..., eq(1)
        mlvecsum 'lnf' 'd2' = formula for ∂ ln ℓⱼ/∂θ₂ⱼ if ..., eq(2)
        ...
        matrix 'g' = ('d1','d2',...)
end
```

The gradient vector you define is

$$\mathbf{g} = \frac{\partial \ln L}{\partial \boldsymbol{\beta}} = \left(\frac{\partial \ln L}{\partial \boldsymbol{\beta}_1}, \frac{\partial \ln L}{\partial \boldsymbol{\beta}_2}, \ldots, \frac{\partial \ln L}{\partial \boldsymbol{\beta}_k} \right)$$

ml provides the coefficient vector as a $1 \times k$ row vector and expects the gradient vector to be returned as a $1 \times k$ row vector.

Many likelihood functions produce group-level log-likelihood values, and in those cases, you can use **mlvecsum** to sum over the groups when computing the derivatives.

The following points enumerate what **ml** supplies and expects of a method-**d1** likelihood evaluator.

- You are passed the vector of coefficients. Use **mleval** to obtain the values of each of the linear predictors of each equation from it; see section 5.2.2.

- The names of the dependent variables (if any) are placed in global macros by **ml**; the first dependent variable is \$ML_y1, the second is \$ML_y2, ... Use \$ML_y1, \$ML_y2, ... just as you would any existing variable in the data.

- You must create any temporary variables for intermediate results as **doubles**. Type

    ```
    gen double name = ...
    ```

    Do not omit the word **double**.

- The estimation subsample is identified by \$ML_samp==1. You may safely ignore this if you do not specify **ml model**'s **nopreserve** option. When your program is called, only relevant data will be in memory. If you do specify **nopreserve**, **mleval** and **mlsum** automatically restrict themselves to the estimation subsample;

it is not necessary to include `if $ML_samp==1` with these commands. However, you will need to restrict other calculation commands to the `$ML_samp==1` subsample.

- `ml` stores the name of the weight variable in `$ML_w`, which contains 1 in every observation if no weights are specified. If you use `mlsum` to produce the log likelihood, you may ignore this because `mlsum` handles that itself.

- The resulting log-likelihood value is to be saved in scalar '`lnf`'. Use `mlsum` to produce this value if possible; see section 6.2.3.

- The resulting gradient vector is to be saved in vector '`g`'. You can use `mlvecsum` to produce it when the likelihood function is produced at the group level. See sections 6.4 and 6.5 for examples.

6.3.3 Method d2

The method-**d2** evaluator calculates the overall log likelihood along with its first and second derivatives. The outline for the program is

```
program myprog
        version 11
        args todo b lnf g H

        tempvar theta1 theta2 ...
        mleval 'theta1' = 'b', eq(1)
        mleval 'theta2' = 'b', eq(2) // if there is a θ_2
        ...

        // if you need to create any intermediate results:
        tempvar tmp1 tmp2 ...
        gen double 'tmp1' = ...
        ...

        mlsum 'lnf' = ...
        if ('todo'==0 | 'lnf'>=.) exit

        tempname d1 d2 ...
        mlvecsum 'lnf' 'd1' = formula for ∂ln ℓ_j/∂θ_1j if ..., eq(1)
        mlvecsum 'lnf' 'd2' = formula for ∂ln ℓ_j/∂θ_2j if ..., eq(2)
        ...
        matrix 'g' = ('d1','d2',...)
        if ('todo'==1 | 'lnf'>=.) exit

        tempname d11 d12 d22 ...
        mlmatsum 'lnf' 'd11' = formula for ∂²ln ℓ_j/∂θ²_1j if ..., eq(1)
        mlmatsum 'lnf' 'd12' = formula for ∂²ln ℓ_j/∂θ_1j∂θ_2j if ..., eq(1,2)
        mlmatsum 'lnf' 'd22' = formula for ∂²ln ℓ_j/∂θ²_2j if ..., eq(2)
        ...
        matrix 'H' = ('d11','d12',... \ 'd12'','d22',...)
end
```

You specify that you are using method **d2** on the `ml model` statement,

```
. ml model d2 myprog ...
```

which you issue after you have defined the likelihood-evaluation program.

The method-**d2** evaluator continues from where the method-**d1** evaluator left off. Taking out the common code, here is how the evaluators differ:

```
program myprog
        version 11
        args todo b lnf g H                      // H is added
        ...
        if ('todo'==1 | 'lnf'==.) exit           // from here down is new
        tempname d11 d12 d22 ...
        mlmatsum 'lnf' 'd11' = formula for ∂²ln ℓⱼ/∂θ²₁ⱼ if ..., eq(1)
        mlmatsum 'lnf' 'd12' = formula for ∂²ln ℓⱼ/∂θ₁ⱼ∂θ₂ⱼ if ..., eq(1,2)
        mlmatsum 'lnf' 'd22' = formula for ∂²ln ℓⱼ/∂θ²₂ⱼ if ..., eq(2)
        ...
        matrix 'negH' = ('d11','d12',... \ 'd12'','d22',...)
end
```

The Hessian matrix you are to define is

$$\mathbf{H} = \frac{\partial^2 \ln L}{\partial \boldsymbol{\beta} \partial \boldsymbol{\beta}'} = \begin{pmatrix} \frac{\partial^2 \ln L}{\partial \beta_1^2} & \frac{\partial^2 \ln L}{\partial \beta_1 \partial \beta_2} & \cdots & \frac{\partial^2 \ln L}{\partial \beta_1 \partial \beta_k} \\ \frac{\partial^2 \ln L}{\partial \beta_2 \partial \beta_1} & \frac{\partial^2 \ln L}{\partial \beta_2^2} & \cdots & \frac{\partial^2 \ln L}{\partial \beta_2 \partial \beta_k} \\ \vdots & \vdots & \ddots & \vdots \\ \frac{\partial^2 \ln L}{\partial \beta_k \partial \beta_1} & \frac{\partial^2 \ln L}{\partial \beta_k \partial \beta_2} & \cdots & \frac{\partial^2 \ln L}{\partial \beta_k^2} \end{pmatrix}$$

ml provides the coefficient vector as a $1 \times k$ row vector and expects the negative Hessian matrix to be returned as a $k \times k$ matrix.

The following points enumerate what **ml** supplies and expects of a method-**d2** likelihood evaluator.

- You are passed the vector of coefficients. Use **mleval** to obtain the values of each of the linear predictors of each equation from it; see section 5.2.2.

- The names of the dependent variables (if any) are placed in global macros by **ml**; the first dependent variable is \$ML_y1, the second is \$ML_y2, ... Use \$ML_y1, \$ML_y2, ... just as you would any existing variable in the data.

- You must create any temporary variables for intermediate results as **doubles**. Type

  ```
  gen double name = ...
  ```

  Do not omit the word **double**.

- The estimation subsample is identified by \$ML_samp==1. You may safely ignore this if you do not specify **ml model**'s **nopreserve** option. When your program is called, only relevant data will be in memory. If you do specify **nopreserve**, **mleval** and **mlsum** automatically restrict themselves to the estimation subsample; it is not necessary to include **if** \$ML_samp==1 with these commands. However, you will need to restrict other calculation commands to the \$ML_samp==1 subsample.

- `ml` stores the name of the weight variable in `$ML_w`, which contains 1 in every observation if no weights are specified. If you use `mlsum` to produce the log likelihood, you may ignore this because `mlsum` handles that itself.

- The resulting log-likelihood value is to be saved in scalar 'lnf'. Use `mlsum` to produce this value if possible; see section 6.2.3.

- The resulting gradient vector is to be saved in vector 'g'. You can use `mlvecsum` to produce it when the likelihood function is produced at the group level. See sections 6.4 and 6.5 for examples.

- The Hessian matrix is to be saved in matrix 'H'. The `mlmatsum` command can be used if the log likelihood meets the linear-form restrictions, and the `mlmatbysum` can be used when the log-likelihood function is defined for groups of observations, as is the case with panel-data models.

6.4 Panel-data likelihoods

In section 6.2, we discussed methods `d0`, `d1`, and `d2` in the context of linear-form likelihood functions for simplicity. However, as we have said before, if you have a linear-form likelihood function, you should instead use methods `lf`, `lf0`, `lf1`, or `lf2` because they are easier to use.

Not all likelihood functions meet the linear-form restrictions, and it is in those cases that methods `d0`, `d1`, and `d2` are required. In this section, we will consider a special case when the linear-form restrictions are violated, that of panel-data estimators. In section 6.5, we will consider other likelihood functions.

In panel data, you have repeated observations for each of a collection of individual persons, firms, etc. Each observation in the data represents one of the observations for one of those individuals. For instance, if we had 5 observations for each of 20 individuals, our dataset would contain $5 \times 20 = 100$ observations in total. This is illustrated in table 6.1.

Table 6.1. Layout for panel data. Variable i identifies the panels, and t identifies the time or instant an observation was made or a measurement was taken.

Obs. no.	i	t	y_{it}	$\mathbf{x}_{it}$
1.	1	1	y_{11}	$\mathbf{x}_{11}$
2.	1	2	y_{12}	$\mathbf{x}_{12}$
3.	1	3	y_{13}	$\mathbf{x}_{13}$
4.	1	4	y_{14}	$\mathbf{x}_{14}$
5.	2	1	y_{21}	$\mathbf{x}_{21}$
6.	2	2	y_{22}	$\mathbf{x}_{22}$
7.	2	3	y_{23}	$\mathbf{x}_{23}$
8.	2	4	y_{24}	$\mathbf{x}_{24}$
9.	3	1	y_{31}	$\mathbf{x}_{31}$
10.	3	2	y_{32}	$\mathbf{x}_{32}$
11.	3	3	y_{33}	$\mathbf{x}_{33}$
12.	3	4	y_{34}	$\mathbf{x}_{34}$

For panel-data estimators, the observations are not independent, but groups of the observations are independent. Panel datasets are usually recorded in the long form, as are the data given in table 6.1. In this dataset, we have independent information on three separate groups i (perhaps i indexes persons), and we have four observations on each of the i's.

Likelihood functions for this type of data will violate the linear-form restrictions because the observations are not independent from others in the same group. We still have independence between groups of observations, however, and in writing our evaluator, it is merely a matter of adjusting our program to exploit this more complicated structure.

To fix ideas, let's consider the particular case of random-effects linear regression.

$$y_{it} = \mathbf{x}_{it}\boldsymbol{\beta} + u_i + e_{it}, \quad u_i \sim N(0, \sigma_u^2), \quad e_{it} \sim N(0, \sigma_e^2)$$

In the above, there are three parameters we need to estimate: $\boldsymbol{\beta}$ (presumably a vector) and σ_u^2 and σ_e^2 (presumably scalars). The likelihood function for random-effects regression is

$$\ln L = \sum_{i=1}^{N} \ln L_i$$

where i indexes group, and the individual log-likelihood values for each group are themselves based on the calculation of a cluster of observations within group:

$$
\ln L_i = \ln \int_{-\infty}^{\infty} f(u_i) \prod_{t=1}^{T_i} f(y_{it}|u_i) \, du_i
$$

$$
= \ln \int_{-\infty}^{\infty} \frac{1}{\sigma_u \sqrt{2\pi}} \exp\left(-\frac{u_i^2}{2\sigma_u^2}\right) \prod_{t=1}^{T_i} \left[\frac{1}{\sigma_e \sqrt{2\pi}} \exp\left\{-\frac{(y_{it} - u_i - \mathbf{x}_{it}\boldsymbol{\beta})^2}{2\sigma_e^2}\right\} \right] du_i
$$

$$
= -\frac{1}{2} \left\{ \frac{\sum_t z_{it}^2 - a_i (\sum_t z_{it})^2}{\sigma_e^2} + \ln(T_i \sigma_u^2/\sigma_e^2 + 1) + T_i \ln(2\pi\sigma_e^2) \right\} \tag{6.3}
$$

where

$$
T_i = \text{number of observations in the } i\text{th group}
$$
$$
z_{it} = y_{it} - \mathbf{x}_{it}\boldsymbol{\beta} \tag{6.4}
$$
$$
a_i = \sigma_u^2/(T_i \sigma_u^2 + \sigma_e^2)
$$

6.4.1   Calculating lnf

The trick to calculating the log-likelihood function in panel datasets is to create a new variable containing the likelihood for each group in the last observation of the group, and then sum only those last observations.

Obs. no.	i	t	y_{it}	$\mathbf{x}_{it}$	New variable	
1.	1	1	y_{11}	$\mathbf{x}_{11}$	.	(missing)
2.	1	2	y_{12}	$\mathbf{x}_{12}$	.	
3.	1	3	y_{13}	$\mathbf{x}_{13}$	.	
4.	1	4	y_{14}	$\mathbf{x}_{14}$	$\ln L_1$	
5.	2	1	y_{21}	$\mathbf{x}_{21}$	.	
6.	2	2	y_{22}	$\mathbf{x}_{22}$	.	
7.	2	3	y_{23}	$\mathbf{x}_{23}$	.	
8.	2	4	y_{24}	$\mathbf{x}_{24}$	$\ln L_2$	
9.	3	1	y_{31}	$\mathbf{x}_{31}$	.	
10.	3	2	y_{32}	$\mathbf{x}_{32}$	.	
11.	3	3	y_{33}	$\mathbf{x}_{33}$	.	
12.	3	4	y_{34}	$\mathbf{x}_{34}$	$\ln L_3$	
					$\ln L$	(sum of nonmissing)

Making the calculation in this way is easy to do. Calculating $\ln L_1$, $\ln L_2$, etc., is easy because most of the terms will be based on sums within i, so we can code

```
by i: generate double newvar = sum(...)
```

although sometimes multiple sums may appear, such as

```
by i: generate double newvar = sum(...)*sum(...)/sum(...)
```

To compute the overall value of the log likelihood, we will still use `mlsum`, but we must remember to sum only the last observation in each group because those are the observations that contain the likelihood contribution for their respective group:

```
tempvar last
by i:  gen byte `last' = (_n==_N)
mlsum `lnf' = ... if `last'
```

In the above sketch of code, we did not bother to ensure that missing values were placed in the intermediate observations, but that does not matter because we explicitly excluded them by coding if `last' at the end of our `mlsum` command.

For random-effects linear regression, the likelihood function is (6.3) with the intermediate calculations as defined in (6.4), and our method-d0 evaluator is

```
                                            begin myrereg_d0.ado
program myrereg_d0
        version 11
        args todo b lnf
        tempvar xb z T S_z2 Sz_2 a last
        tempname s_u s_e
        mleval `xb'  = `b', eq(1)
        mleval `s_u' = `b', eq(2) scalar
        mleval `s_e' = `b', eq(3) scalar
        scalar `s_u' = exp(`s_u')
        scalar `s_e' = exp(`s_e')
        // MY_panel contains the panel ID
        local by $MY_panel
        sort `by'
        local y $ML_y1
        quietly {
                gen double `z' = `y' - `xb'
                by `by': gen `last' = _n==_N
                by `by': gen `T' = _N
                by `by': gen double `S_z2' = sum(`z'^2)
                by `by': gen double `Sz_2' = sum(`z')^2
                gen double `a' = `s_u'^2 / (`T'*`s_u'^2 + `s_e'^2)
                mlsum `lnf'= -.5 *                              ///
                        (                                       ///
                                (`S_z2'-`a'*`Sz_2')/`s_e'^2 +   ///
                                ln(`T'*`s_u'^2/`s_e'^2 + 1) +   ///
                                `T'*ln(2*c(pi)*`s_e'^2)         ///
                        )                                       ///
                        if `last' == 1
        }
end
                                            end myrereg_d0.ado
```

There is a lot of code here, so let's go over it in blocks. We start by retrieving the values from each equation:

$$\theta_{1it} = \mathbf{x}_{it}\boldsymbol{\beta}$$

$$\theta_{2it} = \ln\sigma_u$$

$$\theta_{3it} = \ln\sigma_e$$

We are modeling the variances in the log space, so we must exponentiate the values from the second and third equations:

```
version 11
args todo b lnf
tempvar xb z T S_z2 Sz_2 a last
tempname s_u s_e
mleval 'xb'  = 'b', eq(1)
mleval 's_u' = 'b', eq(2) scalar
mleval 's_e' = 'b', eq(3) scalar
scalar 's_u' = exp('s_u')
scalar 's_u' = exp('s_e')
```

Next we set up some local macros for our own convenience and sort the data by panel. (The technical notes from the Cox model, pages 149–150, regarding the **sort** command apply equally here.) `myrereg_d0` assumes that $MY_panel contains the panel ID variable, so we must remember to set this global macro before running `ml maximize`.

```
// MY_panel contains the panel ID
local by $MY_panel
sort 'by'
local y $ML_y1
```

Inside the **quietly** block, we perform the calculations. The '**last**' variable indicates the last observation of each panel within the estimation subsample. We use this variable to indicate which observations contain contributions to the log likelihood. The '**S_z2**' variable contains the sum of the squared deviations, and '**Sz_2**' contains the square of the summed deviations.

```
quietly {
        gen double 'z' = 'y' - 'xb'
        by 'by': gen 'last' = _n==_N
        by 'by': gen 'T' = _N
        by 'by': gen double 'S_z2' = sum('z'^2)
        by 'by': gen double 'Sz_2' = sum('z')^2
        gen double 'a' = 's_u'^2 / ('T'*'s_u'^2 + 's_e'^2)
        mlsum 'lnf'= -.5 *                               ///
                (                                        ///
                        ('S_z2'-'a'*'Sz_2')/'s_e'^2 +    ///
                        ln('T'*'s_u'^2/'s_e'^2 + 1) +    ///
                        'T'*ln(2*c(pi)*'s_e'^2)          ///
                )                                        ///
                if 'last' == 1
}
```

❏ Technical note

As written, `myrereg_d0` does not accept weights—`$ML_w` does not appear in any of the calculations above, and the `mlsum` command will not work if you specify weights and use an `if` condition to limit its calculations. The solution is to set the log likelihood equal to zero for observations other than the last one in each panel. Then we can omit the `if` condition with `mlsum` and still have it apply weights appropriately. We modify the last few lines of `myrereg_d0` to read

```
        gen double 'a' = 's_u'^2 / ('T'*'s_u'^2 + 's_e'^2)
        tempvar mylnf                              //  <- New
        gen double 'mylnf' = 0                     //  <- New
        replace 'mylnf' = -.5 *                    /// Modified
                (                                  ///
                        ('S_z2'-'a'*'Sz_2')/'s_e'^2 +   ///
                        ln('T'*'s_u'^2/'s_e'^2 + 1) +   ///
                        'T'*ln(2*c(pi)*'s_e'^2)         ///
                )                                  ///
                if 'last' == 1
        mlsum 'lnf' = 'mylnf'                       //  <- New
end
```

There is one more issue to consider when working with weights and panel data. Panel-data estimators typically assume that weights are constant within each panel. You are responsible for verifying that this is true of your dataset; `ml` cannot tell when it is optimizing a panel likelihood. Our method of handling weights with panel data only considers the weight of the last observation in each panel.

❏

(Continued on next page)

▷ **Example**

To verify that `myrereg_d0` works, we use data from Greene (2008):

```
. use http://www.stata-press.com/data/ml4/tablef7-1
. global MY_panel i        // the panel variable
. ml model d0 myrereg_d0 (xb: lnC = lnQ lnPF lf) /lns_u /lns_e
. ml max, nolog
```

```
                                            Number of obs   =          90
                                            Wald chi2(3)    =    11468.30
Log likelihood =  114.72904                 Prob > chi2     =      0.0000
```

lnC	Coef.	Std. Err.	z	P>|z|	[95% Conf. Interval]	
xb						
lnQ	.90531	.0253759	35.68	0.000	.8555742	.9550459
lnPF	.4233757	.013888	30.49	0.000	.3961557	.4505957
lf	-1.064456	.1962307	-5.42	0.000	-1.449061	-.679851
_cons	9.618648	.2066218	46.55	0.000	9.213677	10.02362
lns_u						
_cons	-2.170817	.3026644	-7.17	0.000	-2.764028	-1.577605
lns_e						
_cons	-2.828404	.077319	-36.58	0.000	-2.979947	-2.676862

◁

6.4.2   Calculating g

If we wish to convert our `d0` evaluator into a `d1` evaluator and thus gain speed and precision, we must program the gradient vector calculation. The code to do that is nearly identical to the code for the linear-form case. You use `mlvecsum` to create the sums.

You will discover that gradient calculations in panel-data models come in two flavors: one that varies observation by observation within each group and another that is calculated over the entire group, just as the likelihood value itself.

For the first, you will program the calculation just as you would program a linear-form likelihood:

Obs. no.	i	t	y_{it}	$\mathbf{x}_{it}$	New variable
1.	1	1	y_{11}	$\mathbf{x}_{11}$	g_{11}
2.	1	2	y_{12}	$\mathbf{x}_{12}$	g_{12}
3.	1	3	y_{13}	$\mathbf{x}_{13}$	g_{13}
4.	1	4	y_{14}	$\mathbf{x}_{14}$	g_{14}
5.	2	1	y_{21}	$\mathbf{x}_{21}$	g_{21}
6.	2	2	y_{22}	$\mathbf{x}_{22}$	g_{22}
7.	2	3	y_{23}	$\mathbf{x}_{23}$	g_{23}
8.	2	4	y_{24}	$\mathbf{x}_{24}$	g_{24}
9.	3	1	y_{31}	$\mathbf{x}_{31}$	g_{31}
10.	3	2	y_{32}	$\mathbf{x}_{32}$	g_{32}
11.	3	3	y_{33}	$\mathbf{x}_{33}$	g_{33}
12.	3	4	y_{34}	$\mathbf{x}_{34}$	g_{34}

You obtain the sum using `mlvecsum`:

```
. mlvecsum 'lnf' 'g' = expression based on new variable
```

For any term that applies only to the entire group, you will use the same scheme that you used for calculating the likelihood value:

Obs. no.	i	t	y_{it}	$\mathbf{x}_{it}$	New variable	
1.	1	1	y_{11}	$\mathbf{x}_{11}$	.	(missing)
2.	1	2	y_{12}	$\mathbf{x}_{12}$	.	
3.	1	3	y_{13}	$\mathbf{x}_{13}$	.	
4.	1	4	y_{14}	$\mathbf{x}_{14}$	g_1	
5.	2	1	y_{21}	$\mathbf{x}_{21}$	.	
6.	2	2	y_{22}	$\mathbf{x}_{22}$	.	
7.	2	3	y_{23}	$\mathbf{x}_{23}$	.	
8.	2	4	y_{24}	$\mathbf{x}_{24}$	g_2	
9.	3	1	y_{31}	$\mathbf{x}_{31}$	.	
10.	3	2	y_{32}	$\mathbf{x}_{32}$	.	
11.	3	3	y_{33}	$\mathbf{x}_{33}$	.	
12.	3	4	y_{34}	$\mathbf{x}_{34}$	g_3	

You obtain the sum using `mlvecsum`, but this time you must be careful to sum only the last observation for each group:

```
. mlvecsum 'lnf' 'g' = expression based on new variable if 'last'
```

For instance, the derivatives of our likelihood function for random-effects regression are

$$\frac{\partial \ln L_i}{\partial \mathbf{x}_{it}\boldsymbol{\beta}} = \frac{1}{\sigma_e^2}\left(z_{it} - a_i \sum_{s=1}^{T_i} z_{is}\right) \tag{6.5}$$

$$\frac{\partial \ln L_i}{\partial \ln \sigma_u} = \frac{a_i^2}{\sigma_u^2}\left(\sum_{t=1}^{T_i} z_{it}\right)^2 - T_i a_i \tag{6.6}$$

$$\frac{\partial \ln L_i}{\partial \ln \sigma_e} = \frac{1}{\sigma_e^2}\left\{\sum_{t=1}^{T_i} z_{it}^2 - a_i\left(\sum_{t=1}^{T_i} z_{it}\right)^2\right\} - (T_i - 1) - \frac{a_i^2}{\sigma_u^2}\left(\sum_{t=1}^{T_i} z_{it}\right)^2 - \frac{\sigma_e^2}{\sigma_u^2}a_i \tag{6.7}$$

Consider writing the code to calculate $\partial \ln L / \partial \boldsymbol{\beta}$:

```
. by 'i': gen double 'S_z' = sum('z')
. by 'i': replace 'S_z' = 'S_z'[_N]
. gen double 'dxb' = ('z' - 'a'*'S_z') / 's_e'^2
. mlvecsum 'lnf' 'g1' = 'dxb'
```

There are other ways we could have coded this; perhaps you would prefer

```
. by 'i': gen double 'S_z' = sum('z')
. by 'i': gen double 'dxb' = ('z' - 'a'*'S_z'[_N]) / 's_e'^2
. mlvecsum 'lnf' 'g1' = 'dxb'
```

Either way, think carefully about what is in `dxb`, and you will realize that each observation in `dxb` enters the sum. We can use `dxb` with `mlvecsum` to compute $\partial \ln L / \partial \boldsymbol{\beta}$ because

$$\frac{\partial \ln L_i}{\partial \boldsymbol{\beta}} = \sum_{t=1}^{T_i} \frac{\partial \ln L_i}{\partial \mathbf{x}_{it}\boldsymbol{\beta}}\mathbf{x}_{it}$$

Now consider writing the code to calculate $\partial \ln L / \partial \ln \sigma_e$:

```
. gen double 'de' = ('S_z2'-'a'*'Sz_2')/'s_e'^2
>         - 'T' + 1
>         - 'a'^2*'Sz_2'/'s_u'^2 -
>         - 'a'*'s_e'^2/'s_u'^2
>         if 'last'==1
. mlvecsum 'lnf' 'g3' = 'de' if 'last'==1 , eq(3)
```

All the ingredients to compute $\partial \ln L_i / \partial \ln \sigma_e$ for each group are present in the last observation of the group; thus we use `if 'last'==1`.

The lines that differ between `myrereg_d0` and `myrereg_d1` (our method-d1 evaluator) are

———————————————————————————— fragment `myrereg_d1.ado` ————————

```
program myrereg_d1
        version 11
        args todo b lnf g
        ...
        quietly {
                ...
                by 'by': gen double 'S_z2' = sum('z'^2)
                by 'by': replace    'S_z2' = 'S_z2'[_N]          // new line
                by 'by': gen double 'Sz_2' = sum('z')^2
                by 'by': replace    'Sz_2' = 'Sz_2'[_N]          // new line
                gen double 'a' = ...
                mlsum 'lnf'= -...
                // the following lines are new
                if ('todo'==0 | 'lnf'==.) exit

                // compute the gradient
                tempvar S_z
                tempname dxb du de
                by 'by': gen double 'S_z' = sum('z')
                by 'by': replace 'S_z' = 'S_z'[_N]
                mlvecsum 'lnf' 'dxb' = ('z'-'a'*'S_z')/'s_e'^2   , eq(1)
                mlvecsum 'lnf' 'du' = 'a'^2*'Sz_2'/'s_u'^2       ///
                        -'T'*'a'                                 ///
                        if 'last'==1                             , eq(2)
                mlvecsum 'lnf' 'de' = 'S_z2'/'s_e'^2 -           ///
                        'a'*'Sz_2'/'s_e'^2 -                     ///
                        'a'^2*'Sz_2'/'s_u'^2 -                   ///
                        'T'+1-'a'*'s_e'^2/'s_u'^2                ///
                        if 'last'==1                             , eq(3)
                mat 'g' = ('dxb','du','de')
        }
end
```

———————————————————————————— end `myrereg_d1.ado` ————————

(Continued on next page)

▷ **Example**

Below we verify that the results from `myrereg_d1` agree with those from `myrereg_d0`.

```
. use http://www.stata-press.com/data/ml4/tablef7-1, clear
. global MY_panel i       // the panel variable
. ml model d1 myrereg_d1 (xb: lnC = lnQ lnPF lf) /lns_u /lns_e
. ml max, nolog
```

				Number of obs	=	90
				Wald chi2(3)	=	11468.30
Log likelihood =	114.72904			Prob > chi2	=	0.0000

lnC	Coef.	Std. Err.	z	P>|z|	[95% Conf. Interval]	
xb						
lnQ	.90531	.0253759	35.68	0.000	.8555742	.9550459
lnPF	.4233757	.013888	30.48	0.000	.3961557	.4505957
lf	-1.064456	.1962308	-5.42	0.000	-1.449061	-.6798508
_cons	9.618648	.2066218	46.55	0.000	9.213677	10.02362
lns_u						
_cons	-2.170818	.3026641	-7.17	0.000	-2.764028	-1.577607
lns_e						
_cons	-2.828404	.077319	-36.58	0.000	-2.979947	-2.676862

◁

6.4.3 Calculating H

If we wish to convert our d1 evaluator into a d2 evaluator, we must program the Hessian matrix calculation. As with the gradient, the Hessian calculations in panel-data models come in two flavors. This is further complicated by the fact that `mlmatsum` is not sufficient to perform all the required calculations; `mlmatbysum` completes the tool set for panel-data models.

❑ **Using mlmatbysum to help define H**

For the following discussion, suppose that we have E equations in our model and that the observations are grouped. Let i be the group index and t the observation index within a group. $\mathbf{x}_{eit}$ denotes a vector of values of independent variables for equation e from observation t of group i.

Many models that do not meet the linear-form restrictions have Hessian matrices such that the within-equation submatrices have the form

$$\frac{\partial L}{\partial \boldsymbol{\beta}_1' \partial \boldsymbol{\beta}_1} = \sum_{i=1}^{N} \left\{ \sum_{t=1}^{T_i} d_{it} \mathbf{x}_{1it}' \mathbf{x}_{1it} + \left(\sum_{t=1}^{T_i} a_{it} \right) \left(\sum_{t=1}^{T_i} b_{it} \mathbf{x}_{1it}' \right) \left(\sum_{t=1}^{T_i} b_{it} \mathbf{x}_{1it} \right) \right\}$$

and the between-equation submatrices have the form

$$\frac{\partial L}{\partial \boldsymbol{\beta}_1' \partial \boldsymbol{\beta}_2} = \sum_{i=1}^{N} \left\{ \sum_{t=1}^{T_i} d_{it} \mathbf{x}_{1it}' \mathbf{x}_{2it} + \left(\sum_{t=1}^{T_i} a_{it} \right) \left(\sum_{t=1}^{T_i} b_{it} \mathbf{x}_{1it}' \right) \left(\sum_{t=1}^{T_i} c_{it} \mathbf{x}_{2it} \right) \right\}$$

where a_{it}, b_{it}, c_{it}, and d_{it} are not necessarily the same between derivative formulas. In this case, the overall Hessian matrix may be put together using both `mlmatsum` and `mlmatbysum`. The observation-level outer product matrices

$$\sum_{i=1}^{N} \sum_{t=1}^{T_i} d_{it} \mathbf{x}_{1it}' \mathbf{x}_{1it} \quad \text{and} \quad \sum_{i=1}^{N} \sum_{t=1}^{T_i} d_{it} \mathbf{x}_{1it}' \mathbf{x}_{2it}$$

can be calculated using `mlmatsum`. The group-level outer product matrices

$$\sum_{i=1}^{N} \left(\sum_{t=1}^{T_i} a_{it} \right) \left(\sum_{t=1}^{T_i} b_{it} \mathbf{x}_{1it}' \right) \left(\sum_{t=1}^{T_i} b_{it} \mathbf{x}_{1it} \right)$$

and

$$\sum_{i=1}^{N} \left(\sum_{t=1}^{T_i} a_{it} \right) \left(\sum_{t=1}^{T_i} b_{it} \mathbf{x}_{1it}' \right) \left(\sum_{t=1}^{T_i} c_{it} \mathbf{x}_{2it} \right)$$

can be calculated using `mlmatbysum`.

The syntax for `mlmatbysum` is

`mlmatbysum` *scalarname*$_{\text{lnf}}$ *matrixname varname$_a$ varname$_b$* [*varname$_c$*] [*if*],
   by(*varname*) [eq($\#$[,$\#$])]

Thus obtaining $\partial^2 \ln L / \partial \boldsymbol{\beta} \partial \boldsymbol{\beta}'$ is simply a matter of issuing `mlmatsum` and `mlmatbysum` for each equation pair and joining the results into a single matrix.

In a single-equation system,

```
tempname xx sx
mlmatsum 'lnf' 'xx' = 'd'
mlmatbysum 'lnf' 'sx' 'a' 'b' , by(group_variable)
matrix 'H' = ('xx'+'sx')
```

In a two-equation system,

```
tempname xx11 xx12 xx22 sx11 sx12 sx22 d11 d12 d22
mlmatsum 'lnf' 'xx11' = 'd11', eq(1)
mlmatsum 'lnf' 'xx12' = 'd12', eq(1,2)
mlmatsum 'lnf' 'xx22' = 'd22', eq(2)
mlmatbysum 'lnf' 'sx11' 'a11' 'b11'        , by(group_variable) eq(1)
mlmatbysum 'lnf' 'sx12' 'a12' 'b12' 'c12'  , by(group_variable) eq(1,2)
mlmatbysum 'lnf' 'sx22' 'a22' 'b22'        , by(group_variable) eq(2)
matrix 'd11' = 'xx11'+'sx11'
matrix 'd12' = 'xx12'+'sx12'
matrix 'd22' = 'xx22'+'sx22'
matrix 'H' = ('d11','d12' \ 'd12'','d22')
```

`mlmatbysum` requires '`lnf`' as an argument.

```
   . mlmatbysum 'lnf' 'whatever' var_a var_b , eq(1)
   . mlmatbysum 'lnf' 'whatever' var_a var_b var_c , eq(1,2)
```

This is so `mlmatbysum` can reset '`lnf`' to contain missing if var_b or var_c contains a missing value for some relevant observation. Method-**d2** evaluators are to set '`lnf`' to missing when called with impossible values of the coefficient vector. It is possible that '`lnf`' can be calculated but the Hessian cannot. `mlmatbysum` appropriately handles such cases.

❏

Returning to our random-effects regression model, the first and second derivatives of $\ln L_i$ with respect to $\boldsymbol{\beta}$ are

$$
\frac{\partial \ln L_i}{\partial \boldsymbol{\beta}} = \frac{1}{\sigma_e^2} \sum_{t=1}^{T_i} \left(z_{it} - a_i \sum_{s=1}^{T_i} z_{is} \right) \mathbf{x}_{it}
$$

$$
\frac{\partial^2 \ln L_i}{\partial \boldsymbol{\beta} \partial \boldsymbol{\beta}'} = -\frac{1}{\sigma_e^2} \left\{ \sum_{t=1}^{T_i} \mathbf{x}_{it}' \mathbf{x}_{it} - a_i \left(\sum_{s=1}^{T_i} \mathbf{x}_{is}' \right) \left(\sum_{t=1}^{T_i} \mathbf{x}_{it} \right) \right\} \tag{6.8}
$$

We cannot use `mlmatsum` to calculate $\partial^2 \ln L_i / \partial \boldsymbol{\beta} \partial \boldsymbol{\beta}'$ directly because

$$
\frac{\partial^2 \ln L_i}{\partial \boldsymbol{\beta} \partial \boldsymbol{\beta}'} \neq \sum_{t=1}^{T_i} \frac{\partial^2 \ln L_i}{\partial^2 \mathbf{x}_{it} \boldsymbol{\beta}} \mathbf{x}_{it}' \mathbf{x}_{it}
$$

but we can use it to calculate

$$
\sum_{t=1}^{T_i} \mathbf{x}_{it}' \mathbf{x}_{it}
$$

and we can use `mlmatbysum` to calculate

$$
a_i \left(\sum_{s=1}^{T_i} \mathbf{x}_{is}' \right) \left(\sum_{t=1}^{T_i} \mathbf{x}_{it} \right)
$$

The second derivatives with respect to $\ln \sigma_u$ and $\ln \sigma_e$ are

$$\frac{\partial^2 \ln L_i}{\partial^2 \ln \sigma_u} = -\left\{ \frac{2a_i^2}{\sigma_u^2} \left(\sum_{t=1}^{T_i} z_{it} \right)^2 - \frac{4a_i^3 \sigma_e^2}{\sigma_u^4} \left(\sum_{t=1}^{T_i} z_{it} \right)^2 + \frac{2T_i a_i^2 \sigma_e^2}{\sigma_u^2} \right\}$$

$$\frac{\partial^2 \ln L_i}{\partial^2 \ln \sigma_e} = -\left[\frac{2}{\sigma_e^2} \left\{ \sum_{t=1}^{T_i} z_{it}^2 - a_i \left(\sum_{t=1}^{T_i} z_{it} \right)^2 \right\} \right.$$

$$\left. - \left\{ \frac{2a_i^2}{\sigma_u^2} + \frac{4a_i^3 \sigma_e^2}{\sigma_u^4} \right\} \left(\sum_{t=1}^{T_i} z_{it} \right)^2 + \frac{2a_i \sigma_e^2}{\sigma_u^2} - \frac{2a_i^2 \sigma_e^4}{\sigma_u^4} \right]$$

$$\frac{\partial^2 \ln L_i}{\partial \ln \sigma_e \partial \ln \sigma_u} = -\left\{ \frac{4a_i^3 \sigma_e^2}{\sigma_u^4} \left(\sum_{t=1}^{T_i} z_{it} \right)^2 - \frac{2T_i a_i^2 \sigma_e^2}{\sigma_u^2} \right\}$$

As with the gradient calculations in `myrereg_d1`, the component of the Hessian derived based on these two parameters can be calculated using the `mlmatsum` command (with `if 'last'==1` because these calculations are constant within group).

The second derivatives with respect to $\mathbf{x}_{it}\boldsymbol{\beta}$ and $\ln \sigma_u$ (and $\ln \sigma_e$) are

$$\frac{\partial^2 \ln L_i}{\partial \mathbf{x}_{it}\boldsymbol{\beta}\partial \ln \sigma_u} = -\frac{2a_i^2}{\sigma_u^2} \sum_{t=1}^{T_i} z_{it}$$

$$\frac{\partial^2 \ln L_i}{\partial \mathbf{x}_{it}\boldsymbol{\beta}\partial \ln \sigma_e} = -\left[\frac{2}{\sigma_e^2} \left\{ z_{it} - a_i \sum_{t=1}^{T_i} z_{it} \right\} - \frac{2a_i^2}{\sigma_u^2} \sum_{t=1}^{T_i} z_{it} \right]$$

We can use the above formulas in the `mlmatsum` command to compute the respective components of the Hessian matrix because

$$\frac{\partial^2 \ln L_i}{\partial \boldsymbol{\beta}\partial \ln \sigma_u} = \sum_{t=1}^{T_i} \frac{\partial^2 \ln L_i}{\partial \mathbf{x}_{it}\boldsymbol{\beta}\partial \ln \sigma_u} \mathbf{x}_{it}$$

$$\frac{\partial^2 \ln L_i}{\partial \boldsymbol{\beta}\partial \ln \sigma_e} = \sum_{t=1}^{T_i} \frac{\partial^2 \ln L_i}{\partial \mathbf{x}_{it}\boldsymbol{\beta}\partial \ln \sigma_e} \mathbf{x}_{it}$$

(Continued on next page)

The lines that differ between **myrereg_d1** and **myrereg_d2** (our method-d2 evaluator) are

—————————————————————————————————— fragment myrereg_d2.ado ——————————

```
program myrereg_d2
        version 11
        args todo b lnf g H
        ...
        quietly {
                ...
                mat 'g' = ('dxb','du','de')
                // the following lines are new
                if ('todo'==1 | 'lnf'==.) exit

                // compute the Hessian
                tempname d2xb1 d2xb2 d2xb d2u d2e dxbdu dxbde dude one
                mlmatsum 'lnf' 'd2u' = -(2*'a'^2*'Sz_2'/'s_u'^2 -     ///
                        4*'s_e'^2*'a'^3*'Sz_2'/'s_u'^4 +             ///
                        2*'T'*'a'^2*'s_e'^2/'s_u'^2)                 ///
                        if 'last'==1                       , eq(2)
                mlmatsum 'lnf' 'd2e' =                                ///
                        -(2*('S_z2'-'a'*'Sz_2')/'s_e'^2 -            ///
                        2*'a'^2*'Sz_2'/'s_u'^2 -                     ///
                        4*'a'^3*'Sz_2'*'s_e'^2/'s_u'^4 +            ///
                        2*'a'*'s_e'^2/'s_u'^2 -                      ///
                        2*'a'^2*'s_e'^4/'s_u'^4)                     ///
                        if 'last'==1                       , eq(3)
                mlmatsum 'lnf' 'dude' =                               ///
                        -(4*'a'^3*'Sz_2'*'s_e'^2/'s_u'^4 -          ///
                        2*'T'*'a'^2*'s_e'^2/'s_u'^2)                 ///
                        if 'last'==1                       , eq(2,3)
                mlmatsum 'lnf' 'dxbdu' = -2*'a'^2*'S_z'/'s_u'^2 , eq(1,2)
                mlmatsum 'lnf' 'dxbde' =                              ///
                        -(2*('z'-'a'*'S_z')/'s_e'^2 -               ///
                        2*'a'^2*'S_z'/'s_u'^2                        , eq(1,3)
                // 'a' is constant within panel; and
                // -mlmatbysum- treats missing as 0 for 'a'
                by 'by': replace 'a' = . if !'last'
                mlmatsum 'lnf' 'd2xb2' = 1                      , eq(1)
                gen double 'one' = 1
                mlmatbysum 'lnf' 'd2xb1' 'a' 'one', by($MY_panel) eq(1)
                mat 'd2xb' = -('d2xb2'-'d2xb1')/'s_e'^2
                mat 'H' = (                               ///
                        'd2xb',   'dxbdu', 'dxbde'       \ ///
                        'dxbdu'', 'd2u',   'dude'        \ ///
                        'dxbde'', 'dude',  'd2e'           ///
                )
        }
end
```

—————————————————————————————————————— end myrereg_d2.ado ——————————

▷ **Example**

We can now verify that the results from `myrereg_d2` agree with those of `myrereg_d0` and `myrereg_d1`.

```
. use http://www.stata-press.com/data/ml4/tablef7-1
. global MY_panel i        // the panel variable
. ml model d2 myrereg_d2 (xb: lnC = lnQ lnPF lf) /lns_u /lns_e
. ml max, nolog
```

					Number of obs	=	90
					Wald chi2(3)	=	11468.30
Log likelihood = 114.72904					Prob > chi2	=	0.0000

lnC	Coef.	Std. Err.	z	P>|z|	[95% Conf. Interval]	
xb						
lnQ	.90531	.0253759	35.68	0.000	.8555742	.9550459
lnPF	.4233757	.013888	30.48	0.000	.3961557	.4505957
lf	-1.064456	.1962308	-5.42	0.000	-1.449061	-.6798508
_cons	9.618648	.2066218	46.55	0.000	9.213677	10.02362
lns_u						
_cons	-2.170818	.3026641	-7.17	0.000	-2.764028	-1.577607
lns_e						
_cons	-2.828404	.077319	-36.58	0.000	-2.979947	-2.676862

◁

❏ **Technical note**

Suppose that we wanted to model heteroskedasticity. That is, instead of

$$\theta_{1it} = \mathbf{x}_{it}\boldsymbol{\beta}$$
$$\theta_{2it} = \ln\sigma_u$$
$$\theta_{3it} = \ln\sigma_e$$

we decided to write a likelihood evaluator in which $\ln\sigma_e$ was a linear equation:

$$\theta_{1it} = \mathbf{x}_{1it}\boldsymbol{\beta}_1$$
$$\theta_{2it} = \ln\sigma_u \tag{6.9}$$
$$\theta_{3it} = \ln\sigma_{eit} = \mathbf{x}_{3it}\boldsymbol{\beta}_3$$

We must admit that we were tempted to start with the likelihood as it is given in (6.3); after all, we went through a lot of algebra to complete the squares and integrate out the random effects u_i. The problem is that (6.9) has a substantial effect on the form of the likelihood function. To see why, let's look at the steps we took in deriving (6.3).

$$
\begin{aligned}
\ln L_i &= \ln \int_{-\infty}^{\infty} f(u_i) \prod_{t=1}^{T_i} f(y_{it}|u_i)\, du_i \\
&= \ln \int_{-\infty}^{\infty} \frac{1}{\sigma_u \sqrt{2\pi}} \exp\left(-\frac{u_i^2}{2\sigma_u^2}\right) \prod_{t=1}^{T_i} \left[\frac{1}{\sigma_e \sqrt{2\pi}} \exp\left\{ -\frac{(y_{it} - u_i - \mathbf{x}_{it}\boldsymbol{\beta})^2}{2\sigma_e^2} \right\} \right] du_i \\
&= \ln \int_{-\infty}^{\infty} \frac{1}{\sigma_u \sqrt{2\pi}} \exp\left(-\frac{u_i^2}{2\sigma_u^2}\right) \prod_{t=1}^{T_i} \left[\frac{1}{\sigma_e \sqrt{2\pi}} \exp\left\{ -\frac{(z_{it} - u_i)^2}{2\sigma_e^2} \right\} \right] du_i \\
&= \ln \mathcal{K}_1 + \ln \int_{-\infty}^{\infty} \exp\left\{ -\frac{u_i^2}{2\sigma_u^2} - \sum_{t=1}^{T_i} \frac{(z_{it} - u_i)^2}{2\sigma_e^2} \right\} du_i \qquad (6.10)
\end{aligned}
$$

where

$$
\mathcal{K}_1 = \frac{1}{\sigma_u \sqrt{2\pi}} \prod_{t=1}^{T_i} \frac{1}{\sigma_e \sqrt{2\pi}}
$$

We then complete the square in (6.10) and integrate with respect to u_i to get (6.3).

For the heteroskedastic model, we have σ_{eit} instead of σ_e, thus

$$
\ln L_i = \ln \mathcal{K}_2 + \ln \int_{-\infty}^{\infty} \exp\left\{ -\frac{u_i^2}{2\sigma_u^2} - \sum_{t=1}^{T_i} \frac{(z_{it} - u_i)^2}{2\sigma_{eit}^2} \right\} du_i \qquad (6.11)
$$

where

$$
\mathcal{K}_2 = \frac{1}{\sigma_u \sqrt{2\pi}} \prod_{t=1}^{T_i} \frac{1}{\sigma_{eit} \sqrt{2\pi}}
$$

After we complete the square in (6.11) and integrate with respect to u_i, the log-likelihood function looks like

$$
\ln L_i = -\frac{1}{2} \left\{ \sum_{t=1}^{T_i} \frac{z_{it}^2}{\sigma_{eit}^2} - a_i \left(\sum_{t=1}^{T_i} \frac{z_{it}}{\sigma_{eit}^2} \right)^2 - \ln a_i + \sum_{t=1}^{T_i} \ln(2\pi\sigma_{eit}^2) + \ln \sigma_u^2 \right\}
$$

where

$$
\frac{1}{a_i} = \frac{1}{\sigma_u^2} + \sum_{t=1}^{T_i} \frac{1}{\sigma_{eit}^2}
$$

and our method-d0 evaluator is

```
                                              ─── begin myrereghet_d0.ado ───
    program myrereghet_d0
            version 11
            args todo b lnf
            tempvar xb z T S_z2 Sz_2 a last s_e S_se2
            tempname s_u
            mleval 'xb'  = 'b', eq(1)
            mleval 's_u' = 'b', eq(2) scalar
            mleval 's_e' = 'b', eq(3)
            scalar 's_u' = exp('s_u')
            // MY_panel contains the panel ID
            local by $MY_panel
            sort 'by'
            local y $ML_y1
            quietly {
                    replace 's_e' = exp('s_e')
                    by 'by': gen double 'S_se2' = sum(ln(2*c(pi)*'s_e'^2))
                    gen double 'z' = 'y' - 'xb'
                    by 'by': gen double 'a' = 1 / ( 1/'s_u'^2 + sum(1/'s_e'^2) )
                    by 'by': gen 'last' = _n==_N
                    by 'by': gen 'T' = _N
                    by 'by': gen double 'S_z2' = sum('z'^2/'s_e'^2)
                    by 'by': gen double 'Sz_2' = sum('z'/'s_e'^2)^2
                    mlsum 'lnf'= -.5 *                                    ///
                        (                                                 ///
                                    ('S_z2'-'a'*'Sz_2') -                 ///
                                    ln('a') +                            ///
                                    'S_se2' +                            ///
                                    2*ln('s_u')                          ///
                        )                                                 ///
                        if 'last' == 1
            }
    end
                                              ─── end myrereghet_d0.ado ───
```

Using the example at the end of section 6.4.1, we can verify that `myrereghet_d0` will reproduce the results of the homoskedastic model:

(Continued on next page)

```
. use http://www.stata-press.com/data/ml4/tablef7-1

. global MY_panel i        // the panel variable

. ml model d0 myrereghet_d0 (xb: lnC = lnQ lnPF lf) /lns_u /lns_e

. ml max, nolog
                                           Number of obs   =         90
                                           Wald chi2(3)    =   11468.30
           Log likelihood =   114.72904    Prob > chi2     =     0.0000
```

lnC	Coef.	Std. Err.	z	P>|z|	[95% Conf. Interval]
xb					
lnQ	.90531	.0253759	35.68	0.000	.8555742 .9550459
lnPF	.4233757	.013888	30.48	0.000	.3961557 .4505957
lf	-1.064456	.1962308	-5.42	0.000	-1.449061 -.6798508
_cons	9.618648	.2066218	46.55	0.000	9.213677 10.02362
lns_u					
_cons	-2.170817	.3026644	-7.17	0.000	-2.764028 -1.577605
lns_e					
_cons	-2.828404	.077319	-36.58	0.000	-2.979947 -2.676862

When developing the gradient calculations for a method-**d1** evaluator, you will find that

$$\frac{\partial L}{\partial \boldsymbol{\beta}_3}$$

can be calculated using `mlvecsum`.

For the method-**d2** evaluator, you will find that

$$\frac{\partial^2 L}{\partial \boldsymbol{\beta}_3 \partial \boldsymbol{\beta}_3'}$$

can be calculated using `mlmatsum` and `mlmatbysum`, similarly to (6.8) for $\boldsymbol{\beta}_1$. You will also find that

$$\frac{\partial^2 L}{\partial \boldsymbol{\beta}_1 \partial \boldsymbol{\beta}_3'}$$

can be calculated using `mlmatsum` and `mlmatbysum`.

❑

6.5   Other models that do not meet the linear-form restrictions

Here we discuss other models that do not meet the linear-form restrictions. It is hard to talk about specific challenges you will face when writing an evaluator for these models' likelihood functions, so we will use the Cox proportional hazards model as an example.

Let $\mathbf{x}_j$ be a (fixed) row vector of covariates for a subject. The hazard of failure is assumed to be

$$h_j(t) = h_0(t) \exp(\mathbf{x}_j \boldsymbol{\beta})$$

The Cox proportional hazards model provides estimates of $\boldsymbol{\beta}$ without estimating $h_0(t)$. Given a dataset of N individuals in which each individual is observed from time 0 to t_j, at which time the individual either is observed to fail or is censored, the log-likelihood function is

$$\ln L = \sum_{i=1}^{N_d} \left[\sum_{k \in D_i} \mathbf{x}_k \boldsymbol{\beta} - d_i \ln \left\{ \sum_{j \in R_i} \exp(\mathbf{x}_j \boldsymbol{\beta}) \right\} \right] \tag{6.12}$$

where i indexes the ordered failure times $t_{(i)}$ ($i = 1, \ldots, N_d$), N_d is the number of unique failure times, d_i is the number that fail at $t_{(i)}$, D_i is the set of observations that fail at $t_{(i)}$, and R_i is the set of observations that are at risk at time $t_{(i)}$. R_i is called the risk pool (or risk set) at time $t_{(i)}$.

Actually, (6.12) is just one of several possible log likelihoods used to fit a Cox model. This particular form is attributable to Peto (1972) and Breslow (1974). This likelihood has the following properties:

1. There are likelihood values only when failures occur.

2. The likelihood is a function of all the subjects who have not failed.

As such, it violates the linear-form assumptions. Moreover, it is not an easy likelihood function to calculate in any language.

Let's rewrite the likelihood function as

$$\ln L = \sum_{i=1}^{N_d} \left\{ A_i - d_i \ln(B_i) \right\}$$

where

$$A_i = \sum_{k \in D_i} \mathbf{x}_k \boldsymbol{\beta} \qquad \text{and} \qquad B_i = \sum_{j \in R_i} \exp(\mathbf{x}_j \boldsymbol{\beta})$$

As before, we tend to break a problem into manageable pieces. Specifically, we remove a complicated term from the likelihood function, give it a letter, substitute the letter in the original formula, and write the definition of the letter on the side. Once we have done that, we can think about how to program each of the ingredients. As things stand right now, if we knew A_i, d_i, and B_i, calculating $\ln L$ would be easy. So now we think about how to obtain A_i, d_i, and B_i.

To make things a little more concrete, let's pretend that our data are as given in table 6.2. Here t is time of death or censoring, and d is 1 if death or 0 if censored. A person is in the risk pool until he or she dies or is censored.

Table 6.2. Layout for survival data.

i	t	d	$\mathbf{x}_i$
1	1	1	$\mathbf{x}_1$
2	3	0	$\mathbf{x}_2$
3	5	1	$\mathbf{x}_3$
4	5	1	$\mathbf{x}_4$
5	5	0	$\mathbf{x}_5$
6	6	1	$\mathbf{x}_6$

For calculating B_i, let's think about the pool of still-living persons:

i	t	d	$\mathbf{x}_i$	
1	1	1	$\mathbf{x}_1$	first death every person was alive
2	3	0	$\mathbf{x}_2$	
3	5	1	$\mathbf{x}_3$	second and third deaths;
4	5	1	$\mathbf{x}_4$	$i = 1, 2$ are out of the pool
5	5	0	$\mathbf{x}_5$	
6	6	1	$\mathbf{x}_6$	fourth death; everyone above is out

If you read the above table from bottom to top, the calculation of who is in the pool becomes easy: at the bottom, no one is in the pool, then person 6 is added to the pool, then person 5, and so on. So we could reverse the data and calculate as follows:

i	t	d	$\mathbf{x}_i$	empty pool
6	6	1	$\mathbf{x}_6$	add to pool, $B_6 = \exp(\mathbf{x}_6\boldsymbol{\beta})$
5	5	0	$\mathbf{x}_5$	add to pool, $B_5 = B_6 + \exp(\mathbf{x}_5\boldsymbol{\beta})$
4	5	1	$\mathbf{x}_4$	add to pool, $B_4 = B_5 + \exp(\mathbf{x}_4\boldsymbol{\beta})$
3	5	1	$\mathbf{x}_3$	add to pool, $B_3 = B_4 + \exp(\mathbf{x}_3\boldsymbol{\beta})$
2	3	0	$\mathbf{x}_2$	add to pool, $B_2 = B_3 + \exp(\mathbf{x}_2\boldsymbol{\beta})$
1	1	1	$\mathbf{x}_1$	add to pool, $B_1 = B_2 + \exp(\mathbf{x}_1\boldsymbol{\beta})$

With Stata, this is easy to do with

```
tempvar negt B
gen double 'negt' = -t
sort 'negt' d
gen double 'B' = sum(exp('xb'))
```

where 'xb' is the variable that contains $\mathbf{x}_j\boldsymbol{\beta}$ (we get it using `mleval`).

Calculating A_i is no more difficult:

```
tempvar A
by 'negt' d: gen double 'A' = cond(_n==_N, sum('xb'), .) if d==1
```

We need one more ingredient, d_i, the number of people who die at each time:

```
tempvar sumd
by 'negt' d: gen double 'sumd' = cond(_n==_N, sum(d), .)
```

Now that we have 'A', 'B', and 'sumd', calculating the log-likelihood value is easy:

```
tempvar L last
gen double 'L' = 'A' - 'sumd'*ln('B')
by 'negt' d: gen byte 'last' = (_n==_N & d==1)
mlsum 'lnf' = 'L' if 'last'
```

The if 'last' on the mlsum statement is crucial. By default, mlsum assumes that every observation should contribute to the likelihood value. If there are any missing values among the contributions, mlsum assumes that means the likelihood cannot be calculated at the given value of β. It fills in the overall log likelihood 'lnf' with a missing value, and then later, when ml sees that result, it takes the appropriate action. We purposely have lots of missing values in 'L' because we have contributions to the likelihood only when there is a failure. Thus we need to mark the observations that should contribute to the likelihood and tell mleval about it. We were, we admit, tempted to write

```
gen double 'L' = 'A' - 'sumd'*ln('B')
mlsum 'lnf' = 'L' if !missing('L')
```

because then we would sum the contributions that do exist. That would have been a bad idea. Perhaps, given β, one of the contributions that should exist could not be calculated, meaning that the overall likelihood really could not be calculated at β. We want to be sure that, should that event arise, mlsum sets the overall $\ln L$ in 'lnf' to missing so that ml takes the appropriate action. Therefore, we carefully marked which values should be filled in and specified that on the mlsum statement.

(Continued on next page)

Based on the above, our method-d0 evaluator is

```
                                                    ──── begin mycox1_d0.ado ────
   program mycox1_d0
           version 11
           args todo b lnf
           local t "$ML_y1"                  // $ML_y1 is t
           local d "$ML_y2"                  // $ML_y2 is d
           tempvar xb B A sumd negt last L
           mleval 'xb' = 'b'
           quietly gen double 'negt' = -'t'
           local by "'negt' 'd'"
           local byby "by 'by'"
           sort 'by'
           quietly {
                   gen double 'B' = sum(exp('xb'))
                   'byby': gen double 'A' = cond(_n==_N, sum('xb'), .) if 'd'==1
                   'byby': gen 'sumd' = cond(_n==_N, sum('d'), .)
                   'byby': gen byte 'last' = (_n==_N & 'd' == 1)
                   gen double 'L' = 'A' - 'sumd'*ln('B') if 'last'
                   mlsum 'lnf' = 'L' if 'last'
           }
   end
                                                    ──── end mycox1_d0.ado ────
```

▷ Example

Using `cancer.dta` and `mycox1_d0`, we fit a Cox regression of time of death on drug type and age:

```
. sysuse cancer, clear
(Patient Survival in Drug Trial)
. ml model d0 mycox1_d0 (studytime died = i.drug age, nocons)
. ml maximize
initial:       log likelihood = -99.911448
alternative:   log likelihood = -99.606964
rescale:       log likelihood = -99.101334
Iteration 0:   log likelihood = -99.101334
Iteration 1:   log likelihood = -82.649192
Iteration 2:   log likelihood = -81.729635
Iteration 3:   log likelihood = -81.652728
Iteration 4:   log likelihood = -81.652567
Iteration 5:   log likelihood = -81.652567

                                          Number of obs   =          48
                                          Wald chi2(3)    =       28.33
Log likelihood = -81.652567               Prob > chi2     =      0.0000
```

	Coef.	Std. Err.	z	P>|z|	[95% Conf.	Interval]
drug						
2	-1.71156	.4943639	-3.46	0.001	-2.680495	-.7426242
3	-2.956384	.6557432	-4.51	0.000	-4.241617	-1.671151
age	.11184	.0365789	3.06	0.002	.0401467	.1835333

◁

❏ Technical note

Our program is not as efficient as it might be. Every time our likelihood evaluator is called, it re-sorts the data. In the example above, for instance, `ml` called our evaluator 116 times, which we determined using `ml count` (see appendix A). Clearly, sorting the data just once would save considerable execution time.

Instead of sorting the data within the likelihood evaluator, we could sort the data just once before calling `ml model` and `ml maximize`. We also take this opportunity to include weights according to the formula for the log likelihood based on Breslow's approximation as shown in [ST] **stcox**:

```
                                                  ──── begin mycox2_d0.ado ────
program mycox2_d0
        version 11
        args todo b lnf
        local negt "$ML_y1"            // $ML_y1 is -t
        local d "$ML_y2"              // $ML_y2 is d
        tempvar xb B A sumd last L
        mleval `xb' = `b'
        // data assumed already sorted by `negt' and `d'
        local byby "by `negt' `d'"
        local wxb "$ML_w*`xb'"
        quietly {
                gen double `B' = sum($ML_w*exp(`xb'))
                `byby': gen double `A' = cond(_n==_N, sum(`wxb'), .) if `d'==1
                `byby': gen `sumd' = cond(_n==_N, sum($ML_w*`d'), .)
                `byby': gen byte `last' = (_n==_N & `d' == 1)
                gen double `L' = `A' - `sumd'*ln(`B') if `last'
                mlsum `lnf' = `L' if `last', noweight
        }
end
                                                  ──── end mycox2_d0.ado ────
```

```
. sysuse cancer
(Patient Survival in Drug Trial)

. gen negtime = -studytime

. sort negtime died

. ml model d0 mycox2_d0 (negtime died = i.drug age, nocons)

. ml maximize, nolog
                                        Number of obs   =         48
                                        Wald chi2(3)    =      28.33
Log likelihood = -81.652567            Prob > chi2     =     0.0000
```

	Coef.	Std. Err.	z	P>|z|	[95% Conf. Interval]	
drug						
2	-1.71156	.4943639	-3.46	0.001	-2.680495	-.7426242
3	-2.956384	.6557432	-4.51	0.000	-4.241617	-1.671151
age	.11184	.0365789	3.06	0.002	.0401467	.1835333

We are unsure whether to encourage this. The do-file is only slightly faster because, in the original version, Stata did not actually sort the data 116 times. `sort` is smart, and when told to sort, it looks at the sort markers to determine if the dataset is already sorted. Thus the original code sorted the data once and made `sort` verify that it was sorted 115 times.

Still, there are situations where there is a considerable gain to making calculations once and only once. You can do that, but be aware that you make your do-file more dangerous to use.

❑

❑ **Technical note**

If we were using our evaluator as a subroutine in writing a new estimation command, we could gain the efficiency safely. We would be calling `ml` from inside another ado-file and so have complete control. We could create 'negt' and sort on it before the `ml model` statement and then use 'negt' in the `ml model` statement. The user could not forget to sort the data or mistakenly use time rather than negative time because we would be the ones doing that behind the scenes.

We did this in `mycox`, the estimation command discussed in section 15.5. A listing of `mycox` is shown in appendix C.5.

❑

7 Debugging likelihood evaluators

Things probably will not proceed as smoothly for you as they have for us in the previous chapters. You may write an evaluator and observe the following:

```
. ml model lf myprobit_lf (foreign=mpg weight)
. ml maximize
initial:        log likelihood =      -<inf>  (could not be evaluated)
—Break—
r(1);
```

We are the ones who pressed *Break* because we got tired of watching nothing happen. You might see

```
. ml model lf myprobit_lf (foreign=mpg weight)
. ml maximize
invalid syntax
r(198);
```

or

```
. ml model lf myprobit_lf (foreign=mpg weight)
. ml maximize
initial:        log likelihood =              0
alternative:    log likelihood =              0
rescale:        log likelihood =              0
could not calculate numerical derivatives
flat or discontinuous region encountered
r(430);
```

ml has some capabilities that will help you solve these problems.

7.1 ml check

The mistake we made in the above examples was not typing `ml check` between the `ml model` and `ml maximize` statements. If you are lucky—if your program works—here is what you might see:

```
. sysuse auto
(1978 Automobile Data)

. ml model lf myprobit_lf (foreign = mpg weight)

. ml check

Test 1:  Calling myprobit_lf to check if it computes log likelihood and
         does not alter coefficient vector...
         Passed.

Test 2:  Calling myprobit_lf again to check if the same log likelihood value
         is returned...
         Passed.

Test 3:  Calling myprobit_lf to check if 1st derivatives are computed...
         test not relevant for type lf evaluators.

Test 4:  Calling myprobit_lf again to check if the same 1st derivatives are
         returned...
         test not relevant for type lf evaluators.

Test 5:  Calling myprobit_lf to check if 2nd derivatives are computed...
         test not relevant for type lf evaluators.

Test 6:  Calling myprobit_lf again to check if the same 2nd derivatives are
         returned...
         test not relevant for type lf evaluators.
```

```
Searching for alternate values for the coefficient vector to verify that
myprobit_lf returns different results when fed a different coefficient vector:
Searching...
initial:       log likelihood =      -<inf>  (could not be evaluated)
searching for feasible values +

feasible:      log likelihood = -317.73494
improving initial values +.........
improve:       log likelihood = -116.23567

continuing with tests...
```

```
Test 7:  Calling myprobit_lf to check log likelihood at the new values...
         Passed.

Test 8:  Calling myprobit_lf requesting 1st derivatives at the new values...
         test not relevant for type lf evaluators.

Test 9:  Calling myprobit_lf requesting 2nd derivatives at the new values...
         test not relevant for type lf evaluators.
```

```
                     myprobit_lf HAS PASSED ALL TESTS
```

```
Test 10: Does myprobit_lf produce unanticipated output?
         This is a minor issue.  Stata has been running myprobit_lf with all
         output suppressed.  This time Stata will not suppress the output.  If
         you see any unanticipated output, you need to place quietly in front
         of some of the commands in myprobit_lf.
```
```
                                                        ─ begin execution
                                                        ─ end execution
```

You will probably see something like this:

```
. sysuse auto
(1978 Automobile Data)
. ml model lf myprobit_bad_lf (foreign = mpg weight)
. ml check
Test 1:  Calling myprobit_bad_lf to check if it computes log likelihood and
         does not alter coefficient vector...
         FAILED; myprobit_bad_lf returned error 132.
Here is a trace of its execution:
----------------------------------------------------------------------------
-> myprobit_bad_lf __000001 __000002
            - 'begin'
            = capture noisily version 11: myprobit_bad_lf __000001 __000002
                                    ----------------------- begin myprobit_bad_lf -----
            - version 11
            - args lnf xb
            - quietly replace 'lnf' = ln(normal( 'xb')) if $ML_y1 == 1
            = quietly replace __000001 = ln(normal( __000002)) if foreign == 1
            - quietly replace 'lnf' = ln(normal(-'xb') if $ML_y1 == 0
            = quietly replace __000001 = ln(normal(-__000002) if foreign == 0
too few ')' or ']'
                                    ----------------------- end myprobit_bad_lf -----
            - 'end'
            = set trace off
----------------------------------------------------------------------------

Fix myprobit_bad_lf.
r(132);
```

We left out a closing parenthesis in our program. The problem with our program is not that it altered the coefficient vector; it was just that, while attempting to test whether our program altered the coefficient vector, our program generated an error, so `ml check` backed up and showed us a trace of its execution.

`ml check` can be used with any method of likelihood evaluator.

`ml check` will save you hours of time.

7.2 Using the debug methods

All the evaluator methods that require analytic first or second derivatives have a corresponding debug method. Methods that require first derivatives, namely, `lf1` and `d1`, have accompanying debug methods `lf1debug` and `d1debug`. Second-derivative-based methods `lf2` and `d2` have accompanying debug methods `lf2debug` and `d2debug`. These debug methods verify the accuracy of your analytic derivatives. Like with `lf`, `lf0`, and `d0` methods, they compute all the required derivatives numerically when fitting the model. However, they also request the derivatives from your evaluator program and then compare those derivatives with the ones calculated numerically. A large discrepancy between the numeric and analytic derivatives is an indication that you have either incorrectly taken a derivative of your likelihood function or else made a programming error when implementing it in your program.

Remember, `ml check` does not verify that your derivatives are correct; it just looks for mechanical errors. That does not mean you should not bother with `ml check`—just do not expect `ml check` to catch substantive errors.

If you have written a method-`d1` evaluator, we recommend that you type

```
. ml model d1debug ...
. ml check
. ml maximize
```

Similarly, if you have written a method-`lf2` evaluator, we recommend that you type

```
. ml model lf2debug ...
. ml check
. ml maximize
```

We recommend using the debug methods along with `ml check`. They do not interact—`ml check` will produce no more information than it would if it were called with a regular evaluator method.

To show what can happen, we have written a method-`lf1` evaluator for the Weibull model and purposely programmed the derivatives incorrectly. There are no mechanical errors in our program, however.

```
. sysuse cancer
(Patient Survival in Drug Trial)
. ml model lf1debug myweibull_bad_lf1 (studytime died = i.drug age) ()
. ml check

Test 1:  Calling myweibull_bad_lf1 to check if it computes log likelihood and
         does not alter coefficient vector...
         Passed.
```
   (*output omitted*)
```
Test 3:  Calling myweibull_bad_lf1 to check if 1st derivatives are computed...
         Passed.
```
   (*note that ml check did not notice that the derivatives are wrong!*)
   (*output omitted*)

---
                        myweibull_bad_lf1 HAS PASSED ALL TESTS

   (*output omitted*)

You should check that the derivatives are right.

Use the optimizer to obtain estimates.

The output will include a report comparing analytic to numeric derivatives.
Do not be concerned if your analytic derivatives differ from the numeric ones
in early iterations.

The analytic gradient will differ from the numeric one in early iterations,
then the mreldif() difference should become less than 1e-6 in the middle
iterations, and the difference will increase again in the final iterations as
the gradient goes to zero.

`ml check` even reminds us to check that the derivatives are right. We are already using method `lf1debug`, so we will discover our problem when we type `ml maximize`.

7.2.1   First derivatives

We have proceeded too quickly in our examples in the previous chapters. In real life, the probit and Weibull method-lf1 evaluators we wrote in chapter 5 and the d1 evaluators we wrote in chapter 6 would not have worked the first time we tried. Without a doubt, we would have made an error in producing the gradient vector, either substantively because we took derivatives incorrectly or mechanically in how we programmed the evaluator.

The recipe for success is to start with a working method-lf0 or method-d0 evaluator, make the requisite changes to convert it to a lf1 or d1 evaluator, and then specify method lf1debug or d1debug in the `ml model` statement.

For example, here is what happens when we check the gradient calculations in `myweibull_lf1` using method lf1debug:

```
. sysuse cancer
(Patient Survival in Drug Trial)
. ml model lf1debug myweibull_lf1 (studytime died = i.drug age) ()
. ml max
initial:       log likelihood =        -744
alternative:   log likelihood = -356.14276
rescale:       log likelihood = -200.80201
rescale eq:    log likelihood = -136.69232

lf1debug:  Begin derivative-comparison report ──────────────────────
lf1debug:  mreldif(gradient vectors) =  3.98e-10
lf1debug:  Warning:  evaluator did not compute Hessian matrix
lf1debug:  End derivative-comparison report ────────────────────────
Iteration 0:   log likelihood = -136.69232  (not concave)

lf1debug:  Begin derivative-comparison report ──────────────────────
lf1debug:  mreldif(gradient vectors) =  1.44e-10
lf1debug:  Warning:  evaluator did not compute Hessian matrix
lf1debug:  End derivative-comparison report ────────────────────────
Iteration 1:   log likelihood = -124.11726

lf1debug:  Begin derivative-comparison report ──────────────────────
lf1debug:  mreldif(gradient vectors) =  1.35e-08
lf1debug:  Warning:  evaluator did not compute Hessian matrix
lf1debug:  End derivative-comparison report ────────────────────────
Iteration 2:   log likelihood = -113.90508

lf1debug:  Begin derivative-comparison report ──────────────────────
lf1debug:  mreldif(gradient vectors) =  2.94e-10
lf1debug:  Warning:  evaluator did not compute Hessian matrix
lf1debug:  End derivative-comparison report ────────────────────────
Iteration 3:   log likelihood = -110.30519

lf1debug:  Begin derivative-comparison report ──────────────────────
lf1debug:  mreldif(gradient vectors) =  8.96e-07
lf1debug:  Warning:  evaluator did not compute Hessian matrix
lf1debug:  End derivative-comparison report ────────────────────────
Iteration 4:   log likelihood = -110.26747

lf1debug:  Begin derivative-comparison report ──────────────────────
lf1debug:  mreldif(gradient vectors) =  .0000382
lf1debug:  Warning:  evaluator did not compute Hessian matrix
lf1debug:  End derivative-comparison report ────────────────────────
```

```
Iteration 5:    log likelihood = -110.26736
lf1debug:  Begin derivative-comparison report ───────────────────────
lf1debug:  mreldif(gradient vectors) =   .000039
lf1debug:  Warning:  evaluator did not compute Hessian matrix
lf1debug:  End derivative-comparison report ─────────────────────────
Iteration 6:    log likelihood = -110.26736
                                                  Number of obs   =         48
                                                  Wald chi2(3)    =      35.25
Log likelihood = -110.26736                       Prob > chi2     =     0.0000
```

	Coef.	Std. Err.	z	P>|z|	[95% Conf.	Interval]
eq1						
drug						
2	1.012966	.2903917	3.49	0.000	.4438086	1.582123
3	1.45917	.2821195	5.17	0.000	.9062261	2.012114
age	-.0671728	.0205688	-3.27	0.001	-.1074868	-.0268587
_cons	6.060723	1.152845	5.26	0.000	3.801188	8.320259
eq2						
_cons	.5573333	.1402154	3.97	0.000	.2825162	.8321504

The difference between methods lf1 and lf1debug is that lf1debug provides a derivative comparison report interspersed in the iteration log. For instance, between iterations 1 and 2 we see the following:

```
Iteration 1:    log likelihood = -124.11726
lf1debug:  Begin derivative-comparison report ───────────────────────
lf1debug:  mreldif(gradient vectors) =   1.35e-08
lf1debug:  Warning:  evaluator did not compute Hessian matrix
lf1debug:  End derivative-comparison report ─────────────────────────
Iteration 2:    log likelihood = -113.90508
```

Whenever ml needs first derivatives, method lf1debug calls the likelihood evaluator just as method lf1 would, but method lf1debug also calculates the derivatives numerically. It then displays a comparison of the derivatives calculated both ways and returns to ml the *numerically calculated derivatives*. Thus lf1debug proceeds to a correct answer, even if the analytical derivative calculation is incorrect.

mreldif() in the derivative-comparison report stands for maximum relative difference; this would be easier to understand if we could see the gradient vector. Method lf1debug shows us the vectors as well as the mreldif() if we specify ml maximize's gradient option:

```
. sysuse cancer
(Patient Survival in Drug Trial)
. ml model lf1debug myweibull_lf1 (studytime died = i.drug age) ()
. ml max, gradient
initial:       log likelihood =       -744
trying nonzero initial values +.
alternative:   log likelihood = -356.14276
rescaling entire vector .++.
rescale:       log likelihood = -200.80201
rescaling equations +++++++++++++++++++++++++++++++++++++.
sign reverse +++++++++++++++++++++++++++++++++++++.+++.
rescaling equations ....
rescale eq:    log likelihood = -136.69232
```

```
Iteration 0:
lf1debug:  Begin derivative-comparison report ────────────────────
lf1debug:  mreldif(gradient vectors) =  3.98e-10

lf1debug:  evaluator calculated gradient:
           eq1:       eq1:       eq1:       eq1:       eq1:       eq2:
           1b.        2.         3.
           drug       drug       drug       age        _cons      _cons
r1          0  -.7819348   1.243378   -632.982  -10.67705   2.580612

lf1debug:  numerically calculated gradient (used for stepping):
           eq1:       eq1:       eq1:       eq1:       eq1:       eq2:
           1b.        2.         3.
           drug       drug       drug       age        _cons      _cons
r1          0  -.7819348   1.243378   -632.982  -10.67705   2.580612

lf1debug:  relative differences:
           eq1:       eq1:       eq1:       eq1:       eq1:       eq2:
           1b.        2.         3.
           drug       drug       drug       age        _cons      _cons
r1          0   3.34e-10   3.98e-10   3.55e-10   3.51e-10   3.73e-10

lf1debug:  Warning:  evaluator did not compute Hessian matrix
lf1debug:  End derivative-comparison report ──────────────────────
                                       log likelihood = -136.69232
                                                         (not concave)
```

(Continued on next page)

```
Iteration 1:

lf1debug:  Begin derivative-comparison report ──────────────────────────
lf1debug:  mreldif(gradient vectors) =  1.44e-10

lf1debug:  evaluator calculated gradient:
           eq1:      eq1:      eq1:      eq1:      eq1:      eq2:
           1b.       2.        3.
           drug      drug      drug      age      _cons     _cons
r1            0  12.10675  11.93993  1358.373  25.85834  7.363412

lf1debug:  numerically calculated gradient (used for stepping):
           eq1:      eq1:      eq1:      eq1:      eq1:      eq2:
           1b.       2.        3.
           drug      drug      drug      age      _cons     _cons
r1            0  12.10675  11.93993  1358.373  25.85834  7.363412

lf1debug:  relative differences:
           eq1:      eq1:      eq1:      eq1:      eq1:      eq2:
           1b.       2.        3.
           drug      drug      drug      age      _cons     _cons
r1            0  4.52e-11  6.31e-12  9.77e-12  2.50e-12  1.44e-10

lf1debug:  Warning:  evaluator did not compute Hessian matrix
lf1debug:  End derivative-comparison report ────────────────────────────
                                            log likelihood = -124.11726
────────────────────────────────────────────────────────────────────────
Iteration 2:

lf1debug:  Begin derivative-comparison report ──────────────────────────
lf1debug:  mreldif(gradient vectors) =  1.35e-08

lf1debug:  evaluator calculated gradient:
           eq1:      eq1:      eq1:      eq1:      eq1:      eq2:
           1b.       2.        3.
           drug      drug      drug      age      _cons     _cons
r1            0  -1.769302  -1.718409  -28.53524  -.6443933  -24.31936

lf1debug:  numerically calculated gradient (used for stepping):
           eq1:      eq1:      eq1:      eq1:      eq1:      eq2:
           1b.       2.        3.
           drug      drug      drug      age      _cons     _cons
r1            0  -1.769302  -1.718409  -28.53524  -.6443933  -24.31936

lf1debug:  relative differences:
           eq1:      eq1:      eq1:      eq1:      eq1:      eq2:
           1b.       2.        3.
           drug      drug      drug      age      _cons     _cons
r1            0  4.43e-10  4.67e-10  1.35e-08  4.27e-09  9.37e-13

lf1debug:  Warning:  evaluator did not compute Hessian matrix
lf1debug:  End derivative-comparison report ────────────────────────────
                                            log likelihood = -113.90508
────────────────────────────────────────────────────────────────────────
```

```
Iteration 3:
lf1debug:  Begin derivative-comparison report ─────────────────────────
lf1debug:  mreldif(gradient vectors) =  2.94e-10
lf1debug:  evaluator calculated gradient:
           eq1:        eq1:        eq1:        eq1:        eq1:        eq2:
           1b.         2.          3.
           drug        drug        drug        age        _cons       _cons
r1          0      -.0193315   -.0616992    66.97723    1.170796   -2.054476
lf1debug:  numerically calculated gradient (used for stepping):
           eq1:        eq1:        eq1:        eq1:        eq1:        eq2:
           1b.         2.          3.
           drug        drug        drug        age        _cons       _cons
r1          0      -.0193315   -.0616992    66.97723    1.170796   -2.054476
lf1debug:  relative differences:
           eq1:        eq1:        eq1:        eq1:        eq1:        eq2:
           1b.         2.          3.
           drug        drug        drug        age        _cons       _cons
r1          0       1.25e-10    9.63e-11    2.94e-10    1.49e-10    1.30e-10
lf1debug:  Warning:  evaluator did not compute Hessian matrix
lf1debug:  End derivative-comparison report ───────────────────────────
                                             log likelihood = -110.30519
─────────────────────────────────────────────────────────────────────
Iteration 4:
lf1debug:  Begin derivative-comparison report ─────────────────────────
lf1debug:  mreldif(gradient vectors) =  8.96e-07
lf1debug:  evaluator calculated gradient:
           eq1:        eq1:        eq1:        eq1:        eq1:        eq2:
           1b.         2.          3.
           drug        drug        drug        age        _cons       _cons
r1          0       .0036573    .0010349    5.602418    .0986067   -.0890069
lf1debug:  numerically calculated gradient (used for stepping):
           eq1:        eq1:        eq1:        eq1:        eq1:        eq2:
           1b.         2.          3.
           drug        drug        drug        age        _cons       _cons
r1          0       .0036574    .0010349    5.602424    .0986068   -.089007
lf1debug:  relative differences:
           eq1:        eq1:        eq1:        eq1:        eq1:        eq2:
           1b.         2.          3.
           drug        drug        drug        age        _cons       _cons
r1          0       2.00e-08    2.01e-08    8.96e-07    9.47e-08    8.43e-08
lf1debug:  Warning:  evaluator did not compute Hessian matrix
lf1debug:  End derivative-comparison report ───────────────────────────
                                             log likelihood = -110.26747
─────────────────────────────────────────────────────────────────────
```

(Continued on next page)

```
Iteration 5:

lf1debug:  Begin derivative-comparison report ─────────────────────────────
lf1debug:  mreldif(gradient vectors) =  .0000382
lf1debug:  evaluator calculated gradient:
           eq1:       eq1:       eq1:       eq1:       eq1:       eq2:
           1b.        2.         3.
           drug       drug       drug       age       _cons      _cons
r1          0        .0000179   .0000103   .022547   .0003982   -.0003297
lf1debug:  numerically calculated gradient (used for stepping):
           eq1:       eq1:       eq1:       eq1:       eq1:       eq2:
           1b.        2.         3.
           drug       drug       drug       age       _cons      _cons
r1          0        .000018    .0000104   .0225861  .0003989   -.0003307
lf1debug:  relative differences:
           eq1:       eq1:       eq1:       eq1:       eq1:       eq2:
           1b.        2.         3.
           drug       drug       drug       age       _cons      _cons
r1          0        1.33e-07   1.33e-07   .0000382  6.87e-07   9.81e-07
lf1debug:  Warning:  evaluator did not compute Hessian matrix
lf1debug:  End derivative-comparison report ───────────────────────────────
                                              log likelihood = -110.26736
```

```
Iteration 6:

lf1debug:  Begin derivative-comparison report ─────────────────────────────
lf1debug:  mreldif(gradient vectors) =   .000039
lf1debug:  evaluator calculated gradient:
           eq1:       eq1:       eq1:       eq1:       eq1:       eq2:
           1b.        2.         3.
           drug       drug       drug       age       _cons      _cons
r1          0       -1.33e-07  -1.33e-07  -.0000387  -6.81e-07   9.77e-07
lf1debug:  numerically calculated gradient (used for stepping):
           eq1:       eq1:       eq1:       eq1:       eq1:       eq2:
           1b.        2.         3.
           drug       drug       drug       age       _cons      _cons
r1          0        2.92e-10   1.92e-10   3.42e-07  6.07e-09   -4.94e-09
lf1debug:  relative differences:
           eq1:       eq1:       eq1:       eq1:       eq1:       eq2:
           1b.        2.         3.
           drug       drug       drug       age       _cons      _cons
r1          0        1.33e-07   1.33e-07   .000039   6.87e-07   9.82e-07
lf1debug:  Warning:  evaluator did not compute Hessian matrix
lf1debug:  End derivative-comparison report ───────────────────────────────
                                              log likelihood = -110.26736
```

```
                                         Number of obs   =       48
                                         Wald chi2(3)    =    35.25
        Log likelihood = -110.26736      Prob > chi2     =   0.0000
```

	Coef.	Std. Err.	z	P>|z|	[95% Conf. Interval]	
eq1						
drug						
2	1.012966	.2903917	3.49	0.000	.4438086	1.582123
3	1.45917	.2821195	5.17	0.000	.9062261	2.012114
age	-.0671728	.0205688	-3.27	0.001	-.1074868	-.0268587
_cons	6.060723	1.152845	5.26	0.000	3.801188	8.320259
eq2						
_cons	.5573333	.1402154	3.97	0.000	.2825162	.8321504

When we did not specify the `gradient` option, we saw the following reported just before the `Iteration 1` line:

```
lf1debug:  Begin derivative-comparison report ——————————————————
lf1debug:  mreldif(gradient vectors) =  1.44e-10
lf1debug:  Warning:  evaluator did not compute Hessian matrix
lf1debug:  End derivative-comparison report ——————————————————
Iteration 1:   log likelihood = -124.11726
```

With the `gradient` option, we now see

```
Iteration 1:

lf1debug:  Begin derivative-comparison report ——————————————————
lf1debug:  mreldif(gradient vectors) =  1.44e-10

lf1debug:  evaluator calculated gradient:
        eq1:      eq1:      eq1:      eq1:      eq1:      eq2:
        1b.       2.        3.
        drug      drug      drug      age       _cons     _cons
r1         0  12.10675  11.93993  1358.373  25.85834  7.363412

lf1debug:  numerically calculated gradient (used for stepping):
        eq1:      eq1:      eq1:      eq1:      eq1:      eq2:
        1b.       2.        3.
        drug      drug      drug      age       _cons     _cons
r1         0  12.10675  11.93993  1358.373  25.85834  7.363412

lf1debug:  relative differences:
        eq1:      eq1:      eq1:      eq1:      eq1:      eq2:
        1b.       2.        3.
        drug      drug      drug      age       _cons     _cons
r1         0  4.52e-11  6.31e-12  9.77e-12  2.50e-12  1.44e-10

lf1debug:  Warning:  evaluator did not compute Hessian matrix
lf1debug:  End derivative-comparison report ——————————————————
                                  log likelihood = -124.11726
```

We have the same `mreldif()` of 1.44e–10, but now we also see the gradient vector element by element. `mreldif()` is the maximum of the relative differences calculated element by element, where relative difference is defined as $|g_i - n_i|/(|n_i| + 1)$, g_i is the element of the gradient vector `myweibull_d1` calculated, and n_i is the numerically calculated gradient value. We can safely ignore the warning regarding computation of the Hessian matrix because we are working with an `lf1` evaluator, not an `lf2` one.

Specifying the `gradient` option can be useful when the calculated gradient value is suspect. Here `myweibull_lf1` works, and the reported `mreldif()`s are typical:

```
iteration 0:   3.98e–10
iteration 1:   1.44e–10
iteration 2:   1.35e–08
iteration 3:   2.94e–10
iteration 4:   8.96e–07
iteration 5:   3.83e–05
iteration 6:   3.91e–05
```

What is being revealed here is not the accuracy of `myweibull_lf1` but that of the numerical approximation. Here they are initially good, remain good through the next few iterations, and then become poor near the end. They end up poor because we are measuring error as a relative difference and the derivatives themselves are going to zero as the algorithm converges. In other cases, you might see the initial relative differences be somewhat larger than those in the middle, though if you have programmed the analytic derivatives correctly, at some point the relative differences should become relatively small (1e–8 or smaller).

The purpose of the output, however, is not to test the accuracy of `ml`'s numeric derivative calculator but to verify that the analytic derivatives are in rough agreement with them. If they are, we can safely assume that the analytic derivatives are correct.

To demonstrate what you might see when the analytic derivatives are somehow wrong, we will use `myweibull_bad_lf1`, a modified copy of `myweibull_lf1`. We changed the line that reads

```
replace 'g2' = 'd' - 'R'*'p'*('M'-'d')
```

to read

```
replace 'g2' = 'd' + 'R'*'p'*('M'-'d')
```

and so introduce a rather typical sign error. The result of fitting the model with method `lf1debug` is

```
. sysuse cancer
(Patient Survival in Drug Trial)

. ml model lf1debug myweibull_bad_lf1 (studytime died = i.drug age) ()

. ml max

initial:       log likelihood =        -744
alternative:   log likelihood = -356.14276
rescale:       log likelihood = -200.80201
rescale eq:    log likelihood = -136.69232

lf1debug:  Begin derivative-comparison report ──────────────────────
lf1debug:  mreldif(gradient vectors) = .9407374
lf1debug:  Warning:  evaluator did not compute Hessian matrix
lf1debug:  End derivative-comparison report ────────────────────────
Iteration 0:   log likelihood = -136.69232  (not concave)

lf1debug:  Begin derivative-comparison report ──────────────────────
lf1debug:  mreldif(gradient vectors) = .8496778
lf1debug:  Warning:  evaluator did not compute Hessian matrix
lf1debug:  End derivative-comparison report ────────────────────────
Iteration 1:   log likelihood = -124.11726

lf1debug:  Begin derivative-comparison report ──────────────────────
lf1debug:  mreldif(gradient vectors) = 1.267058
lf1debug:  Warning:  evaluator did not compute Hessian matrix
lf1debug:  End derivative-comparison report ────────────────────────
Iteration 2:   log likelihood = -113.90508

lf1debug:  Begin derivative-comparison report ──────────────────────
lf1debug:  mreldif(gradient vectors) = 1.016209
lf1debug:  Warning:  evaluator did not compute Hessian matrix
lf1debug:  End derivative-comparison report ────────────────────────
Iteration 3:   log likelihood = -110.30519

lf1debug:  Begin derivative-comparison report ──────────────────────
lf1debug:  mreldif(gradient vectors) = .9855602
lf1debug:  Warning:  evaluator did not compute Hessian matrix
lf1debug:  End derivative-comparison report ────────────────────────
Iteration 4:   log likelihood = -110.26747

lf1debug:  Begin derivative-comparison report ──────────────────────
lf1debug:  mreldif(gradient vectors) = .9841323
lf1debug:  Warning:  evaluator did not compute Hessian matrix
lf1debug:  End derivative-comparison report ────────────────────────
Iteration 5:   log likelihood = -110.26736

lf1debug:  Begin derivative-comparison report ──────────────────────
lf1debug:  mreldif(gradient vectors) =  .984127
lf1debug:  Warning:  evaluator did not compute Hessian matrix
lf1debug:  End derivative-comparison report ────────────────────────
Iteration 6:   log likelihood = -110.26736
```

(Continued on next page)

```
                                                    Number of obs    =          48
                                                    Wald chi2(3)     =      35.25
        Log likelihood = -110.26736                 Prob > chi2      =     0.0000
```

	Coef.	Std. Err.	z	P>|z|	[95% Conf.	Interval]
eq1						
drug						
2	1.012966	.2903917	3.49	0.000	.4438086	1.582123
3	1.45917	.2821195	5.17	0.000	.9062261	2.012114
age	-.0671728	.0205688	-3.27	0.001	-.1074868	-.0268587
_cons	6.060723	1.152845	5.26	0.000	3.801188	8.320259
eq2						
_cons	.5573333	.1402154	3.97	0.000	.2825162	.8321504

First, despite the error in `myweibull_bad_lf1`, the overall model is still correctly fit because method `lf1debug` used the numerically calculated derivatives. The log, however, reveals that our calculation of the derivatives has problems:

```
lf1debug:  Begin derivative-comparison report ————————————————————
lf1debug:  mreldif(gradient vectors) =  .9407374
lf1debug:  Warning:  evaluator did not compute Hessian matrix
lf1debug:  End derivative-comparison report ————————————————————
Iteration 0:   log likelihood = -136.69232  (not concave)
lf1debug:  Begin derivative-comparison report ————————————————————
lf1debug:  mreldif(gradient vectors) =  .8496778
lf1debug:  Warning:  evaluator did not compute Hessian matrix
lf1debug:  End derivative-comparison report ————————————————————
Iteration 1:   log likelihood = -124.11726
```

The `mreldif()`s we now observe are

iteration 0:   0.94
iteration 1:   0.85
iteration 2:   1.27
iteration 3:   1.02
iteration 4:   0.99
iteration 5:   0.98
iteration 6:   0.98

Specifying the `gradient` option allows us to spot that the problem arises in calculating the derivative of the second equation. We can easily see this when we look at the relative differences of the gradient elements in the iteration log:

```
lf1debug:   Begin derivative-comparison report ————————————————————
lf1debug:   mreldif(gradient vectors) = .8496778

lf1debug:   evaluator calculated gradient:
           eq1:        eq1:        eq1:        eq1:        eq1:        eq2:
           1b.          2.          3.
           drug        drug        drug         age       _cons       _cons
r1            0    12.10675    11.93993    1358.373    25.85834    54.63659

lf1debug:   numerically calculated gradient (used for stepping):
           eq1:        eq1:        eq1:        eq1:        eq1:        eq2:
           1b.          2.          3.
           drug        drug        drug         age       _cons       _cons
r1            0    12.10675    11.93993    1358.373    25.85834    7.363412

lf1debug:   relative differences:
           eq1:        eq1:        eq1:        eq1:        eq1:        eq2:
           1b.          2.          3.
           drug        drug        drug         age       _cons       _cons
r1            0    4.52e-11    6.31e-12    9.77e-12    2.50e-12    .8496778
```

Note the large value for eq2:_cons and the small values for eq1. This tells us that the derivatives for eq2 are either incorrectly derived or badly coded but that the eq1 derivatives are correct.

7.2.2 Second derivatives

As mentioned in the previous section, in real life the probit and Weibull evaluators we wrote in chapters 5 and 6 would not have worked the first time we tried. Without a doubt, we would have made an error in producing the Hessian, either substantively because we took the derivatives incorrectly or mechanically in how we programmed the evaluator.

Similar to starting with a derivative-free evaluator and then programming and testing our first derivatives, here we start with a working method-lf1 or method-d1 evaluator and make the changes necessary to convert it to method-lf2 or method-d2. This time we specify method-lf2debug or method-d2debug in the ml model statement. They are just like methods lf2 and d2 except that they check the analytical derivatives against derivatives calculated numerically.

Here is the result of using method d2debug to check the Hessian calculations in myprobit_d2:

(Continued on next page)

```
. sysuse auto
(1978 Automobile Data)

. ml model d2debug myprobit_d2 (foreign = mpg weight)

. ml max

initial:       log likelihood = -51.292891
alternative:   log likelihood = -45.055272
rescale:       log likelihood = -45.055272
d2debug:  Begin derivative-comparison report ──────────────────────
d2debug:  mreldif(gradient vectors) =   .0002445
d2debug:  mreldif(Hessian matrices) =    .000014
d2debug:  End derivative-comparison report ────────────────────────
Iteration 0:   log likelihood = -45.055272

d2debug:  Begin derivative-comparison report ──────────────────────
d2debug:  mreldif(gradient vectors) =   9.12e-10
d2debug:  mreldif(Hessian matrices) =   .0027612
d2debug:  End derivative-comparison report ────────────────────────
Iteration 1:   log likelihood = -27.908441

d2debug:  Begin derivative-comparison report ──────────────────────
d2debug:  mreldif(gradient vectors) =   3.19e-09
d2debug:  mreldif(Hessian matrices) =   1.21e-07
d2debug:  End derivative-comparison report ────────────────────────
Iteration 2:   log likelihood = -26.863584

d2debug:  Begin derivative-comparison report ──────────────────────
d2debug:  mreldif(gradient vectors) =   2.20e-06
d2debug:  mreldif(Hessian matrices) =   2.58e-08
d2debug:  End derivative-comparison report ────────────────────────
Iteration 3:   log likelihood = -26.844207

d2debug:  Begin derivative-comparison report ──────────────────────
d2debug:  mreldif(gradient vectors) =   .0003472
d2debug:  mreldif(Hessian matrices) =   6.72e-09
d2debug:  End derivative-comparison report ────────────────────────
Iteration 4:   log likelihood = -26.844189

d2debug:  Begin derivative-comparison report ──────────────────────
d2debug:  mreldif(gradient vectors) =   .0003653
d2debug:  mreldif(Hessian matrices) =   7.08e-09
d2debug:  End derivative-comparison report ────────────────────────
Iteration 5:   log likelihood = -26.844189
                                          Number of obs   =         74
                                          Wald chi2(2)    =      20.75
Log likelihood = -26.844189               Prob > chi2     =     0.0000
```

foreign	Coef.	Std. Err.	z	P>|z|	[95% Conf. Interval]	
mpg	-.1039503	.0515689	-2.02	0.044	-.2050235	-.0028772
weight	-.0023355	.000566	-4.13	0.000	-.003445	-.0012261
_cons	8.275464	2.554141	3.24	0.001	3.269439	13.28149

The difference between methods **d2** and **d2debug** is that method **d2debug** provides a derivative comparison report interspersed in the iteration log. For instance, before iteration 1, we see the following:

```
d2debug:  Begin derivative-comparison report ───────────────────────
d2debug:  mreldif(gradient vectors) =  9.12e-10
d2debug:  mreldif(Hessian matrices) =  .0027612
d2debug:  End derivative-comparison report ─────────────────────────
Iteration 1:   log likelihood = -27.908441
```

Whenever `ml` needs derivatives, method `d2debug` calls the likelihood evaluator just as method `d2` would, but method `d2debug` also calculates the derivatives numerically. It then displays a comparison of the derivatives calculated both ways and then returns to `ml` the *numerically calculated derivatives*. Thus `d2debug` proceeds to a correct answer, even if an analytical derivative calculation (first or second) is incorrect.

`d2debug` reports comparisons for both first and second derivatives. The issues concerning comparing first derivatives are the same as for `d1debug`—see section 7.2.1; it is the value of the `mreldif()` in the middle iterations that matters.

Concerning second derivatives, it is the comparison in the last few iterations that matters. The numeric second derivatives `d2debug` calculates can be poor—very poor— in early and middle iterations:

iteration 0:	1.40e–05
iteration 1:	2.76e–03
iteration 2:	1.21e–07
iteration 3:	2.58e–08
iteration 4:	6.72e–09
iteration 5:	7.08e–09

That the numerical derivatives are poor in early iterations does not matter; by the last iteration, the numerical derivative will be accurate, and that is the one you should use to compare. Remember, the purpose of the output is not to test the accuracy of `ml`'s numeric derivative calculator but to verify that the analytic derivatives roughly agree with the numeric derivatives. If they do, we can safely assume that the analytic derivatives are correct.

To demonstrate what you might see when the analytic derivatives are somehow wrong, suppose our probit evaluator is `myprobit_bad_d2`. This is a copy of `myprobit_d2` in which the line that reads

```
mlmatsum 'lnf' 'H' = -'g1'*('g1'+'xb'), eq(1)
```

was modified to read

```
tempvar h
gen double 'h' = 'g1'*('g1'+'xb') if $ML_y1 == 1
replace    'h' = -'g1'*('g1'-'xb') if $ML_y1 == 0
mlmatsum 'lnf' 'H' = -'h', eq(1)
```

With `d2debug`, our overall results would still be correct because `d2debug` uses the numeric derivatives (first and second), not the analytic values we calculate. But the `mreldif()`s for the Hessian are

```
iteration 0:    3.581
iteration 1:    21.72
iteration 2:    45.15
iteration 3:    48.94
iteration 4:    46.73
iteration 5:    46.73
```

Clearly our analytic Hessian calculation is incorrect. Whenever the `mreldif()` value is greater than 10^{-6} in the last iteration, you should suspect problems.

When using d2debug, you can specify `ml maximize`'s `hessian` option to see the analytically calculated Hessian matrix compared with the numerically calculated Hessian matrix, or specify `gradient` to see a comparison of gradient vectors, or both.

7.3   ml trace

Between `ml check` and methods `d1debug` and `d2debug`, you should be able to find and fix most problems.

Even so, there may be a bug in your code that appears only rarely because it is in a part of the code that is rarely executed. The way we typically find such bugs is to **set trace on** and then review a log of the output.

Do not do this with `ml` because the trace will contain a trace of `ml`'s code, too, and there is a lot of it. Instead, type **ml trace on**. That will trace the execution of your program only:

```
. sysuse auto
(1978 Automobile Data)
. ml model lf myprobit_lf (foreign = mpg weight)
. ml trace on                  // trace the evaluator
. ml maximize
-> myprobit_lf __000001 __000002
                    - 'begin'
                    = capture noisily version 11: myprobit_lf __000001 __000002
                    ─────────────────────────────────────── begin myprobit_lf ──
                      - version 11
                      - args lnfj xb
                      - quietly replace 'lnfj' = ln(normal( 'xb')) if $ML_y1 == 1
                      = quietly replace __000001 = ln(normal( __000002)) if foreign
>   == 1
                      - quietly replace 'lnfj' = ln(normal(-'xb')) if $ML_y1 == 0
                      = quietly replace __000001 = ln(normal(-__000002)) if foreign
>   == 0
                    ─────────────────────────────────────────── end myprobit_lf ──
                    - 'end'
                    = set trace off
initial:        log likelihood = -51.292891
-> myprobit_lf __000001 __000002
                    - 'begin'
                    = capture noisily version 11: myprobit_lf __000001 __000002
                    ─────────────────────────────────────── begin myprobit_lf ──
                      - version 11
                      - args lnfj xb
                      - quietly replace 'lnfj' = ln(normal( 'xb')) if $ML_y1 == 1
                      = quietly replace __000001 = ln(normal( __000002)) if foreign
>   == 1
                      - quietly replace 'lnfj' = ln(normal(-'xb')) if $ML_y1 == 0
                      = quietly replace __000001 = ln(normal(-__000002)) if foreign
>   == 0
                    ─────────────────────────────────────────── end myprobit_lf ──
                    - 'end'
                    = set trace off
    (output omitted )
```

You may type `ml trace off` later, but that is seldom necessary because the next time you issue an `ml model` statement, trace is turned off when the previous problem is cleared.

8　Setting initial values

Typically, but not always, you can get away without specifying initial values. In the examples so far in this book, we have typed

```
. ml model ...
. ml maximize
```

and not bothered at all with initial values. We have been able to do that because `ml maximize` is pretty smart about coming up with initial values on its own. By default,

1. `ml maximize` assumes an initial value vector $\boldsymbol{\beta}_0 = (0, 0, \ldots, 0)$.

2. If $\ln L(\boldsymbol{\beta}_0)$ can be evaluated, `ml maximize` tries to improve the initial values deterministically by rescaling $\boldsymbol{\beta}_0$, overall and equation by equation.

   If $\ln L(\boldsymbol{\beta}_0)$ cannot be evaluated,

   a. `ml maximize` uses a pseudorandom algorithm to hunt for a $\boldsymbol{\beta}_0$ vector that will allow $\ln L(\boldsymbol{\beta}_0)$ to be evaluated and, once found,

   b. `ml maximize` tries to improve the initial values deterministically by rescaling $\boldsymbol{\beta}_0$, overall and equation by equation.

`ml maximize` does all of this *before* attempting to maximize the function using the standard techniques.

Actually, `ml maximize` does not do all of this. It calls `ml search` to do all of this. There are good reasons to invoke `ml search` yourself, however, and there are other ways to set the initial values, too:

1. Invoking `ml search` does a more thorough job of setting initial values. When `ml search` is called by `ml maximize`, it tries to be quick and find something that is merely good enough.

2. `ml plot` graphs slices of the likelihood function (one parameter at a time) and resets the parameter to the value corresponding to the maximum of the slice. `ml plot` is great entertainment, too.

3. You can also use `ml init`. This is a method of last resort when you are working interactively, but programmers often use this method because they are willing to perform extra work to obtain good initial values, especially because it is faster than using either of the above techniques.

You may combine the three methods above. If you know that 5 is a good value for one of the parameters, you can use `ml init` to set it. You can then use `ml search` to grope around from there for even better values—`ml search` cannot make things worse. Then you can use `ml plot` to examine another of the parameters that you know is likely to cause problems and so set it to a reasonable value. You can then use `ml search` again to improve things. If you kept working like this, you could dispense with the maximization step altogether.

8.1   ml search

The syntax of `ml search` is

`ml search` $\big[\,\big[\,/\,\big]$ *eqname* $\big[\,:\,\big]$ $\#_{lb}$ $\#_{ub}\big]$ $\big[\,\big[\,/\,\big]$ *eqname* $\big[\,:\,\big]$ $\#_{lb}$ $\#_{ub}\big]$ $\big[\ldots\big]$ $\big[\,,$
        `repeat(`$\#$`)` `nolog` `trace` `restart` `norescale`$\big]$

`ml search` looks for better (or feasible) initial values. By feasible, we mean values of the parameters for which the log likelihood can be calculated. For instance, in the log-likelihood function

$$\ln \ell_j = \ln \phi \left(\frac{y_j - \mu_j}{\sigma} \right) - \ln \sigma \tag{8.1}$$

any starting value with $\sigma \leq 0$ is not feasible.

The simplest syntax for `ml search` is

    . ml search

You can also follow `ml search` with the name of an equation and two numbers, such as

    . ml search eq1 -3 3

and `ml search` will restrict the search to those bounds. Using the log likelihood in (8.1), suppose that we type

    . ml model lf mynormal_lf (mu: y = x1 x2) /sigma

It would make sense—but not be necessary—to restrict `ml search` to positive and reasonable values for σ:

    . ml search sigma .1 5

You specify bounds for equations, not individual parameters. That means you do not specify bounds for the coefficient on, say, `x1`; rather, you specify bounds for the equation `mu`. In this example, if `y` varied between 1 and 10, it would make sense to specify

    . ml search sigma .1 5 mu 1 10

In practice, it is seldom worth the trouble to specify the bounds because `ml search` comes up with bounds on its own by probing the likelihood function.

`ml search` defines initial values on the basis of equations by setting the intercept (coefficient on `_cons`) to numbers randomly chosen within the range and setting the remaining coefficients to zero. If the equation has no intercept, a number c within the range is chosen, and the coefficients are then set from the regression $c = \mathbf{x}\boldsymbol{\beta}_0 + e$.

`ml search` can be issued repeatedly.

Using ml search

We strongly recommend that you use `ml search`. You can specify bounds for the search, but typically you need only type

```
. ml search
```

We recommend that you use `ml search`, even if you specify initial values (which you do with `ml init`, as we will explain later):

```
. ml model lf mynormal_lf (mu: mpg=weight displ) /sigma
. ml init /sigma = 3
. ml search
```

If you use `ml search`, any previous starting values will be taken as a suggestion, so it would probably be better to type

```
. ml model lf mynormal_lf (mu: mpg=weight displ) /sigma
. ml search /sigma 1 10
```

and so restrict the search for `/sigma` to the range $1 < \sigma < 10$ rather than suggesting $\sigma = 3$ and leaving the range at $-\infty < \sigma < \infty$. Of course, we could do both, but the suggestion is not worth nearly as much as the bounds.

Not specifying an equation does not mean the equation is not searched; it means just that no special bounds are specified for the search.

Determining equation names

Remember, you do not have to specify bounds, but the search will be more efficient if you do. You specify the bounds in terms of the equations, not the individual parameters.

`ml model` labels equations as `eq1`, `eq2`, ... if you do not specify otherwise. You can name an equation by coding "(*eqname*: ...)" when you specify the equation. If you are confused, you can type `ml query` any time after the `ml model` statement to find out what you have defined and where you are. Suppose that we set up a problem by typing

```
. sysuse auto
(1978 Automobile Data)
. ml model lf myprobit_lf (foreign = mpg weight)
```

Then `ml query` will report

```
. ml query
Settings for moptimize() ──────────────────────────────────────────
Version:                                        1.00
Evaluator
    Type:                                       lf
    Function:                                   myprobit_lf
Dependent variable:                             foreign
Equation 1:
    predictor     1:                            mpg
    predictor     2:                            weight
                                                _cons

  (output omitted)

Starting values
    Coefficient values
        1:                                      0
        2:                                      0
        3:                                      0
    Function value:                             .
Current status
    Coefficient values:                         unknown
    Function value:                             .
    Converged:                                  no
```

Our equation is called `eq1`. Next let's search over the restricted range $(-3, 3)$:

```
. ml search eq1 -3 3
initial:       log likelihood = -51.292891
alternative:   log likelihood = -45.055272
rescale:       log likelihood = -45.055272
```

If we now do an `ml query`, we see

```
. ml query
Settings for moptimize() ──────────────────────────────────────────
Version:                                        1.00
Evaluator
    Type:                                       lf
    Function:                                   myprobit_lf
Dependent variable:                             foreign
Equation 1:
    predictor     1:                            mpg
    predictor     2:                            weight
                                                _cons

  (output omitted)

Starting values
    Coefficient values
        1:                                      0
        2:                                      0
        3:                                      -.5
    Function value:                             -45.0552724
```

```
Current status
    Coefficient values
        1:                              0
        2:                              0
        3:                              -.5
    Function value:                     -45.0552724
    Converged:                          no
```

ml remembers the search bounds we specified. If we simply type `ml search`, the $(-3, 3)$ bounds are assumed.

8.2   ml plot

The basic syntax of `ml plot` is

ml p̲l̲o̲t̲ [*eqname:*] *name* [*#* [*#* [*#*]]]

`ml plot` graphically helps fill in initial values. In practice, this command is more cute than useful because `ml search` works so well. Still, `ml plot` focuses on individual parameters, not on the overall equation as with `ml search`, and that can sometimes be useful.

`ml plot` graphs the likelihood function for one parameter at a time and then replaces the value of the parameter according to the maximum value of the likelihood function. These graphs are cuts of the likelihood function—they are not graphs of the profile-likelihood function.

`ml plot` is not an alternative to `ml search`; you can use them both, in any order, and repeatedly. We recommend that you use `ml search` first and then try `ml plot`. This way, `ml plot` is at least starting with feasible values.

The simplest syntax is to type `ml plot` followed by a coefficient name:

```
. ml plot _cons
. ml plot /sigma
. ml plot eq2:_cons
. ml plot eq1:foreign
```

When the name of the equation is not explicitly supplied, the first equation, whatever the name, is assumed (thus `_cons` probably means `eq1:_cons`).

For instance, after setting up a model,

```
. sysuse auto
(1978 Automobile Data)
. ml model lf myprobit_lf (foreign = mpg weight)
```

we might type the following, which results in the plot in figure 8.1.

```
. ml plot _cons
              reset _cons =          -.5  (was           0)
         log likelihood = -45.055272  (was -51.292891)
```

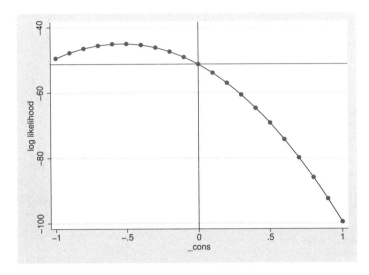

Figure 8.1. Probit log likelihood versus the constant term.

By default, `ml plot` graphs 21 values of the likelihood within ±1 of the current value of the parameter. Rather than displaying a graph, `ml plot` might present you with a message:

```
. ml plot _cons
(all values are missing)
```

This could mean that

1. The current value for the coefficient on `_cons` and all values within ±1 of the current value are infeasible, so you need to find a better range.

2. The current value for the coefficient on `_cons` is infeasible, and the range is too wide. Remember, `ml plot` graphs 21 points; perhaps values within ±1/100 of the range would be feasible.

That is why we recommend that you first use `ml search`.

You can also type `ml plot` followed by the name of a parameter and then a number. For example, the following results in the plot in figure 8.2.

```
. ml plot weight .01
            keeping weight =              0
            log likelihood = -45.055272
```

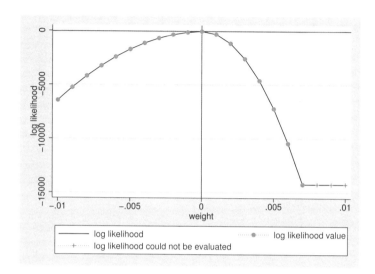

Figure 8.2. Probit log likelihood versus the `weight` coefficient.

Here we graph ±0.01 around the current value of the coefficient on `weight`. If we specified two numbers, such as `ml plot mpg -.5 1`, we would graph the range $[\beta - 0.5, \beta + 1]$ where β would be the current value of the coefficient on `mpg`. If we typed a third number, that many points (plus 1) would be graphed.

`ml` maintains the current value of the initial-value vector β_0. β_0 is set to zero when we issue an `ml model` statement, and thereafter other routines can update it. Each time we perform an `ml search`, β_0 is updated. Each time we use `ml plot`, β_0 is updated if a better value for the parameter is found. Thus you may use `ml search` and `ml plot` together. The current value of β_0 is reported by `ml query`.

8.3   ml init

The syntax of `ml init` is

`ml init { [ eqname: ]name=# | /eqname=# } [ ... ]`

`ml init # [ # ... ], copy`

`ml init matname [ , skip copy ]`

ml init allows you to specify initial values explicitly. ml init without arguments resets
the parameter vector to contain all zeros.

Using ml init

ml init can be issued not at all, once, or repeatedly.

To specify initial values, type the coefficient's name, an equal sign, and the value to
which it should be set. For instance,

```
. ml model lf myprobit_lf (foreign=mpg weight)
. ml init mpg=1.5 _cons=-.5
```

More than one coefficient may be specified and not all coefficients need be specified.
Initial values for omitted coefficients remain unchanged.

In multiple-equation models, specify the equation name and a colon to identify the
coefficient:

```
. ml model lf myprog (y1=x1 x2 x3) (y2=x1 x4) /sigma1 /sigma2
. ml init eq1:_cons=5 eq2:_cons=3
```

Here the first two equations are named eq1 and eq2 because we did not supply our own
equation names in the ml model statement. When you omit the equation name in the
ml init statement, as in

```
. ml init x1=3.14
```

Stata assumes that you are referring to the first equation.

Also, when you set the coefficient on the constant, you can refer either to *eqname*:_cons or to /*eqname*—they mean the same thing:

```
. ml init /sigma1=3 sigma2:_cons=4
```

You can even use the slash notation with equations that have more than just a constant.
The line

```
. ml init eq1:_cons=5 eq2:_cons=3
```

could just as well be specified

```
. ml init /eq1=5 /eq2=3
```

In the rare instance when you wish to specify initial values for all parameters, you
can type them one after the other:

```
. ml init 5 7 2 9  3 2 1  1  1, copy
```

In most such cases, however, you would have a vector containing the values, and it is
easier to use the syntax described next.

Loading initial values from vectors

You can obtain initial values from row vectors (matrices). This is rarely done, and this feature is included for programmers who want to obtain initial values from some other estimator.

```
. regress y1 x2 x3
. matrix b0 = e(b)
. ml model lf myprog (y1=x1 x2 x3) (y2=x1 x4) /sigma1 /sigma2
. ml init b0
```

By default, values are copied from the specified vector by name; it does not matter that the vector obtained from `regress` is in a different order than that used by `ml`. In this case, the initial values were applied to the first equation because the vector `b0` fetched from `regress` did not include equation names; if we had set the equation names, the vector could have applied to the second equation.

Our use of the name `b0` has no special significance in the above example; we could just as well have called it `initvec` or anything else.

9 Interactive maximization

You type `ml maximize` to obtain estimates once you have issued the `ml model` statement. The process can be as short as

```
. ml model ...
. ml maximize
```

but we recommend

```
. ml model ...
. ml check
. ml search
. ml maximize
```

9.1 The iteration log

If you type `ml maximize`, you will see something like

```
. ml maximize
initial:       log likelihood =       -744
alternative:   log likelihood = -356.14276
rescale:       log likelihood = -200.80201
rescale eq:    log likelihood = -136.69232
Iteration 0:   log likelihood = -136.69232  (not concave)
Iteration 1:   log likelihood = -124.13044
Iteration 2:   log likelihood = -113.99049
...
```

The first four lines of the output really have to do with `ml search`. The first line, `initial`, reports the log likelihood at the initial values β_0. Here the log likelihood is -744, but it might be reported as `-<inf>`, meaning that the likelihood could not be evaluated at β_0. That would not cause problems. The next line, `alternative`, reports the result of a random search for an alternative β_0.

`ml` then takes the better of the two. `rescale` reports the results of rescaling the entire β_0 vector—that is, searching for a scalar c such that $\ln L(c\beta_0) > \ln L(\beta_0)$. `rescale eq` then repeats the search for c, equation by equation.

At this point, maximization has not even begun: starting values are being improved, and already we have improved the log likelihood from -744 to -136.7.

The goal is to find a β_0 so that the chosen optimization algorithm has a reasonable chance of converging to the maximum.

At iteration 0, the maximization process really starts, and note that iteration 0 reports an opening log likelihood that is equal to where the searcher left off.

The purpose of an iteration is to take $\boldsymbol{\beta}_i$ and produce a better value $\boldsymbol{\beta}_{i+1}$, where i is the iteration number. Iteration 0 takes $\boldsymbol{\beta}_0$ and produces $\boldsymbol{\beta}_1$.

An iteration is the calculation of a new direction vector. When `ml` reports an iteration, it is reporting $\ln L(\boldsymbol{\beta}_i)$, the value received. The iteration calculates a direction vector, $\mathbf{d}_i$, based on the function at that point, where $\mathbf{d}_i$ is some function of the gradient and Hessian (or an alternative to the Hessian when using the BHHH, DFP, or BFGS algorithms). The iteration then steps in that direction until the log likelihood ceases to improve. $\boldsymbol{\beta}_{i+1}$ is defined as $\boldsymbol{\beta}_i + s\mathbf{d}_i$, where s is the stepsize taken. The process then repeats.

Thus the derivatives are calculated once per iteration. Between iterations, log-likelihood evaluations are made.

9.2   Pressing the Break key

You can press *Break* while `ml` is iterating:

```
. ml maximize
initial:       log likelihood =        -744
alternative:   log likelihood = -356.14276
rescale:       log likelihood = -200.80201
rescale eq:    log likelihood = -136.69232
Iteration 0:   log likelihood = -136.69232  (not concave)
Iteration 1:   log likelihood = -124.13044
Iteration 2:   log likelihood = -113.99049
Iteration 3:   log likelihood = -110.30732
—Break—
r(1);
```

Nothing is lost; you can pick up right from where you left off:

```
. ml maximize
initial:       log likelihood = -110.26748
rescale:       log likelihood = -110.26748
rescale eq:    log likelihood = -110.26748
Iteration 0:   log likelihood = -110.26748
Iteration 1:   log likelihood = -110.26736
Iteration 2:   log likelihood = -110.26736
(estimation output appears)
```

The iteration numbers change, but note that the opening log-likelihood values are the same as the values just before you pressed *Break*.

In between, you can use `ml query` to obtain a summary of where you are:

```
. ml query
Settings for moptimize() ─────────────────────────────────────────
Version:                                 1.00
Evaluator
    Type:                                d1
    Function:                            myweibullb_d1
Dependent variables
    1:                                   studytime
    2:                                   died
Equation 1:
    predictor    1:                      1b.drug
    predictor    2:                      2.drug
    predictor    3:                      3.drug
    predictor    4:                      age
                                         _cons
Equation 2:
    predictors:                          <none>
                                         _cons
User-defined information:                <none>
Weights:                                 <none>
Optimization technique:                  nr
Singular H method:                       m-marquardt
Convergence
    Maximum iterations:                  16000
    Tolerance for the function value:    1.0000e-07
    Tolerance for the parameters:        1.0000e-06
    Tolerance for the scaled gradient:   1.0000e-05
    Warning:                             on
Iteration log
    Value ID:                            log likelihood
    Trace dots:                          off
    Trace value:                         on
    Trace tol:                           off
    Trace step:                          off
    Trace coefficients:                  off
    Trace gradient:                      off
    Trace Hessian:                       off
Constraints:                             yes
Search for starting values:              off
Starting values
    Coefficient values
        1:                               0
        2:                               0
        3:                               0
        4:                               0
        5:                               4
        6:                               -.25
    Function value:                      -136.692316
```

```
Current status
    Coefficient values
        1:                                          0
        2:                                          1.01373004
        3:                                          1.4600912
        4:                                          -.067167153
        5:                                          6.05907233
        6:                                          .558621114
    Function value:                                 -110.307324
    Converged:                                      no
```

Or you can type `ml report`:

```
. ml report
Current coefficient vector:
           eq1:      eq1:      eq1:      eq1:      eq1:      eq2:
           1b.        2.        3.
           drug      drug      drug       age     _cons     _cons
r1            0   1.01373  1.460091  -.0671672  6.059072  .5586211
Value of log likelihood function = -110.26748
Gradient vector (length =  5.77133):
           eq1:      eq1:      eq1:      eq1:      eq1:      eq2:
           1b.        2.        3.
           drug      drug      drug       age     _cons     _cons
r1            0  .0029079  .0004267  5.769678  .1014669  -.0936064
```

You can even use any of the methods for setting initial values—`ml search`, `ml plot`, or `ml init`—to attempt to improve the parameter vector before continuing with the maximization by typing `ml maximize`.

9.3   Maximizing difficult likelihood functions

When you have a difficult likelihood function, specify `ml maximize`'s `difficult` option. The `difficult` option varies how a direction is found when the negative Hessian, $-\mathbf{H}$, cannot be inverted and thus the usual direction calculation $\mathbf{d} = \mathbf{g}(-\mathbf{H})^{-1}$ cannot be calculated. `ml` flags such instances by mentioning that the likelihood is "not concave":

```
. ml maximize
initial:       log likelihood =       -744
alternative:   log likelihood = -356.14276
rescale:       log likelihood = -200.80201
rescale eq:    log likelihood = -136.69232
Iteration 0:   log likelihood = -136.69232  (not concave)
Iteration 1:   log likelihood = -124.13044
Iteration 2:   log likelihood = -113.99049
...
```

`ml`'s usual solution is to mix a little steepest ascent into the calculation, which works well enough in most cases.

When you specify `difficult`, `ml` goes through more work. First, `ml` computes the eigenvalues of $-\mathbf{H}$ and then, for the part of the orthogonal subspace where the eigenvalues are negative or small positive numbers, `ml` uses steepest ascent. In the other subspace, `ml` uses a regular Newton–Raphson step.

A "not concave" message does not necessarily indicate problems. It can be particularly maddening when `ml` reports "not concave" iteration after iteration as it crawls toward a solution. It is only a problem if `ml` reports "not concave" at the last iteration. The following log has "not concave" messages along the way and at the end:

```
. ml maximize
initial:        log likelihood =  -2702.155
alternative:    log likelihood =  -2702.155
rescale:        log likelihood = -2315.2187
rescale eq:     log likelihood = -2220.5288
Iteration 0:    log likelihood = -2220.5288  (not concave)
Iteration 1:    log likelihood = -2152.5211  (not concave)
Iteration 2:    log likelihood = -2141.3916  (not concave)
Iteration 3:    log likelihood = -2140.7298  (not concave)
Iteration 4:    log likelihood = -2139.5807  (not concave)
Iteration 5:    log likelihood = -2139.0407  (not concave)
(output omitted )
Iteration 40:   log likelihood = -2138.0806  (not concave)
Iteration 41:   log likelihood = -2138.0803  (not concave)
Iteration 42:   log likelihood = -2138.0801  (not concave)
Iteration 43:   log likelihood = -2138.0799  (not concave)
(results presented; output omitted)
```

Specifying `difficult` sometimes helps:

```
. ml maximize, difficult
initial:        log likelihood =  -2702.155
alternative:    log likelihood =  -2702.155
rescale:        log likelihood = -2315.2187
rescale eq:     log likelihood = -2220.5288
Iteration 0:    log likelihood = -2220.5288  (not concave)
Iteration 1:    log likelihood = -2143.5267  (not concave)
Iteration 2:    log likelihood = -2138.2358
Iteration 3:    log likelihood = -2137.9355
Iteration 4:    log likelihood = -2137.8971
Iteration 5:    log likelihood = -2137.8969
(results presented; output omitted)
```

Here `difficult` not only sped estimation but solved the problem of nonconcavity at the final iteration. Note that the final log likelihood without `difficult` was -2138.1, which is less than -2137.9. The difference may seem small, but it was enough to place the final results on a concave portion of the likelihood function.

Usually, however, specifying `difficult` just slows things down.

10 Final results

Once you have typed `ml maximize` and obtained the results, you can do anything Stata usually allows after estimation. That means you can use `test` to perform hypothesis tests, `predict` and `predictnl` to obtain predicted values, and so on.

10.1 Graphing convergence

A special feature of `ml` is that, once estimation output has been produced, you can graph the convergence by typing `ml graph`. Typing the following produces the output shown in figure 10.1.

```
. sysuse cancer
(Patient Survival in Drug Trial)
. ml model lf1 myweibull_lf1 (studytime died = i.drug age) ()
. ml maximize, nolog
```

					Number of obs Wald chi2(3) Prob > chi2	= = =	48 35.25 0.0000

Log likelihood = -110.26736

		Coef.	Std. Err.	z	P>|z|	[95% Conf. Interval]	
eq1							
	drug						
	2	1.012966	.2903917	3.49	0.000	.4438086	1.582123
	3	1.45917	.2821195	5.17	0.000	.9062261	2.012114
	age	-.0671728	.0205688	-3.27	0.001	-.1074868	-.0268587
	_cons	6.060723	1.152845	5.26	0.000	3.801188	8.320259
eq2							
	_cons	.5573333	.1402154	3.97	0.000	.2825163	.8321504

```
. ml graph
```

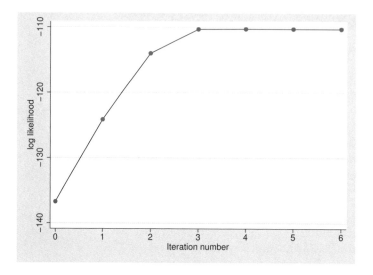

Figure 10.1. Weibull log-likelihood values by iteration.

Given how Stata's optimizers work, for smooth likelihood functions the graph may jump around at first, but once it gets close to the maximum, it should move smoothly and with smaller and smaller steps toward it; see section 1.4. This graph shows that the process has truly converged.

The basic syntax of `ml graph` is

ml <u>graph</u> $\left[\,\#\,\right]$

$\#$ specifies the number of iterations to be graphed, counted from the last. For example, `ml graph 5` would graph the last five iterations. `ml graph` keeps track of the last 20 iterations, sufficient for determining convergence.

10.2   Redisplaying output

Once `ml maximize` has displayed results, you can use `ml display` to redisplay them.

```
. ml display
```

```
                                              Number of obs   =         48
                                              Wald chi2(3)    =      35.25
        Log likelihood = -110.26736           Prob > chi2     =     0.0000
```

		Coef.	Std. Err.	z	P>|z|	[95% Conf.	Interval]
eq1							
	drug						
	2	1.012966	.2903917	3.49	0.000	.4438086	1.582123
	3	1.45917	.2821195	5.17	0.000	.9062261	2.012114
	age	-.0671728	.0205688	-3.27	0.001	-.1074868	-.0268587
	_cons	6.060723	1.152845	5.26	0.000	3.801188	8.320259
eq2							
	_cons	.5573333	.1402154	3.97	0.000	.2825163	.8321504

The syntax of ml display is

ml <u>di</u>splay [, <u>noh</u>eader <u>nof</u>ootnote level(#) <u>f</u>irst neq(#) showeqns <u>pl</u>us

         <u>noc</u>nsreport <u>noo</u>mitted vsquish <u>noe</u>mptycells <u>base</u>levels

         allbaselevels cformat(% *fmt*) pformat(% *fmt*) sformat(% *fmt*)

         <u>coef</u>legend *eform_option*]

where *eform_option* is <u>ef</u>orm(*string*), <u>ef</u>orm, hr, shr, <u>ir</u>r, or, or <u>rr</u>r.

noheader suppresses the display of the header above the coefficient table. In the non-
svy context, the header displays the final log-likelihood (criterion) value, the number
of observations, and the model significance test. In the svy context, the header
displays the number of observations, the estimate of the population size, and the
model significance test.

nofootnote suppresses the footnote display below the coefficient table. The footnote
displays a warning if the model fit did not converge within the specified number of
iterations. Use ml footnote to display the warning if 1) you add to the coefficient
table using the plus option or 2) you have your own footnotes and want the warning
to be last.

level(#) is the standard confidence-level option. It specifies the confidence level, as a
percentage, for confidence intervals of the coefficients. The default is level(95) or
as set by set level; see [U] **20.7 Specifying the width of confidence intervals**.

first displays a coefficient table reporting results for the first equation only, and the report makes it appear that the first equation is the only equation. This option is used
by programmers who estimate ancillary parameters in the second and subsequent
equations and who wish to report the values of such parameters themselves.

neq(#) is an alternative to first. neq(#) displays a coefficient table reporting results
for the first # equations. This option is used by programmers who estimate ancillary
parameters in the #+1 and subsequent equations and who wish to report the values
of such parameters themselves.

showeqns is a seldom-used option that displays the equation names in the coefficient
table. ml display uses the numbers stored in e(k_eq) and e(k_aux) to determine
how to display the coefficient table. e(k_eq) identifies the number of equations,
and e(k_aux) identifies how many of these are for ancillary parameters. The first
option is implied when showeqns is not specified and all but the first equation are
for ancillary parameters.

plus displays the coefficient table, but rather than ending the table in a line of dashes,
ends it in dashes–plus-sign–dashes. This is so that programmers can write additional
display code to add more results to the table and make it appear as if the combined
result is one table. Programmers typically specify plus with the first or neq()
option. plus implies nofootnote.

nocnsreport suppresses the display of constraints above the coefficient table. This
option is ignored if constraints were not used to fit the model.

noomitted specifies that variables that were omitted because of collinearity not be
displayed. The default is to include in the table any variables omitted because of
collinearity and to label them as "(omitted)".

vsquish specifies that the blank space separating factor-variable terms or time-series–
operated variables from other variables in the model be suppressed.

noemptycells specifies that empty cells for interactions of factor variables not be displayed. The default is to include in the table interaction cells that do not occur in
the estimation sample and to label them as "(empty)".

baselevels and allbaselevels control whether the base levels of factor variables
and interactions are displayed. The default is to exclude from the table all base
categories.

  baselevels specifies that base levels be reported for factor variables and for interactions whose bases cannot be inferred from their component factor variables.

  allbaselevels specifies that all base levels of factor variables and interactions be
  reported.

cformat(%fmt) specifies how to format coefficients, standard errors, and confidence
limits in the coefficient table.

pformat(%fmt) specifies how to format *p*-values in the coefficient table.

sformat(%fmt) specifies how to format test statistics in the coefficient table.

coeflegend specifies that the legend of the coefficients and how to specify them in an
expression be displayed rather than the coefficient table.

eform_option: `eform(`*string*`)`, `eform`, `hr`, `shr`, `irr`, `or`, and `rrr` display the coefficient
   table in exponentiated form: for each coefficient, $\exp(b)$ rather than b is displayed,
   and standard errors and confidence intervals are transformed. Display of the intercept, if any, is suppressed. *string* is the table header that will be displayed above the
   transformed coefficients and must be 11 characters or shorter in length—for example, `eform("Odds ratio")`. The options `eform`, `hr`, `shr`, `irr`, `or`, and `rrr` provide a
   default *string* equivalent to "`exp(b)`", "`Haz. Ratio`", "`SHR`", "`IRR`", "`Odds Ratio`",
   and "`RRR`", respectively. These options may not be combined.

`ml display` looks at `e(k_eform)` to determine how many equations are affected by
   an *eform_option*; by default, only the first equation is affected.

11 Mata-based likelihood evaluators

Mata is a full-featured matrix programming language built into Stata. It is designed for linear algebra and other advanced mathematical computation. Many complicated statistical models are written using matrix notation, and Mata makes programming those models in Stata easy. In this chapter, we illustrate how to write likelihood-evaluator programs in Mata while continuing to use Stata's `ml` commands to control the process. Advanced programmers who wish to work entirely within Mata can use the `moptimize()` suite of Mata functions; see [M-5] **moptimize( )**. This chapter assumes that you are already familiar with the basics of Mata and that you know how to write likelihood-evaluator programs using Stata's ado-language.

11.1 Introductory examples

The easiest way to see how likelihood evaluators are written in Mata is to compare a couple of them with evaluators that we have already discussed, which were written in ado-code.

11.1.1 The probit model

In section 4.2.1, we used the following method-`lf` likelihood evaluator for the probit model:

```
                                            begin myprobit_lf.ado
program myprobit_lf
        version 11
        args lnfj xb
        quietly replace `lnfj' = ln(normal( `xb')) if $ML_y1 == 1
        quietly replace `lnfj' = ln(normal(-`xb')) if $ML_y1 == 0
end
                                            end myprobit_lf.ado
```

To fit a probit model of `died` on `i.drug`, and `age`, we would type

```
. ml model lf myprobit_lf (died = i.drug age)
. ml maximize
```

To implement the same model using a Mata-based likelihood evaluator, we create the following do-file:

―――――――――――――――――――――――――――― begin `myprobit_lf_mata.do` ―――――

```
version 11
sysuse cancer, clear

mata:
void myprobit_lf(transmorphic scalar ML, real rowvector b,
                 real colvector lnfj)
{
        real vector depvar, xb
        depvar = moptimize_util_depvar(ML, 1)
        xb     = moptimize_util_xb(ML, b, 1)

        lnfj = depvar:*ln(normal(xb)) + (1:-depvar):*ln(normal(-xb))
}
end

ml model lf myprobit_lf() (died = i.drug age)
ml maximize
```

―――――――――――――――――――――――――――――― end `myprobit_lf_mata.do` ―――――

We wrote our evaluator and `ml` commands in a do-file so that we could define our Mata function `myprobit_lf`. We could type the function interactively, but that is error-prone, and if you type as badly as we do, it would take you several tries to type it in correctly. For most purposes, defining the Mata function in a do-file is the easiest alternative. Other alternatives include creating a Mata `.mo` file (see [M-3] **mata mosave**) and including the Mata code in the ado-file (see [M-1] **ado**), which is most practical if you are creating an ado-file, as described in chapter 13.

Method-`lf` likelihood evaluators in Mata always receive three arguments. The first argument is a structure that `ml` uses to record information about the current problem; we refer to this structure as the "problem handle". If you include declarations for the arguments to your function, the problem handle must be declared a `transmorphic scalar`; we named ours ML. We named the second argument b; this holds the vector of coefficients at which we are to evaluate the likelihood function and is passed as a `real rowvector`. Unlike `lf` evaluators written in ado-code, which receive the values of linear combinations of the independent variables and coefficients, `ml` always passes the raw vector of coefficients to evaluators written in Mata. The third argument is a column vector we called `lnfj` that is to contain the observation-level log-likelihood values; it is analogous to the `lnfj` temporary variable passed to ado-based `lf` evaluators.

The functions `moptimize_util_depvar()` and `moptimize_util_xb()` are utilities that aid in implementing likelihood evaluators. The first of these requires us to pass the problem handle as well as an equation number. In return, we obtain a vector containing the dependent variable for the kth equation, where k is the second argument passed to `moptimize_util_depvar()`. The second function, `moptimize_util_xb()` is analogous to the `mleval` command used in ado-based evaluators. We pass to it the problem handle, coefficient vector, and an equation number, and it returns the linear combination (model parameter) for that equation. At the end of our function, we replace the elements of vector `lnfj` with the current log-likelihood values.

The final portion of our do-file uses `ml model` to set up our problem and then calls `ml maximize` to fit it. The only difference between this call to `ml model` and calls used when likelihood evaluators are written as ado-files is that here we followed the name of our Mata-based likelihood evaluator with parentheses. The () after the name indicates to `ml` that we have a Mata-based evaluator. All other features of `ml model` and, in fact, all other `ml` subcommands work identically whether the evaluator is ado based or Mata based.

Executing our do-file, we obtain

```
. version 11
. sysuse cancer, clear
(Patient Survival in Drug Trial)
. mata:
------------------------------------------------ mata (type end to exit) ------
: void myprobit_lf(transmorphic scalar ML, real rowvector b,
>                  real colvector lnfj)
> {
>         real vector depvar, xb
>         depvar = moptimize_util_depvar(ML, 1)
>         xb      = moptimize_util_xb(ML, b, 1)
>
>         lnfj = depvar:*ln(normal(xb)) + (1:-depvar):*ln(normal(-xb))
> }
: end
-------------------------------------------------------------------------------

. ml model lf myprobit_lf() (died = i.drug age)
. ml maximize
initial:       log likelihood = -33.271065
alternative:   log likelihood = -31.427839
rescale:       log likelihood = -31.424556
Iteration 0:   log likelihood = -31.424556
Iteration 1:   log likelihood = -21.883453
Iteration 2:   log likelihood = -21.710899
Iteration 3:   log likelihood = -21.710799
Iteration 4:   log likelihood = -21.710799
```

Number of obs	=	48
Wald chi2(3)	=	13.58
Prob > chi2	=	0.0035

Log likelihood = -21.710799

died	Coef.	Std. Err.	z	P>|z|	[95% Conf. Interval]	
drug						
2	-1.941275	.597057	-3.25	0.001	-3.111485	-.7710644
3	-1.792924	.5845829	-3.07	0.002	-2.938686	-.6471626
age	.0666733	.0410666	1.62	0.104	-.0138159	.1471624
_cons	-2.044876	2.284283	-0.90	0.371	-6.521988	2.432235

Our results match those we obtained in section 4.2.1.

11.1.2   The Weibull model

In section 5.4.3, we wrote a method-lf2 likelihood evaluator for the Weibull model. Here we write a do-file that implements the likelihood evaluator in Mata:

```
                                        begin myweibull_lf2_mata.do
version 11
sysuse cancer, clear
mata:
void myweibull_lf2(transmorphic scalar ML, real scalar todo,
                   real rowvector b, real colvector lnfj,
                   real matrix S, real matrix H)
{
        real scalar lgam, p
        real vector leta, t, d, M, R
        real matrix H11, H12, H22

        leta = moptimize_util_xb(ML, b, 1)
        lgam = moptimize_util_xb(ML, b, 2)
        t    = moptimize_util_depvar(ML, 1)
        d    = moptimize_util_depvar(ML, 2)

        p = exp(lgam)
        M = (t:*exp(-1*leta)):^p
        R = ln(t) - leta
        lnfj = -M + d:*(lgam:-leta + (p-1):*R)
        if (todo == 0) return

        S = J(rows(leta), 2, .)
        S[.,1] = p*(M - d)
        S[.,2] = d - p*R:*(M - d)
        if (todo == 1) return

        H11 = moptimize_util_matsum(ML, 1, 1, p^2 :*M, lnfj[1])
        H12 = moptimize_util_matsum(ML, 1, 2, -p*(M - d + p*R:*M), lnfj[1])
        H22 = moptimize_util_matsum(ML, 2, 2, p*R:*(p*R:*M + M - d), lnfj[1])
        H = -1*(H11, H12 \ H12', H22)
}
end

ml model lf2 myweibull_lf2() (lneta: studytime died = i.drug age) /lngamma
ml maximize
                                          end myweibull_lf2_mata.do
```

Method-lf2 Mata-based likelihood evaluators are passed six arguments: the problem handle, a scalar that functions just like the ado-based `todo` parameter, the coefficient vector, a vector that we are to fill in with the observation-level log-likelihood values, a matrix to hold the observation-level scores, and a matrix to contain the Hessian of the overall log-likelihood function.

The first part of our program again uses the utility functions `moptimize_util_xb()` and `moptimize_util_depvar()` to obtain the model parameters and dependent variables. We then create several vectors containing intermediate calculations, analogous to our use of temporary variables in the ado-version of our evaluator program. We use these vectors to fill in `lnfj` with the log likelihood. We check `todo` and return from the function if `ml` does not want our evaluator to compute derivatives (`todo = 0`).

If `todo` is nonzero, we next compute the scores. Our model has two equations, so our matrix of scores will have two columns and several rows equal to the number of observations in the dataset. Because we want to fill in the matrix S one column at a time, we must first initialize it to have the correct dimensions. With that done, we then fill in the two columns with the two equations' scores. If `todo` equals one, we need not compute the Hessian, and in that case, we return from the function.

The last part of our code computes the Hessian of the overall log likelihood. We use the function `moptimize_util_matsum()` to aid us in those computations; that function is equivalent to the `mlmatsum` command. The first argument is again the problem handle, and the second and third arguments are the two equations for which we are computing the second derivative. The fourth argument represents the second derivative $\partial^2 \ln L/\partial\theta_i\theta_j$. The fifth argument is the variable we use to store the log likelihood; `moptimize_util_matsum()` will set this to missing if the Hessian cannot be computed.

Executing our do-file, we obtain

```
. version 11
. sysuse cancer, clear
(Patient Survival in Drug Trial)
. mata:
---------------------------------------------------- mata (type end to exit) ----------
: void myweibull_lf2(transmorphic scalar ML, real scalar todo,
>                     real rowvector b, real colvector lnfj,
>                     real matrix S, real matrix H)
> {
>         real scalar lgam, p
>         real vector leta, t, d, M, R
>         real matrix H11, H12, H22
>
>         leta = moptimize_util_xb(ML, b, 1)
>         lgam = moptimize_util_xb(ML, b, 2)
>         t    = moptimize_util_depvar(ML, 1)
>         d    = moptimize_util_depvar(ML, 2)
>
>         p = exp(lgam)
>         M = (t:*exp(-1*leta)):^p
>         R = ln(t) - leta
>         lnfj = -M + d:*(lgam:-leta + (p-1):*R)
>         if (todo == 0) return
>
>         S = J(rows(leta), 2, .)
>         S[.,1] = p*(M - d)
>         S[.,2] = d - p*R:*(M - d)
>         if (todo == 1) return
>
>         H11 = moptimize_util_matsum(ML, 1, 1, p^2 :*M, lnfj[1])
>         H12 = moptimize_util_matsum(ML, 1, 2, -p*(M - d + p*R:*M), lnfj[1])
>         H22 = moptimize_util_matsum(ML, 2, 2, p*R:*(p*R:*M + M - d), lnfj[1])
>         H = -1*(H11, H12 \ H12', H22)
> }
: end
----------------------------------------------------------------------------------------
. ml model lf2 myweibull_lf2() (lneta: studytime died = i.drug age) /lngamma
```

```
. ml maximize
initial:        log likelihood =        -744
alternative:    log likelihood = -356.14276
rescale:        log likelihood = -200.80201
rescale eq:     log likelihood = -136.69232
Iteration 0:    log likelihood = -136.69232  (not concave)
Iteration 1:    log likelihood = -124.13044
Iteration 2:    log likelihood = -113.99047
Iteration 3:    log likelihood = -110.30732
Iteration 4:    log likelihood = -110.26748
Iteration 5:    log likelihood = -110.26736
Iteration 6:    log likelihood = -110.26736
```

		Number of obs	=	48
		Wald chi2(3)	=	35.25
Log likelihood = -110.26736		Prob > chi2	=	0.0000

	Coef.	Std. Err.	z	P>|z|	[95% Conf.	Interval]
lneta						
drug						
2	1.012966	.2903917	3.49	0.000	.4438086	1.582123
3	1.45917	.2821195	5.17	0.000	.9062261	2.012114
age	-.0671728	.0205688	-3.27	0.001	-.1074868	-.0268587
_cons	6.060723	1.152845	5.26	0.000	3.801188	8.320259
lngamma						
_cons	.5573333	.1402154	3.97	0.000	.2825163	.8321504

These results match those reported in section 5.4.3.

11.2   Evaluator function prototypes

We have seen that method-lf evaluators receive three arguments, while method-lf2 evaluators receive six. In this section, we outline the arguments required for each type of likelihood evaluator written in Mata.

Method-lf evaluators

void *lfeval*(transmorphic scalar M, real rowvector b, real rowvector *lnfj*)

> *input*:
> M: problem handle
> b: $1 \times k$ coefficient vector
> *output*:
> *lnfj*: $N \times 1$ log-likelihood vector

Notes:

- *lfeval* is the name of your evaluator function.

- k is the number of coefficients in the model, and N is the number of observations in the estimation subsample accounting for if, in, and missing values.

- *lnfj* is to be filled in with the observation-level log-likelihood values.

- Unlike ado-based method-lf evaluators, you are passed the coefficient vector b. Use moptimize_util_xb to extract the θ probability model parameters.

- To define your model using ml model, you type

    ml model lf *lfeval*() ...

  You must include the parentheses after the name of your Mata-based evaluator.

lf-family evaluators

void *lfxeval*(transmorphic scalar M, real scalar *todo*, real rowvector b,
        real rowvector *lnfj*, real matrix S, real matrix H)

> *input*:
> M: problem handle
> *todo*:   scalar *todo* = 0, 1, or 2
> b: $1 \times k$ coefficient vector
> *output*:
> *lnfj*: $N \times 1$ log-likelihood vector
> S: $N \times E$ matrix of scores
> H: $k \times k$ Hessian matrix

Notes:

- *lfxeval* is the name of your evaluator function.

- Your function must accept all six arguments even if you are writing a method-lf0 or method-lf1 evaluator. Simply ignore the S and H arguments in your program if you are not required to fill them in.

- If you are implementing a method-lf1 or method-lf2 evaluator and $todo = 1$, you must fill in the $N \times E$ matrix S with the observation-level scores, where E is the number of equations in the model. The columns in S are ordered in the same way you specified the equations in your ml model statement. The (s,t) element of S is to contain $\partial \ln \ell_s / \partial \theta_{ts}$, where $s = 1, \ldots, N$ indexes observations and $t = 1, \ldots, E$ indexes equations.

- If you are implementing a method-lf2 evaluator and $todo = 2$, you must fill in the $E \times E$ matrix H with the Hessian matrix of the overall log-likelihood function. The (s,t) element of H is to contain $\partial^2 \ln L / \partial \beta_s \partial \beta_t$, where $s = 1, \ldots, k$ and $t = 1, \ldots, k$ index coefficients.

- To define your model using ml model, you type

      ml model lf0 *lfxeval()* ...

  or

      ml model lf1 *lfxeval()* ...

  or

      ml model lf2 *lfxeval()* ...

  depending on whether you have implemented an evaluator of method lf0, lf1, or lf2, respectively. You must include the parentheses after the name of your evaluator.

d-family evaluators

    void *deval*(transmorphic scalar M, real scalar *todo*, real rowvector b,
            real scalar *lnf*, real rowvector g, real matrix H)

	input:	
	M:	problem handle
	todo:	scalar $todo = 0$, 1, or 2
	b:	$1 \times k$ coefficient vector
	output:	
	lnf:	scalar containing log likelihood
	g:	$1 \times k$ gradient vector
	H:	$k \times k$ Hessian matrix

Notes:

- *deval* is the name of your evaluator function.

- Your function must accept all six arguments even if you are writing a method-d0 or method-d1 evaluator. Simply ignore the g and H arguments in your program if you are not required to fill them in.

- If you are implementing a method-**d1** or method-**d2** evaluator and $todo = 1$, you must fill in the $1 \times k$ vector g with the gradient of the overall log-likelihood function. Element s of g is to contain $\partial \ln L / \partial \beta_s$, where $s = 1, \ldots, k$ indexes coefficients.

- If you are implementing a method-**lf2** evaluator and $todo = 2$, you must fill in the $E \times E$ matrix H with the Hessian matrix of the overall log-likelihood function. The (s, t) element of H is to contain $\partial^2 \ln L / \partial \beta_s \partial \beta_t$, where $s = 1, \ldots, k$ and $t = 1, \ldots, k$ index coefficients.

- To define your model using **ml model**, you type

      ml model d0 *deval()* ...

  or

      ml model d1 *deval()* ...

  or

      ml model d2 *deval()* ...

  depending on whether you have implemented an evaluator of method **d0**, **d1**, or **d2**, respectively. You must include the parentheses after the name of your evaluator.

11.3   Utilities

When writing ado-file–based likelihood evaluators, utilities like `mleval`, `mlvecsum`, and `mlmatsum` are often useful, as are the macros `$ML_y1`, `$ML_y2`, and so on. Analogous functions exist for Mata-based likelihood evaluators. Here we focus on those functions that have analogues in the ado-language. For other functions that may be of interest, see [M-5] **moptimize( )**.

Table 11.1 lists the most popular utilities used in ado-based likelihood evaluators and their corresponding Mata functions. We then briefly discuss each function in turn, using the same notation as in section 11.2.

Table 11.1. Ado-based and Mata-based utilities

Ado-code	Mata function
`$y1, $y2, ...`	`moptimize_util_depvar()`
`mleval`	`moptimize_util_xb()`
	`moptimize_util_eq_indices()`
`mlsum`	`moptimize_util_sum()`
`mlvecsum`	`moptimize_util_vecsum()`
`mlmatsum`	`moptimize_util_matsum()`
`mlmatbysum`	`moptimize_util_matbysum()`

Dependent variables

In ado-based likelihood evaluators, the names of the dependent variables are stored in macros $ML_y1, $ML_y2, and so on. You can use these macros in, for example, `generate` statements to refer to the model's dependent variables.

In Mata-based likelihood evaluators, if you program

```
y1 = moptimize_util_depvar(M, 1)
```

then `y1` will be a column vector containing the observations on the first dependent variable. This vector is automatically restricted to the estimation subsample, so you need not do extra work to handle any `if` or `in` restrictions specified in the `ml model`. In this example, M is the problem handle, and the second argument represents the equation for which you want the dependent variable.

Obtaining model parameters

For all ado-based likelihood-evaluator types other than `lf`, you use the `mleval` command to obtain the θ model parameters from 'b'. With Mata-based likelihood evaluators, you instead use the `moptimize_util_xb()` function. Like the name suggests, this function computes $\theta_j = \mathbf{X}_j \beta_j$. For example, the code

```
xj = moptimize_util_xb(M, b, j)
```

creates column vector `xj` that contains the model parameters for the jth equation evaluated at the coefficient vector b. As with all the utility functions documented in this section, `moptimize_util_xb()` automatically restricts calculations to the relevant estimation sample.

When working with scalar parameters such as variances in homoskedastic models, you could specify the `scalar` option to `mleval`. Obtaining the same result in Mata is a two-step process. For example, say that your `ml model` statement is

```
ml model myeval() (mpg = gear_ratio turn) /lnsigma /theta
```

The command

```
j = moptimize_util_eq_indices(M, 1)
```

will create j = [1, 3]. The first element of j indicates that the coefficients for the first equation start at position 1 of b. The second element indicates that the coefficients for the first equation end at position 3 of b. That is, b[1], b[2], and b[3] are the coefficients for the first equation (corresponding to the variables `mpg` and `gear_ratio` and the constant, respectively).

The command

```
j = moptimize_util_eq_indices(M, 2)
```

will set j = $[4, 4]$, indicating that the scalar lnsigma is stored in the fourth element of b. Because j is a 1×2 vector, you can use range subscripts to extract the relevant portion of b. Hence, here you would program

```
j = moptimize_util_eq_indices(M, 2)
sigma = exp(b[|j|])
```

and work with the scalar sigma. Finally,

```
j = moptimize_util_eq_indices(M, 3)
```

will set j = $[5, 5]$, allowing you to extract the scalar theta as the fifth element of b.

Summing individual or group-level log likelihoods

With d-family evaluators, you return the overall log-likelihood function. As we saw with panel-data regression and the Cox model, we obtain a log likelihood for a group of observations; summing those groups' log likelihoods yields the overall log likelihood. In the simple case of linear regression, the overall log likelihood is simply the sum of each observation's individual log likelihood. For ado-based likelihood evaluators, the mlsum command is used to assist in computing the overall log likelihood.

With Mata-based evaluators, you use the moptimize_util_sum() function. For example,

```
lnf = moptimize_util_sum(M, v)
```

will sum the vector v to obtain the overall log likelihood, applying weights if specified, restricting the summation to the estimation subsample, and verifying that v contains no missing values. If v does contain missing values, moptimize_util_sum() returns a missing value, indicating that the overall log likelihood cannot be computed at the current value of the coefficient vector.

In fact, moptimize_util_sum() does not check that the number of rows in v equals the number of observations. For panel-data and other group-level log likelihoods, you could create v to have a number of rows equal to the number of panels and groups. However, if you do this, you cannot specify weights during estimation.

Calculating the gradient vector

The tool mlvecsum can be used in ado-based likelihood functions to assist in calculating the gradient vector $\partial \ln L / \partial \beta_i$ when the likelihood function meets the linear-form restrictions. The corresponding Mata utility is moptimize_util_vecsum(). The format for calling this utility is

```
gi = moptimize_util_vecsum(M, i, s, lnf)
```

M is again the problem handle. i is the equation number for which you are computing the gradient and corresponds to the eq() option of mlvecsum. s is a column vector that contains $\partial \ln \ell_j / \partial \theta_{ij}$. Argument *lnf* is the name of the scalar that stores the overall log likelihood of your model. Like `mlvecsum`, if `moptimize_util_vecsum()` cannot compute the gradient vector, likely because of missing values in s, it will set *lnf* to missing to indicate to `ml` that the current value of the coefficient vector is invalid. Weights are handled automatically, and the sum is restricted to the estimation subsample.

Calculating the Hessian

In section 5.2.5, we discussed the command `mlmatsum` to compute the Hessian matrix of a likelihood function that satisfies the linear-form restrictions. The corresponding Mata utility is `moptimize_util_matsum()` and has the syntax

```
Hmn = moptimize_util_matsum(M, m, n, d, lnf)
```

M is again the problem handle. Arguments m and n refer to the two equations whose cross-partial derivative you are computing and correspond to the two arguments in the `eq()` option of `mlmatsum`. Argument d is a vector containing $\partial^2 \ln \ell_j / \partial \theta_m \partial \theta_n$. Argument *lnf* is again the name of the scalar that stores the overall log likelihood of your model. If `moptimize_util_matsum()` cannot compute the Hessian, it will set *lnf* to missing. Weights are handled automatically, and the sum is restricted to the estimation subsample.

In section 6.4.3, we discussed using the `mlmatbysum` command to facilitate computation of the Hessian of models that have panel- or group-level log-likelihood functions. Terms on the (block) diagonal of the Hessian matrix have the form

$$\mathbf{H}_m = \sum_{i=1}^{N} \left(\sum_{t=1}^{T_i} a_{it} \right) \left(\sum_{t=1}^{T_i} b_{it} \mathbf{x}'_{mit} \right) \left(\sum_{t=1}^{T_i} b_{it} \mathbf{x}_{mit} \right)$$

where m refers to the equation whose submatrix of the Hessian you are computing. To form $\mathbf{H}_m$ using `mlmatbysum`, you would use

```
mlmatbysum 'lnf' 'Hm' varname_a varname_b, by(byvar) eq(m, m)
```

To perform the same operation using `moptimize_util_matbysum()`, you would code

```
moptimize_init_by(M, "byvar")
Hm = moptimize_util_matbysum(M, m, a, b, lnf)
```

M is again the problem handle, m is the equation number, a and b are column vectors that correspond to *varname_a* and *varname_b*, respectively, and *lnf* is the scalar that holds the overall log likelihood. *lnf* will be set to missing if the calculation cannot be completed. Weights and the estimation subsample are taken into account.

Before calling `moptimize_util_matbysum()`, you must call `moptimize_init_by()` to declare the panel or group identifier variable. As with virtually all `moptimize*()` functions, the first argument you pass to `moptimize_init_by()` is the problem handle. Technically, you need to declare the `byvar` only once to fit your model, but the easiest thing to do is just include the function call inside your likelihood evaluator. That will redeclare your *byvar* every time your likelihood is evaluated, but `moptimize_init_by()` is so fast that you will not notice a difference.

Terms on the off-diagonal of the Hessian matrix have the form

$$\mathbf{H}_{mn} = \sum_{i=1}^{N} \left(\sum_{t=1}^{T_i} a_{it} \right) \left(\sum_{t=1}^{T_i} b_{it} \mathbf{x}'_{1it} \right) \left(\sum_{t=1}^{T_i} c_{it} \mathbf{x}_{2it} \right)$$

To compute a matrix like that with `mlmatbysum`, you would use

```
mlmatbysum 'lnf' 'Hm' varname_a varname_b varname_c, by(byvar) eq(m, n)
```

The corresponding `moptimize_util_matbysum()` call is

```
Hmn = moptimize_util_matbysum(M, m, n, a, b, c, lnf)
```

Instead of specifying a single equation (m) as for elements on the diagonal of the Hessian, here you must specify two equations (m and n), and you must specify c, a row vector corresponding to c_{it} in the formula above and *varname_c* in `mlmatbysum`. If you have not previously called `moptimize_init_by()` within your likelihood evaluator, you will need to do so before using `moptimize_util_matbysum()`.

11.4 Random-effects linear regression

In section 6.4, we discussed likelihood functions for panel data, using the random-effects linear regression model as an example. Here we replicate that model using a Mata-based likelihood-evaluator function; this example allows us to use every Mata function in table 11.1. We will first present our Mata-based evaluator function in pieces to aid discussion. You may find it useful to bookmark section 6.4 so that you can flip back and forth to compare ado-code and Mata code.

❑ **Technical note**

In the following example, so that we can focus on the core code, we do not use variable declarations within the body of our Mata function. In general, however, using declarations has at least two advantages: first, debugging is simplified because Mata can tell whether we are, for example, trying to use a matrix as a scalar; and second, Mata executes slightly faster when it knows beforehand what will be stored in a variable. See [M-2] **declarations** and `mata set matastrict` in [M-3] **mata set**. By default, `matastrict` is set to `off`, meaning declarations are not required.

❑

11.4.1   Calculating lnf

The first part of our function, used to compute the likelihood function, is

```
void myrereg_d2_mata(transmorphic scalar M, real scalar todo,
                     real rowvector b, real scalar lnf,
                     real rowvector g, real matrix H)
{
        xb  = moptimize_util_xb(M, b, 1)
        j   = moptimize_util_eq_indices(M, 2)
        s_u = exp(b[|j|])
        j   = moptimize_util_eq_indices(M, 3)
        s_e = exp(b[|j|])

        y = moptimize_util_depvar(M, 1)
        z = y - xb

        // Create view of panels and set up panel-level processing
        st_view(panvar = ., ., st_global("MY_panel"))
        paninfo = panelsetup(panvar, 1)
        npanels = panelstats(paninfo)[1]

        lnfj = J(npanels, 1, 0)
        for(i=1; i <= npanels; ++i)
                zi = panelsubmatrix(z, i, paninfo)
                S_z2 = sum(zi:^2)
                Sz_2 = sum(zi)^2
                T = rows(zi)
                a = s_u^2 / (T*s_u^2 + s_e^2)
                lnfj[i] = - 0.5 * (                           ///
                        (S_z2 - a*Sz_2) / s_e^2 +             ///
                        ln(T*s_u^2/s_e^2 + 1) +               ///
                        T*ln(2*pi()*s_e^2) )

        lnf = moptimize_util_sum(M, lnfj)
        if ((todo == 0) | (lnf == .)) return
```

Following the prototype in section 11.2 for a d-family evaluator, we name our function `myrereg_d2_mata()` and make it accept the requisite arguments. We then use `moptimize_util_xb()` to obtain the first equation (θ_1) of our model. Using the procedure discussed in section 11.3 for extracting scalar parameters from b, we then fill in `s_u` and `s_e` with the standard deviations of the panel-level and idiosyncratic error terms. We use `moptimize_util_depvar()` to retrieve our dependent variable and create the vector z to contain the residuals.

When writing likelihoods for panel-data models, you should be familiar with Mata's `panelsetup()` and related utilities; see [M-5] **panelsetup( )**. These utilities allow you to set up your data so that they can be processed one panel at a time. Particularly useful in this application is the **panelsubmatrix()** function, which allows us to obtain from a data matrix only those rows that correspond to the panel we are considering.

Because the likelihood function exists only at the panel level, we set up vector `lnfj` to have several rows equal to the number of panels in the dataset. We then loop over each panel of the dataset, calculating the panel-level likelihood as in (6.3). We use `moptimize_util_sum()` to sum the elements of `lnfj` to obtain the overall log likelihood, taking advantage of the fact that this function will work with a vector even if the vector does not have its number of rows equal to the number of observations in the estimation sample. Finally, we check to see if `ml` expects us to compute the gradient or Hessian and exit if we are not required to do so.

11.4.2   Calculating g

Our code for computing the gradient is

```
dxb = J(rows(panvar), 1, 0)
du  = 0
de  = 0
for(i=1; i <= npanels; ++i) {
        zi = panelsubmatrix(z, i, paninfo)
        T = rows(zi)
        a = s_u^2 / (T*s_u^2 + s_e^2)
        S_zi  = sum(zi)
        S_z2  = sum(zi:^2)
        Sz_2  = sum(zi)^2
        dxb[|paninfo[i,1] \ paninfo[i,2]|] = (zi :- a*S_zi)/s_e^2
        du = du + a^2*Sz_2 / s_u^2 - T*a
        de = de + S_z2 / s_e^2 - a*Sz_2 / s_e^2 -         ///
                a^2*Sz_2 / s_u^2 - T + 1 -                ///
                a*s_e^2 / s_u^2
        if ((du == .) | (de == .)) {
                lnf = .
                break
        }
}
dxb = moptimize_util_vecsum(M, 1, dxb, lnf)
g = (dxb, du, de)
if ((todo == 1) | (lnf == .)) return
```

Examining (6.5), we see that $\partial \ln L_i / \partial \mathbf{x}_{it}\beta$ is observation specific, while from (6.6) and (6.7), we see that the derivatives with respect to the variance components are panel specific. We therefore create the vector `dxb` to have as its number of rows the number of observations in the estimation sample, which is easily gleaned by looking at the length of a vector that we know contains a row for each observation in the sample. We initialize `du` and `de` to be scalar zero; we will add each panel's contribution to those elements of the gradient ourselves.

As in computing the log-likelihood function, we loop over each panel individually. We use range subscripting (see [M-2] **subscripts**) to fill in at once all the elements of `dxb` corresponding to the current panel. The function `panelsetup()` returns a $P \times 2$ matrix, where P is the number of panels. The first element of the ith row contains the number of the first row of data corresponding to panel i and the second element contains the number of the last row corresponding to panel i, making range subscripting simple. We add the current panel's contributions to the gradients with respect to the variance components. Importantly, because we are not using an `moptimize()` utility function to do the sums, we must check for ourselves that the gradient is valid (does not contain missing values); if it is not valid, we set `lnf` to missing and exit. We use `moptimize_util_vecsum()` to sum the contributions to the gradient with respect to $\mathbf{x}'_{it}\beta$. Finally, we assemble the gradient vector and exit if we are done.

11.4.3   Calculating H

The computations for the Hessian matrix proceed in much the same way as those for the gradient, though of course the formulas are more complicated. Our code is virtually a direct translation into Mata of the ado-code from section 6.4.3. The Mata code is longer because the ado-based mlmatsum and mlmatbysum commands allow you to specify expressions, while the Mata utilities moptimize_util_matsum() and moptimize_util_matbysum() require you to create your own vectors before calling them.

Our code for computing the Hessian is

```
// Compute Hessian
avec  = J(rows(z), 1, 0)
d2u   = J(npanels, 1, 0)
d2e   = J(npanels, 1, 0)
dude  = J(npanels, 1, 0)
dxbdu = J(rows(z), 1, 0)
dxbde = J(rows(z), 1, 0)
for(i=1; i<= npanels; ++i) {
        zi = panelsubmatrix(z, i, paninfo)
        S_zi = sum(zi)
        S_z2 = sum(zi:^2)
        Sz_2 = sum(zi)^2
        T = rows(zi)
        a = s_u^2 / (T*s_u^2 + s_e^2)

        avec[paninfo[i,2]] = a
        d2u[i] = -(2*a^2*Sz_2 / s_u^2 -               ///
                4*s_e^2*a^3*Sz_2 / s_u^4 +            ///
                2*T*a^2*s_e^2 / s_u^2)
        d2e[i] = -(2*(S_z2 - a*Sz_2) / s_e^2 -        ///
                2*a^2*Sz_2 / s_u^2 -                  ///
                4*a^3*Sz_2*s_e^2 / s_u^4 +            ///
                2*a*s_e^2 / s_u^2 -                   ///
                2*a^2*s_e^4 / s_u^4)
        dude[i] = -(4*a^3*S_zi^2*s_e^2 / s_u^4 -      ///
                2*T*a^2*s_e^2 / s_u^2)
        dxbdu[|paninfo[i,1] \ paninfo[i,2]|] =        ///
                J(rows(zi), 1, -2*a^2*S_zi / s_u^2)
        dxbde[|paninfo[i,1] \ paninfo[i,2]|] =        ///
                -(2*(zi :- a*S_zi)/s_e^2 :-           ///
                2*a^2*S_zi / s_u^2)
}
d2u = moptimize_util_matsum(M, 2, 2, d2u, lnf)
d2e = moptimize_util_matsum(M, 3, 3, d2e, lnf)
dude = moptimize_util_matsum(M, 2, 3, dude, lnf)
dxbdu = moptimize_util_matsum(M, 1, 2, dxbdu, lnf)
dxbde = moptimize_util_matsum(M, 1, 3, dxbde, lnf)
moptimize_init_by(M, st_global("MY_panel"))
d2xb1 = moptimize_util_matbysum(M, 1, avec, J(rows(z), 1, 1), lnf)
d2xb2 = moptimize_util_matsum(M, 1, 1, J(rows(z), 1, 1), lnf)
d2xb  = -(d2xb2 - d2xb1) / s_e^2

H = (d2xb,    dxbdu, dxbde \               ///
    dxbdu', d2u,   dude \                  ///
    dxbde', dude,  d2e)
```

We created d2u, d2e, and dude to have their numbers of rows equal to the number of panels because they only have one value per panel. The component $a_i = \sigma_u^2/(T_i\sigma_u^2 + \sigma_e^2)$ has only one value per panel, but because we use the moptimize_util_matbysum() function, we need our avec vector to have the same length as the estimation sample. (We will admit we did not realize that until after discovering that our first draft of the program reported a conformability error.)

11.4.4 Results at last

Executing our complete do-file, we obtain

```
. version 11
. use http://www.stata-press.com/data/ml4/tablef7-1, clear
. mata:
─────────────────────────────────────────────── mata (type end to exit) ───────
: void myrereg_d2_mata(transmorphic scalar M, real scalar todo,
>                      real rowvector b, real scalar lnf,
>                      real rowvector g, real matrix H)
> {
>        real    colvector      xb
>        real    matrix         j
>        real    colvector      s_u
>        real    colvector      s_e
>        real    colvector      y
>        real    colvector      z
>        real    matrix         panvar
>        real    matrix         paninfo
>        real    scalar         npanels
>        real    colvector      lnfj
>        real    scalar         i
>        real    matrix         zi
>        real    matrix         S_z2
>        real    matrix         Sz_2
>        real    matrix         T
>        real    colvector      a
>        real    colvector      dxb
>        real    colvector      du
>        real    colvector      de
>        real    colvector      S_zi
>        real    colvector      avec
>        real    colvector      d2u
>        real    colvector      d2e
>        real    colvector      dude
>        real    colvector      dxbdu
>        real    colvector      dxbde
>        real    colvector      d2xb1
>        real    colvector      d2xb2
>        real    colvector      d2xb
>
>        xb  = moptimize_util_xb(M, b, 1)
>        j   = moptimize_util_eq_indices(M, 2)
>        s_u = exp(b[|j|])
>        j   = moptimize_util_eq_indices(M, 3)
>        s_e = exp(b[|j|])
>
```

```
>               y = moptimize_util_depvar(M, 1)
>               z = y - xb
>
>               // Create view of panels and set up panel-level processing
>               st_view(panvar = ., ., st_global("MY_panel"))
>               paninfo = panelsetup(panvar, 1)
>               npanels = panelstats(paninfo)[1]
>
>               lnfj = J(npanels, 1, 0)
>               for(i=1; i <= npanels; ++i) {
>                       zi = panelsubmatrix(z, i, paninfo)
>                       S_z2 = sum(zi:^2)
>                       Sz_2 = sum(zi)^2
>                       T = rows(zi)
>                       a = s_u^2 / (T*s_u^2 + s_e^2)
>                       lnfj[i] = - 0.5 * (                              ///
>                               (S_z2 - a*Sz_2) / s_e^2 +               ///
>                               ln(T*s_u^2/s_e^2 + 1) +                 ///
>                               T*ln(2*pi()*s_e^2) )
>               }
>               lnf = moptimize_util_sum(M, lnfj)
>               if ((todo == 0) | (lnf == .)) return
>
>               // Compute the gradient
>               dxb = J(rows(panvar), 1, 0)
>               du  = 0
>               de  = 0
>               for(i=1; i <= npanels; ++i) {
>                       zi = panelsubmatrix(z, i, paninfo)
>                       T = rows(zi)
>                       a = s_u^2 / (T*s_u^2 + s_e^2)
>                       S_zi  = sum(zi)
>                       S_z2  = sum(zi:^2)
>                       Sz_2  = sum(zi)^2
>                       dxb[|paninfo[i,1] \ paninfo[i,2]|] = (zi :- a*S_zi)/s_e^2
>                       du = du + a^2*Sz_2 / s_u^2 - T*a
>                       de = de + S_z2 / s_e^2 - a*Sz_2 / s_e^2 -       ///
>                               a^2*Sz_2 / s_u^2 - T + 1 -             ///
>                               a*s_e^2 / s_u^2
>                       if ((du == .) | (de == .)) {
>                               lnf = .
>                               break
>                       }
>
>               }
>               dxb = moptimize_util_vecsum(M, 1, dxb, lnf)
>               g = (dxb, du, de)
>               if ((todo == 1) | (lnf == .)) return
>
>               // Compute Hessian
>               avec  = J(rows(z), 1, 0)
>               d2u   = J(npanels, 1, 0)
>               d2e   = J(npanels, 1, 0)
>               dude  = J(npanels, 1, 0)
>               dxbdu = J(rows(z), 1, 0)
>               dxbde = J(rows(z), 1, 0)
>               for(i=1; i<= npanels; ++i) {
>                       zi = panelsubmatrix(z, i, paninfo)
>                       S_zi  = sum(zi)
```

```
>                         S_z2 = sum(zi:^2)
>                         Sz_2 = sum(zi)^2
>                         T = rows(zi)
>                         a = s_u^2 / (T*s_u^2 + s_e^2)
>
>                         avec[paninfo[i,2]] = a
>                         d2u[i] =                                      ///
>                             -(2*a^2*Sz_2 / s_u^2 -                    ///
>                               4*s_e^2*a^3*Sz_2 / s_u^4 +             ///
>                               2*T*a^2*s_e^2 / s_u^2)
>                         d2e[i] =                                      ///
>                             -(2*(S_z2 - a*Sz_2) / s_e^2 -            ///
>                               2*a^2*Sz_2 / s_u^2 -                   ///
>                               4*a^3*Sz_2*s_e^2 / s_u^4 +            ///
>                               2*a*s_e^2 / s_u^2 -                    ///
>                               2*a^2*s_e^4 / s_u^4)
>                         dude[i] =                                     ///
>                             -(4*a^3*S_zi^2*s_e^2 / s_u^4 -          ///
>                               2*T*a^2*s_e^2 / s_u^2)
>                         dxbdu[|paninfo[i,1] \ paninfo[i,2]|] =       ///
>                             J(rows(zi), 1, -2*a^2*S_zi / s_u^2)
>                         dxbde[|paninfo[i,1] \ paninfo[i,2]|] =       ///
>                             -(2*(zi :- a*S_zi)/s_e^2 :-             ///
>                               2*a^2*S_zi / s_u^2)
>                 }
>         d2u = moptimize_util_matsum(M, 2, 2, d2u, lnf)
>         d2e = moptimize_util_matsum(M, 3, 3, d2e, lnf)
>         dude = moptimize_util_matsum(M, 2, 3, dude, lnf)
>         dxbdu = moptimize_util_matsum(M, 1, 2, dxbdu, lnf)
>         dxbde = moptimize_util_matsum(M, 1, 3, dxbde, lnf)
>         moptimize_init_by(M, st_global("MY_panel"))
>         d2xb1 = moptimize_util_matbysum(M, 1, avec, J(rows(z), 1, 1), lnf)
>         d2xb2 = moptimize_util_matsum(M, 1, 1, J(rows(z), 1, 1), lnf)
>         d2xb  = -(d2xb2 - d2xb1) / s_e^2
>
>         H = (d2xb,   dxbdu, dxbde \             ///
>              dxbdu', d2u,   dude  \             ///
>              dxbde', dude,  d2e)
> }
: end
```

```
.
. global MY_panel i
. sort $MY_panel
. ml model d2 myrereg_d2_mata() (xb: lnC = lnQ lnPF lf) /lns_u /lns_e
. ml maximize
initial:       log likelihood = -615.46011
alternative:   log likelihood = -315.38194
rescale:       log likelihood =  -242.2031
rescale eq:    log likelihood = -111.43007
Iteration 0:   log likelihood = -111.43007   (not concave)
Iteration 1:   log likelihood = -40.001214   (not concave)
Iteration 2:   log likelihood =  49.947998   (not concave)
Iteration 3:   log likelihood =  71.902601   (not concave)
Iteration 4:   log likelihood =  89.534248
Iteration 5:   log likelihood =  110.25833
Iteration 6:   log likelihood =  111.13119
```

```
Iteration 7:    log likelihood =  112.87175
Iteration 8:    log likelihood =  114.72075
Iteration 9:    log likelihood =  114.72904
Iteration 10:   log likelihood =  114.72904
```

```
                                           Number of obs   =          90
                                           Wald chi2(3)    =    11468.30
Log likelihood =  114.72904                Prob > chi2     =      0.0000
```

lnC	Coef.	Std. Err.	z	P>|z|	[95% Conf. Interval]	
xb						
lnQ	.90531	.0253759	35.68	0.000	.8555742	.9550459
lnPF	.4233757	.013888	30.48	0.000	.3961557	.4505957
lf	-1.064456	.1962308	-5.42	0.000	-1.449061	-.6798508
_cons	9.618648	.2066218	46.55	0.000	9.213677	10.02362
lns_u						
_cons	-2.170818	.3026641	-7.17	0.000	-2.764028	-1.577607
lns_e						
_cons	-2.828404	.077319	-36.58	0.000	-2.979947	-2.676862

12 Writing do-files to maximize likelihoods

There are two reasons to maximize likelihood functions:

1. You need to fit the model for a particular analysis or just a few analyses.

2. You need the estimator around constantly because you use it all the time.

We recommend using do-files in case 1 and ado-files (covered in the next chapter) in case 2.

12.1 The structure of a do-file

We organize our do-files differently depending on the problem, but we always start the same way:

```
─────────────────────────────────────── begin outline.do ───────────
clear
capture program drop myprog
program myprog
        version 11
        args ...
        ...
end
use sampledata
ml model ... myprog ...
─────────────────────────────────────── end outline.do ───────────
```

Then, working interactively, we can type

```
. do outline
. ml check
. ml search
. ml max
```

Typically, `ml check` detects problems, but this structure makes it easy to try again. We can edit our do-file and then start all over by typing `do outline`. We do not actually name the file `outline.do`, of course. At this stage, we typically give it some very short name: `it.do` is our favorite. Then we can type `do it`.

We put the loading of some sample data and the `ml model` statement in the do-file. The remaining `ml` commands we need to type are short; we just did not want to be forever retyping things like

```
. ml model d2 myweibull_d2 (lneta: studytime died = i.drug age) /lngamma
```

while we are debugging our code.

12.2   Putting the do-file into production

Once we have our evaluator working, we modify our do-file and rename it:

```
─────────────────────────────────────── begin readytouse.do ───────
capture program drop myprog
program myprog
        version 11
        args ...
        ...
end
capture program drop model
program model
        version 11
        ml model method myprog (lneta: '0') /lngamma
end
──────────────────────────────────── end readytouse.do ───────
```

The changes we made were as follows:

1. We removed the **clear** at the top; we no longer want to eliminate the data in memory. Now we can load the data and run the do-file in any order.

2. We removed the "**use** *sampledata*" and the hardcoded `ml model` statement.

3. We substituted in their place a new program—**model**, a two-line program we wrote to make it easy to type the `ml model` statement.

We previously imagined that model statements such as

```
. ml model d2 myweibull_d2 (lneta: studytime died = i.drug age) /lngamma
```

were reasonable for this estimator. Some of what we need to type will always be the same:

```
. ml model d2 myweibull_d2 (lneta: ...) /lngamma
```

To make it easier for us to use the estimator, we code the fixed part into a new program:

```
program model
        version 11
        ml model d2 myweibull_d2 (lneta: '0') /lngamma
end
```

'0' is just Stata jargon for "what the user typed goes here". So with `model`, we can type things like

> . model studytime died = i.drug age

With *readytouse*.`do` in place in our analysis do-file, we can code things like

```
                                                    begin analysis.do
run readytouse
use realdata
model studytime died = i.drug age
ml search
ml max
                                                    end analysis.do
```

13 Writing ado-files to maximize likelihoods

When you find yourself writing or editing do-files for fitting a specific model to data regularly, it might be worth the effort to write an ado-file that will fit the model for you. Such a program is commonly called an estimation command.

This chapter discusses writing estimation commands. Chapter 14 discusses how to accommodate survey data analysis in your existing `ml`–based estimation commands. Other example estimation commands are presented in chapter 15.

13.1 Writing estimation commands

The term *estimation command* has a special meaning in Stata, because it implies several properties:

- Estimation commands share a common syntax.

- You can, at any time, review the previous estimates by typing the estimation command without arguments. This is known as *replay* in Stata parlance.

- Estimation commands share a common output format.

- You can obtain the variance–covariance matrix of the estimators by typing `estat vce` after estimation.

- You can restrict later commands to the estimation subsample by adding the suffix `if e(sample)`, for example, `summarize ... if e(sample)`.

- You can obtain predictions and the standard error of the linear prediction using `predict`. You can use `predictnl` for nonstandard predictions and their standard errors.

- You can refer to coefficients and standard errors in expressions (such as with `generate`) by typing `_b[name]` or `[eqname]_b[name]` and `_se[name]` or `[eqname]_se[name]`.

- You can perform tests on the estimated parameters using `test` (Wald test of linear hypotheses), `testnl` (Wald test of nonlinear hypotheses), and `lrtest` (likelihood-ratio tests). You can obtain point estimates and confidence intervals of linear and nonlinear functions of the estimated parameters using `lincom` and `nlcom`, respectively.

Except for the first two bulleted items, ml satisfies all these requirements; you do not
have to do anything more. ml's syntax is different from that of other estimation commands (first bullet), and you cannot type ml without arguments to redisplay results
(second bullet), but you can type ml display to redisplay results, so the second viola-
tion is only a technicality.

Convenience of syntax is the main reason to package an ml estimator into an ado-file.
Rather than typing

```
. do such-and-such (to load the evaluator)
. ml model lf mynormal_lf (mpg = foreign weight displacement) () if age>20,
> vce(robust)
. ml search
. ml maximize
```

it would be more convenient to simply type

```
. mynormal0 mpg foreign weight displacement if age>20, vce(robust)
```

Creating such an ado-file would be easy:

─── begin mynormal0.ado ───────────

```
program mynormal0
        version 11
        syntax varlist [if] [in] [, vce(passthru) ]
        gettoken y rhs : varlist
        ml model lf mynormal_lf ('y' = 'rhs') () 'if' 'in', 'vce'
        ml maximize
end
```

── end mynormal0.ado ───────────

─── begin mynormal_lf.ado ───────────

```
program mynormal_lf
        version 11
        args lnf mu lnsigma
        quietly replace 'lnf' = ln(normalden($ML_y1,'mu',exp('lnsigma')))
end
```

── end mynormal_lf.ado ───────────

The mynormal0 command works, although it does not really meet our standards for an
estimation command. Typing mynormal0 without arguments will not redisplay results,
although it would not be difficult to incorporate that feature. Moreover, this command
does not allow us to use factor-variable notation when specifying our variables.

The point is that mynormal0 works and is fairly simple, and it may work well enough
for your purposes. We are about to go to considerable effort to solve what are really
minor criticisms. We want to add every conceivable option, and we want to make the
program bulletproof. Such work is justified only if you or others plan on using the
resulting command often.

In the following, we do not mention how to write your estimation command so that it
will allow the use of Stata's by *varlist:* prefix. If you are going to use the outline for an
estimation command detailed in section 13.3 and want to allow by, see the *Combining
byable(recall) with byable(onecall)* section in [P] **byable**.

13.2   The standard estimation-command outline

The outline for any estimation command—not just those written with `ml`—is

─────────────────────────────────────── begin *newcmd*.`ado` ───────────

```
program newcmd, eclass sortpreserve
        version 11
        local options "display-options"
        if replay() {
                if ("'e(cmd)'" != "newcmd") error 301
                syntax [, 'options']
        }
        else {
                syntax ... [, estimation-options 'options' ]

                // mark the estimation subsample
                marksample touse

                // obtain the estimation results
                ...
                ereturn local cmd "newcmd"
        }
        display estimation results respecting display options
end
```

─── end *newcmd*.`ado` ───────────

The idea behind this outline is as follows. Estimation commands should

- Redisplay results when typed without arguments.

- Compute and display results when typed with arguments.

Variations on the outline are acceptable, and a favorite one is

─────────────────────────────────────── begin *newcmd*.`ado` ───────────

```
program newcmd
        version 11
        if replay() {
                if ("'e(cmd)'" != "newcmd") error 301
                Replay '0'
        }
        else    Estimate '0'
end
program Estimate, eclass sortpreserve
        syntax ... [, estimation-options display-options ]

        // mark the estimation subsample
        marksample touse

        // obtain the estimation results
        ...
        ereturn local cmd "newcmd"

        Replay, ...
end
program Replay
        syntax [, display-options ]
        display estimation results respecting the display options
end
```

─── end *newcmd*.`ado` ───────────

The names `Estimate` and `Replay` are actually used for the subroutine names; there is no possibility of conflict, even if other ado-files use the same names, because the subroutines are local to their ado-file. (We do not bother to restate `version 11` in the subroutines `Estimate` and `Replay`. The subroutines can be called only by *newcmd*, and *newcmd* is already declared to be `version 11`.)

13.3   Outline for estimation commands using ml

The outline for a new estimation command implemented with `ml` is

```
───────────────────────────────────────────── begin newcmd.ado ─────────
program newcmd
        version 11
        if replay() {
                if ("'e(cmd)'" != "newcmd") error 301
                Replay '0'
        }
        else    Estimate '0'
end

program Estimate, eclass sortpreserve
        syntax varlist(fv) [if] [in] [fweight] [,              ///
                Level(cilevel)                                ///
                other-estimation-options                      ///
                other-display-options-if-any                  ///
        ]

        // mark the estimation subsample
        marksample touse

        // perform estimation using ml
        ml model method newcmd_ll ...                         ///
                ['weight''exp'] if 'touse',                   ///
                other-estimation-options                      ///
                maximize
        ereturn local cmd "newcmd"

        Replay, 'level' other-display-options-if-any
end

program Replay
        syntax [, Level(cilevel) other-display-options-if-any ]
        ml display, level('level') other-display-options-if-any
end
─────────────────────────────────────────────── end newcmd.ado ─────────


───────────────────────────────────────────── begin newcmd_ll.ado ─────────
program newcmd_ll
        version 11
        args ...
        ...
end
─────────────────────────────────────────────── end newcmd_ll.ado ─────────
```

An estimation command implemented in terms of `ml` requires two ado-files:

newcmd`.ado`	The estimation command the user sees
newcmd`_ll.ado`	The secret subroutine (the evaluator) `ml` calls

The evaluator must be in a separate ado-file because, otherwise, `ml` could not find it.

We recommend that you give the secret subroutine the name formed by adding `_ll` as a suffix to the name of the new estimation command *newcmd*: *newcmd*`_ll`. No user will ever have to type this name, nor would we want the user to type it, and placing an underscore in the name pretty much ensures that. The standard ending `_ll` reminds you that it is a log-likelihood evaluator. Using the name of the command reminds you which ado-file this subroutine goes with.

For instance, if you planned on calling your new estimation command `mynormal`, then we would recommend that you name the likelihood evaluator `mynormal_ll.ado`. For this book, we decided to use the method as a suffix; this facilitates having different copies of likelihood evaluators for a given model (for example, `mynormal_lf.ado` and `mynormal_lf2.ado`).

13.4 Using ml in noninteractive mode

There are no `ml search`, `ml init`, or `ml maximize` commands. We use only `ml model` and then, later, `ml display`.

`ml model`, it turns out, has a `maximize` option that causes `ml model` to issue the `ml maximize` command for us. It also has a `search()` option corresponding to `ml search`, an `init()` option corresponding to `ml init`, and many other options.

The `ml` system has two modes of operation: interactive mode, which we have used in the previous chapters, and noninteractive mode, which we want to use now. When you specify `ml model`'s `maximize` option, you turn off interactive mode. The `ml model` command proceeds from definition to final solution.

Turning off interactive mode is desirable in this context. We wish to use `ml` as a subroutine, and we do not want the user to know. If we used `ml` in interactive mode— if we coded the `ml model` and `ml maximize` statements separately—then if the user pressed *Break* at the wrong time, `ml` would leave behind traces of the incomplete maximization. Also in interactive mode, `ml` temporarily adds variables to the dataset, but these variables are not official temporary variables as issued by the `tempvar` command. Instead, `ml` gets rid of them when the `ml maximize` command successfully completes. These variables could be left behind were the user to press *Break*.

In noninteractive mode, `ml` uses real temporary variables and arranges that all traces of itself disappear no matter how things work out.

Noninteractive mode is turned on—interactive mode is switched off—when we specify `ml model`'s `maximize` option.

▷ **Example**

The following is our first draft of an estimation command for linear regression.
`mynormal1` has all the properties listed in section 13.1.

```
                                               begin mynormal1.ado
program mynormal1
        version 11
        if replay() {
                if ('""'e(cmd)'"' != "mynormal1") error 301
                Replay '0'
        }
        else    Estimate '0'
end
program Estimate, eclass sortpreserve
        syntax varlist(fv) [if] [in] [,        ///
                vce(passthru)                  ///
                Level(cilevel)                 /// -Replay- option
        ]
        // check syntax
        gettoken lhs rhs : varlist
        // mark the estimation sample
        marksample touse
        // fit the full model
        ml model lf2 mynormal_lf2                      ///
                (mu: 'lhs' = 'rhs')                    ///
                /lnsigma                               ///
                if 'touse',                            ///
                'vce'                                  ///
                maximize
        ereturn local cmd mynormal1
        Replay , level('level')
end
program Replay
        syntax [, Level(cilevel) ]
        ml display , level('level')
end
                                               end mynormal1.ado
```

◁

13.5   Advice

Experience has taught us that developing new commands is best done in steps. For
example, we advocate the following when developing an estimation command using `ml`:

1. Derive the log-likelihood function from your probability model.

2. Write a likelihood evaluator that calculates the log-likelihood values and, optionally, its derivatives. We recommend method `lf` whenever possible; otherwise, start
   with method `d0`.

3. Interactively test this likelihood evaluator using the debugging utilities described
   in chapter 7.

4. Use the outline from section 13.3 to write a first draft for your estimation command.

5. Test your first draft for correctness and ensure that it has all the properties expected of an estimation command. Put these tests into a certification script. For more information on writing certification scripts, see Gould (2001).

6. When adding new features to your estimation command, add one option at a time, test each new option, and add the new tests to the certification script.

For the rest of this chapter, we will give some detailed advice based on our experience writing estimation commands. We will build each piece of advice into our estimation command for linear regression.

13.5.1   Syntax

Keep the syntax simple. Remember, you have to parse it and the user has to type it. Making the syntax unnecessarily complicated serves nobody's interest.

Many models are just single equations with one or more "dependent" variables. In that case, we recommend that the syntax be

cmd varlist $\big[\,if\,\big]$ $\big[\,in\,\big]$ $\big[\,weight\,\big]$ $\big[\,,\ \ldots\,\big]$

No extraneous characters are placed between the dependent and independent variables. This syntax is easy to parse and use. If there is one dependent variable, then the outline is

```
syntax varlist(fv) [fweight pweight] [if] [in] [, ...]
gettoken lhs rhs : varlist
...
ml model ... ('lhs'='rhs') ...
```

This approach works, even if there is really more than one equation but the subsequent equations concern ancillary parameters such as the variance of the residuals in linear regression. The `ml model` statement would then read

```
ml model ... ('lhs'='rhs') /lnsigma ...
```

Including the specification `(fv)` with `syntax varlist` indicates that users can include factor-variable notation in the *varlist* they specify. Typically, the dependent variable is not allowed to contain factor-variable notation. To ensure that the user does not specify such a dependent variable, you can use the `_fv_check_depvar` utility. Programming

```
_fv_check_depvar 'lhs'
```

will check for factor variables and interactions in `'lhs'` and issue an error message if factor variables or interactions are present. In general, you will not need to do any additional work to handle factor variables because `ml` does the rest automatically.

If there are two dependent variables (for example, in Stata's `intreg` command), the outline becomes

```
syntax varlist(min=2 fv) [fweight pweight] [if] [in] [, ... ]
gettoken lhs1 rhs : varlist
gettoken lhs2 rhs : rhs
_fv_check_depvar 'lhs1' 'lhs2', k(2)
...
ml model ... ('lhs1' 'lhs2' = 'rhs') ...
```

The option `k(2)` to `_fv_check_depvar` indicates that we have two dependent variables.

Many multiple-equation models, such as Stata's `heckman` command, lend themselves to the second equation being specified as a required option. In those cases, the outline (with one dependent variable) becomes

```
syntax varlist(fv) [fweight pweight] [if] [in], SECond(varlist)
[... ]
gettoken lhs rhs : varlist
_fv_check_varlist 'lhs'
...
ml model ... ('lhs' = 'rhs') ('second')  ...
```

Above, we specified the independent variables for the second equation in the required option `second()`, which could be minimally abbreviated `sec()`. The name of the option, of course, would be your choice.

▷ Example

We will add another option called `hetero(`*varlist*`)` to our estimation command. This new option will allow the user to specify variables to model heteroskedasticity in the errors. Because it is so easy, we will also allow *varlist* to contain factor-variable notation. By default, `mynormal2` will be a linear regression model with a constant error variance (homoskedastic errors).

Here is the ado-file for the next iteration of our command, which we call `mynormal2`:

```
                                              begin mynormal2.ado
program mynormal2
        version 11
        if replay() {
                if ('"'e(cmd)'"' != "mynormal2") error 301
                Replay '0'
        }
        else    Estimate '0'
end

program Estimate, eclass sortpreserve
        syntax varlist(fv) [if] [in] [,        ///
                vce(passthru)                  ///
                HETero(varlist fv)             /// <- NEW
                Level(cilevel)                 /// -Replay- options
        ]
```

```
        // check syntax
        gettoken lhs rhs : varlist
        _fv_check_depvar 'lhs'

        // mark the estimation sample
        marksample touse

        // fit the full model
        ml model lf2 mynormal_lf2                   ///
                (mu: 'lhs' = 'rhs')                 ///
                (lnsigma: 'hetero')                 /// <- NEW
                if 'touse',                         ///
                'vce'                               ///
                maximize

        ereturn local cmd mynormal2

        Replay , level('level')
end

program Replay
        syntax [, Level(cilevel) ]
        ml display , level('level')
end
```
────────────────────────── end `mynormal2.ado` ──────────────

◁

13.5.2   Estimation subsample

In your code, you may need to make calculations on the data. If so, you must ensure that you make them on the same data that `ml` uses, that is, the estimation subsample.

You should identify that sample early in your code and then restrict your attention to it. In the standard Stata syntax case, this is easy:

```
syntax varlist(fv) [fweight pweight] [if] [in] [, ... ]
marksample touse
```

Thereafter, you code `if 'touse'` on the end of your Stata commands. `marksample` creates a temporary variable—a variable that you thereafter refer to in single quotes—that contains 1 if the observation is to be used and 0 otherwise.

If you obtain a second equation from an option, account for missing values in it as well:

```
syntax varlist(fv) [fweight pweight] [if] [in], SECond(varlist fv) [, ... ]
marksample touse
markout 'touse' 'second'
```

`markout` takes the `'touse'` variable created by `marksample` and sets elements in it to zero in observations where the second, third, ... specified variables are missing.

We must also allow for the possibility that the user specified `vce(cluster` *varname*`)` to request a robust variance calculation that does not assume observations within each cluster are independent but only that observations in different clusters are independent. In general, we need to restrict our calculations to those observations for which the cluster variable does not contain missing values.

The easiest way to find out whether the user specified vce(cluster *varname*) is to use the _vce_parse utility:

```
syntax ... , [ vce(passthru) ... ]

marksample touse
...
_vce_parse 'touse', opt(Robust opg oim) argopt(CLuster) : , 'vce'
```

With this syntax, _vce_parse will scan local macro 'vce' and see if the user specified vce(cluster *varname*); we capitalized the C and L in CLuster to indicate that cluster can be abbreviated as cl (or clu, clus, ...). If the user did specify a cluster variable, then _vce_parse will mark out our touse variable if *varname* contains missing values. For more information on _vce_parse, type in help _vce_parse.

Once we have correctly identified the estimation subsample, there is no reason for ml to go to the work of doing it again. We can specify our ml model statement as

```
ml model ... if 'touse', missing ...
```

ml model's missing option literally means that missing values among the dependent and independent variable are okay, so they will not be marked out from the estimation sample. In fact, we have already marked such missing values out, so we are merely saving time. We are also ensuring that, if we have marked the estimation subsample incorrectly, ml will use the same incorrect sample and that will probably cause the estimation to go violently wrong. Thus such a bug would not go undiscovered.

▷ Example

mynormal3 uses markout to identify missing values in the hetero() option and the cluster variable (if specified), and it also uses the missing option in the ml model statement.

Here are the differences between `mynormal2` and `mynormal3`.

```
────────────────────────────────────────────── fragment mynormal3.ado ──────────
program mynormal3
        version 11
        if replay() {
                if (`"`e(cmd)'"' != "mynormal3") error 301
                Replay `0'
        }
        else    Estimate `0'
end
program Estimate, eclass sortpreserve
        ...
        // mark the estimation sample
        marksample touse
        markout `touse' `hetero'                      // <- NEW
        _vce_parse `touse', opt(Robust opg oim)       /// <- NEW
                argopt(CLuster): , `vce'              // <- NEW
        // fit the full model
        ml model lf2 mynormal_lf2                     ///
                ...                                   ///
                missing                               /// <- NEW
                maximize
        ereturn local cmd mynormal3
        Replay , level(`level')
end
program Replay
        ...
end
────────────────────────────────────────────── end mynormal3.ado ──────────
```

Using `auto.dta` and `mynormal3`, we fit the regression of mileage (in miles per gallon) on vehicle weight and engine displacement. We also use an indicator variable to model heteroskedasticity between foreign and domestic vehicles.

(Continued on next page)

```
. sysuse auto
(1978 Automobile Data)

. mynormal3 mpg weight displacement, hetero(foreign)
initial:       log likelihood =     -<inf>  (could not be evaluated)
feasible:      log likelihood = -6441.8168
rescale:       log likelihood = -368.12496
rescale eq:    log likelihood =  -257.3948
Iteration 0:   log likelihood =  -257.3948  (not concave)
Iteration 1:   log likelihood = -240.09029  (not concave)
Iteration 2:   log likelihood = -229.23871  (not concave)
Iteration 3:   log likelihood = -222.61255  (not concave)
Iteration 4:   log likelihood = -215.68284  (not concave)
Iteration 5:   log likelihood = -207.35199
Iteration 6:   log likelihood = -186.51002
Iteration 7:   log likelihood = -183.74152
Iteration 8:   log likelihood = -183.71457
Iteration 9:   log likelihood = -183.71454
Iteration 10:  log likelihood = -183.71454
                                            Number of obs   =         74
                                            Wald chi2(2)    =     202.12
Log likelihood = -183.71454                 Prob > chi2     =     0.0000
```

mpg	Coef.	Std. Err.	z	P>|z|	[95% Conf. Interval]	
mu						
weight	-.0059578	.000831	-7.17	0.000	-.0075865	-.004329
displacement	.0000592	.0068563	0.01	0.993	-.0133789	.0134974
_cons	39.4995	1.597494	24.73	0.000	36.36847	42.63053
lnsigma						
foreign	.824055	.1810665	4.55	0.000	.4691712	1.178939
_cons	.8187011	.0982573	8.33	0.000	.6261204	1.011282

◁

❏ Technical note

For speed, if you implement your command using an `lf`- or a `d`-family evaluator, the most important option you can add to the `ml model` statement is `nopreserve`. The way `ml` ordinarily works with these evaluators is to preserve the data and drop the irrelevant observations prior to optimization. Dropping irrelevant observations allows the evaluator to be written sloppily and still produce correct results. The data are automatically restored following estimation.

If you specify `nopreserve`, you must manage your estimation subsample carefully, and that means you must write your evaluator to respect `$ML_samp`. The speed advantages in doing so are considerable.

We recommend that during development, you do not specify `nopreserve`. Once you have completed your code, add the `nopreserve` option and then test it carefully against the version without the option.

To use `nopreserve` in `mynormal3`, you would modify the `ml model` statement to read

```
ml model lf2 mynormal_lf2                           ///
         (mu: 'lhs' = 'rhs')                        ///
         (lnsigma: 'hetero')                        ///
         if 'touse',                                ///
         'vce'                                      ///
         missing                                    ///
         nopreserve                                 /// <- NEW
         maximize
```

❏

13.5.3 Parsing with help from mlopts

`mlopts` is an official Stata command that will assist in checking syntax and parsing estimation commands that use `ml`.

The problem `mlopts` solves is this: you are about to write a new estimation command, and doubtlessly, it is going to allow options:

`mynewcmd` *varlist* [*if*] ... [, *options*]

Among those options, you would like to allow all the standard maximization options, such as `iterate(#)`, `tolerance(#)`, and `trace`. Handling these options, once parsed, is easy; you merely need to include them on the `ml model` statement. The problem is that even you, the programmer, probably do not know what all these options are (see appendix A), and even if you do, separating these options from the options unique to `mynewcmd` will be work.

That is where `mlopts` helps. `mlopts` takes the options specified by the user of your command and separates them into two lists: the standard ones that you will want to pass along to `ml model`, and all the rest.

There are two ways to use `mlopts`. In the first, you code

```
        syntax varlist(fv) ... [, *]
        mlopts std mine, 'options'
```

In the second, you code

```
        syntax varlist(fv) ... [, myopts *]
        mlopts std, 'options'
```

In both cases, the standard maximization options, such as `tolerance(#)` and `trace`, are placed in local macro 'std'. In the first case, any other options are placed in 'mine', from which you will have to then arrange that they be parsed. In the second case, your options already are parsed. Most programmers opt for the second syntax.

We will show you an example but first, let's get the official documentation out of the way.

Syntax

mlopts *mlopts* [*rest*] [, *options_to_parse*]

Description

The options specified in *options_to_parse* will be separated into two groups and returned in local macros specified by the caller. The names of the local macros will be *mlopts* and *rest*.

mlopts is required and will contain those options that the ml command recognizes as *maximize* options; see appendix A (pages 297–301). Options returned in *mlopts* are

<u>trace</u> <u>grad</u>ient <u>hess</u>ian showstep <u>iter</u>ate(#) <u>tol</u>erance(#)

<u>ltol</u>erance(#) <u>gtol</u>erance(#) <u>nrtol</u>erance(#) <u>nonrtol</u>erance

shownrtolerance <u>dif</u>ficult <u>constr</u>aints(*clist* | *matname*)

vce(oim | <u>o</u>pg | <u>r</u>obust) <u>tech</u>nique(nr | bhhh | dfp | bfgs)

rest is optional and, if specified, will contain all other options that were not returned in *mlopts*. If *rest* is not supplied, any extra options will cause an error.

Saved results

mlopts saves the following in s():

Macros
    s(constraints)   contents of constraints() option
    s(technique)   contents of technique() option

▷ Example

In mynormal4, we use mlopts to retrieve the *maximize* options and pass them on to the ml model statement. We also added the nolog and noconstant options; nolog suppresses the iteration log from being displayed, and noconstant suppresses the constant term (intercept) in the regression equation.

Here are the differences between mynormal3 and mynormal4.

```
────────────────────────────────── fragment mynormal4.ado ──────────
        program mynormal4
                version 11
                if replay() {
                        if ('"'e(cmd)'"' != "mynormal4") error 301
                        Replay '0'
                }
                else    Estimate '0'
        end
        program Estimate, eclass sortpreserve
                syntax varlist(fv) [if] [in] [,     ///
                        vce(passthru)               ///
                        noLOg                       /// <- NEW
                        noCONStant                  /// <- NEW
                        ...                         ///
                        Level(cilevel)              /// -Replay- option
                        *                           /// <- NEW for -mlopts-
                ]
                // check syntax
                mlopts mlopts, 'options'                    // <- NEW
                gettoken lhs rhs : varlist
                _fv_check_depvar 'lhs'
                ...
                // fit the full model
                ml model lf2 mynormal_lf2                   ///
                        (mu: 'lhs' = 'rhs', 'constant')     /// <- CHANGED
                        (lnsigma: 'hetero')                 ///
                        if 'touse',                         ///
                        'vce'                               ///
                        'log'                               /// <- NEW
                        'mlopts'                            /// <- NEW
                        ...
                ereturn local cmd mynormal4
                Replay , level('level')
        end
        program Replay
                ...
        end
─────────────────────────────────────── end mynormal4.ado ──────────
```

We can now request the outer product of the gradients variance estimates by specifying the vce(opg) option to be automatically processed by mlopts. We also use the noconstant option to directly estimate the intercept for both foreign and domestic vehicles.

(Continued on next page)

```
. sysuse auto
(1978 Automobile Data)

. gen domestic = foreign==0

. mynormal4 mpg foreign domestic weight displacement, vce(opg) noconstant nolog
initial:       log likelihood =      -<inf>  (could not be evaluated)
feasible:      log likelihood = -6441.8168
rescale:       log likelihood = -368.12496
rescale eq:    log likelihood =  -257.3948
                                                Number of obs   =          74
                                                Wald chi2(4)    =     1854.07
Log likelihood = -194.16368                     Prob > chi2     =      0.0000
```

mpg	Coef.	OPG Std. Err.	z	P>|z|	[95% Conf. Interval]	
mu						
foreign	40.24731	2.619865	15.36	0.000	35.11247	45.38216
domestic	41.84794	3.294567	12.70	0.000	35.39071	48.30518
weight	-.0067745	.001664	-4.07	0.000	-.0100358	-.0035132
displacement	.0019286	.0137793	0.14	0.889	-.0250783	.0289356
lnsigma						
_cons	1.204895	.0671298	17.95	0.000	1.073323	1.336467

◁

13.5.4 Weights

We recommend that your estimation commands allow frequency and sampling weights, if they can. For sampling weights, the issues are the same as for allowing robust estimates of variance. There is no issue for method-lf evaluators, `pweights` are not possible with method-d0 evaluators, and method-lf1 and method-lf2 evaluators must return equation-level scores for `pweights` to be allowed (see page 83).

If you allow weights, you should code

```
syntax ... [fweight pweight] ...
...
if "`weight'" != "" {
        local wgt "[`weight'`exp']"
}
```

The idea here is to define the macro 'wgt' so that weights are more convenient to use. Then you can code the `ml model` statement as

```
. ml model ... `wgt' if `touse' ...
```

because 'wgt' either substitutes to nothing or substitutes to [fweight=*exp*] or [pweight=*exp*].

▷ **Example**

We incorporated `fweights` and `pweights` into `mynormal5`.

Here are the differences between `mynormal4` and `mynormal5`:

```
———————————————————————————————— fragment mynormal5.ado ————————
program mynormal5
        version 11
        if replay() {
                if (`"`e(cmd)'"' != "mynormal5") error 301
                Replay `0'
        }
        else    Estimate `0'
end
program Estimate, eclass sortpreserve
        syntax varlist(fv) [if] [in]            ///
                [fweight pweight] [,            /// <- NEW
                noLOg                           /// -ml model- options
                ...                             ///
        ]

        ...
        if "`weight'" != "" {                          // NEW block
                local wgt "[`weight'`exp']"
        }
        // mark the estimation sample
        marksample touse
        markout `touse' `hetero'
        _vce_parse `touse', opt(Robust oim opg)        ///
                argopt(CLuster): `wgt' , `vce'         //  <- CHANGED
        ...
        // fit the full model
        ml model lf2 mynormal_lf2                      ///
                (mu: `lhs' = `rhs', `constant')        ///
                (lnsigma: `hetero')                    ///
                `wgt' if `touse',                      /// <- CHANGED
                `vce'                                  ///
                ...
        ereturn local cmd mynormal5
        Replay , level(`level')
end
program Replay
        ...
end
———————————————————————————————— end mynormal5.ado ————————
```

◁

13.5.5   Constant-only model

We recommend adopting a two-step procedure for obtaining estimates:

```
display as txt _n "Fitting constant-only model:"
ml model ..., maximize ...
display as txt _n "Fitting full model:"
ml model ..., maximize continue ...
```

The idea here is first to fit the constant-only model and then to proceed to fit the full model. `continue` specifies that `ml` is to report a likelihood-ratio test based on the previous `ml` results in place of the Wald test. There are two reasons why we recommend this two-step procedure with the `continue` option:

1. Users want to see a likelihood-ratio test reported, if it is relevant. The two-step procedure, with the `continue` option used at the second step, will accomplish this.

2. Estimation for many commonly used models proceeds more smoothly if you use the coefficients from the constant-only fit as starting values.

With some likelihood functions, you can obtain a solution for the constant-only model analytically. In that case, you can supply the point estimates using the `init()` option and the log-likelihood value using the `lf0()` option (not to be confused with evaluator method `lf0`). The outline then becomes

```
obtain analytic solution
display as txt _n "Fitting full model:"
ml model ..., maximize lf0(...) init(...) ...
```

In both outlines above, we would also specify the `search(off)` option, but that does not matter as far as reporting likelihood-ratio tests is concerned. We would specify `search(off)` because it is unlikely that `ml search` will improve on these starting values; searching would be a waste of time. However, do not specify `search(off)` if the constant-only model involves setting any coefficient on _cons to zero in any equation or if some equations have no _cons.

The likelihood-ratio test is inappropriate in some circumstances. It would be inappropriate for clustered data or when robust variance estimates are being computed; see [SVY] **survey**. The likelihood-ratio test might also be inappropriate if the option `constraints()` is specified, depending on the constraints.

In any case, `ml` watches for these problems and reports the Wald test, even if you specify the `continue` option. In the last case, Wald tests are always reported when the `constraints()` option is specified because it is too difficult to determine whether the likelihood-ratio test is appropriate.

When fitting the constant-only model, you need not bother honoring the user's request for any type of robust variance estimates. Calculating robust variance estimates takes additional time, and doing so does not affect the likelihood-ratio test or the point estimates that are in turn used as starting values for the full model. Only for the full model do you need to compute robust variance estimates if requested by the user.

▷ Example

Below we modify `mynormal5` to create `mynormal6` that follows the two-step procedure outlined above. `mynormal6` will first fit the constant-only model as long as option `noconstant` is not specified. Also, we add a `nolrtest` option to go back to reporting

the Wald test in place of the new likelihood-ratio test. We do this mainly to show that
we can.

Here are the differences between `mynormal5` and `mynormal6`:

```
———————————————————————————————————— fragment mynormal6.ado ————————

program mynormal6
        version 11
        if replay() {
                if ('"'e(cmd)'"' != "mynormal6") error 301
                Replay '0'
        }
        else    Estimate '0'
end
program Estimate, eclass sortpreserve
        syntax varlist(fv) [if] [in]                ///
                [fweight pweight] [,                ///
                ...                                 ///
                noLRTEST                            /// <- NEW
                Level(cilevel)                      /// -Replay- option
                *                                   /// -mlopts- options
        ]
        // check syntax
        mlopts mlopts, 'options'
        local cns 's(constraints)'                      // <- NEW
        ...
if "'log'" != ""
local qui quietly


        ...
        if "'constant'" == "" {                               // NEW block
                'qui' di as txt _n "Fitting constant-only model:"
                ml model lf mynormal_lf                         ///
                        (mu: 'lhs' = )                          ///
                        (lnsigma: 'hetero')                     ///
                        'wgt' if 'touse',                       ///
                        'log'                                   ///
                        'mlopts'                                ///
                        nocnsnotes                              ///
                        missing                                 ///
                        maximize
                if "'lrtest'" == "" {
                        local contin continue search(off)
                }
                else {
                        tempname b0
                        mat 'b0' = e(b)
                        local contin init('b0') search(off)
                }
                'qui' di as txt _n "Fitting full model:"
        }
        // fit the full model
        ml model lf2 mynormal_lf2                       ///
                (mu: 'lhs' = 'rhs', 'constant')         ///
                (lnsigma: 'hetero')                     ///
                'wgt' if 'touse',                       ///
                'vce'                                   ///
```

```
                        ...                              ///
                        'contin'                         /// <- NEW
                        missing                          ///
                        maximize

            ereturn local cmd mynormal6
            Replay , level('level')
    end

    program Replay
            ...
    end
```
——— end mynormal6.ado —————————

Using `auto.dta` and `mynormal6`, we fit a regression model and see the value of the
likelihood-ratio test statistic for the coefficients in the regression equation.

```
. sysuse auto, clear
(1978 Automobile Data)

. mynormal6 mpg foreign weight displacement

Fitting constant-only model:

initial:      log likelihood =      -<inf>  (could not be evaluated)
feasible:     log likelihood = -6441.8168
rescale:      log likelihood = -368.12496
rescale eq:   log likelihood =  -257.3948
Iteration 0:  log likelihood =  -257.3948
Iteration 1:  log likelihood = -253.97778
Iteration 2:  log likelihood = -234.74755
Iteration 3:  log likelihood = -234.39642
Iteration 4:  log likelihood = -234.39434
Iteration 5:  log likelihood = -234.39434

Fitting full model:

Iteration 0:  log likelihood = -234.39434  (not concave)
Iteration 1:  log likelihood = -221.82805  (not concave)
Iteration 2:  log likelihood = -218.56082
Iteration 3:  log likelihood = -195.60215
Iteration 4:  log likelihood = -194.16762
Iteration 5:  log likelihood = -194.16368
Iteration 6:  log likelihood = -194.16368

                                        Number of obs   =          74
                                        LR chi2(3)      =       80.46
Log likelihood = -194.16368             Prob > chi2     =      0.0000
```

mpg	Coef.	Std. Err.	z	P>|z|	[95% Conf. Interval]	
mu						
foreign	-1.600631	1.083131	-1.48	0.139	-3.72353	.5222673
weight	-.0067745	.0011346	-5.97	0.000	-.0089982	-.0045508
displacement	.0019286	.0097942	0.20	0.844	-.0172676	.0211249
_cons	41.84795	2.286289	18.30	0.000	37.3669	46.32899
lnsigma						
_cons	1.204895	.0821995	14.66	0.000	1.043787	1.366003

◁

13.5.6   Initial values

Previously, we did not bother with supplying `ml` with initial values; `ml` is pretty good about filling them in for us. In estimation commands, however, we recommend that you take the effort to produce good starting values.

Mostly, `ml search` does a good job, and if randomly it does not, the interactive `ml` user can type `ml search` yet again. In programming a canned estimator, do not depend on randomness working in your favor.

Produce starting values that you know are good. For most models, it is adequate to produce known-to-be-good starting values for the constant-only model, obtain the constant-only estimates, and then proceed from there to fitting the full model:

```
produce good starting values for constant-only model
display as txt _n "Fitting constant-only model:"
ml model ..., maximize init(...) search(off) ...
display as txt _n "Fitting full model:"
ml model ..., maximize continue search(off) ...
```

It would be acceptable to remove the `search(off)` from the first `ml model` statement. If, however, your starting values really are good, you will want to leave the `search(off)` in place because otherwise you will just be wasting computer time.

If you had questionable starting values, in addition to removing `search(off)` you might be tempted to add, say, `repeat(5)` and thus allow `ml search` a better opportunity to improve the values. In general, we do not do this because, if randomly searching really does improve things, then we are going to have to explain to users why they got one iteration log one time and a different iteration log another. We would also be worried that randomness may not work in our favor next time. Work harder on solving the problem of initial values, because randomly searching really is a waste of computer time.

Notice our very different attitude when programming an estimator for our own use and writing an estimation command. When we are just using `ml` interactively, we seldom bother with initial values. We use `ml search`, and that is safe because we are monitoring results. When we write an estimation command, users may not anticipate problems and therefore not watch for them. Because the estimator is a command, they will assume it works, and we therefore recommend performing the work necessary to meet their expectations.

▷ **Example**

The maximum likelihood estimators of the mean, μ, and variance, σ^2, for the normal distribution are

$$\widehat{\mu} = \overline{y} = \frac{1}{N} \sum_{j=1}^{N} y_j \tag{13.1}$$

$$\widehat{\sigma}^2 = \frac{1}{N} \sum_{j=1}^{N} (y_j - \overline{y})^2 \tag{13.2}$$

`mynormal7` uses `summarize` to compute the initial values according to (13.1) and (13.2) for the constant-only linear regression model.

When weights are supplied, we define the `'awgt'` macro for initial value calculations.

```
local awgt "[aw'exp']"
```

We use `aweights` because `summarize` does not allow `pweights` but, when used with `aweights`, reproduces the correct mean and variance calculations for `pweights`.

Here are the differences between `mynormal6` and `mynormal7`:

——————————————————————————— fragment mynormal7.ado ———————

```
program mynormal7
        version 11
        if replay() {
                if ('""'e(cmd)'"' != "mynormal7") error 301
                Replay '0'
        }
        else    Estimate '0'
end
program Estimate, eclass sortpreserve
        ...
        if "'weight'" != "" {
                local wgt "['weight''exp']"
                // for initial value calculations
                if "'weight'" == "pweight" {              // <-NEW block
                        local awgt "[aw'exp']"
                }
                else    local awgt "'wgt'"

        }
        ...
        if "'constant'" == "" {
                // initial values                    -- NEW block
                quietly sum 'lhs' 'awgt' if 'touse'
                local mean = r(mean)
                local lnsd = ln(r(sd))+ln((r(N)-1)/r(N))/2
                local initopt init(/mu='mean' /lnsigma='lnsd') search(off)

                'qui' di as txt _n "Fitting constant-only model:"
                ml model lf mynormal_lf                    ///
                        ...                                ///
                        'mlopts'                           ///
                        'initopt'                          /// <- NEW
                        nocnsnotes                         ///
                        missing                            ///
                        maximize
                ...
        }
```

```
        ...
        ereturn local cmd mynormal7
        Replay , level('level')
end

program Replay
        ...
end
```
—————————————————————————————————— end mynormal7.ado ——————————

In the following, we use `auto.dta` and `mynormal7` to fit a linear regression. It seems that `ml` performed an extra iteration when fitting the constant-only model, even though `mynormal7` passed it values calculated using the analytical solution. Actually, `ml` took one iteration to verify that it converged; `Iteration 0` is displayed to show the value of the log-likelihood function just before optimization begins in earnest.

```
. sysuse auto
(1978 Automobile Data)

. mynormal7 mpg foreign weight displacement

Fitting constant-only model:

Iteration 0:   log likelihood = -234.39434
Iteration 1:   log likelihood = -234.39434

Fitting full model:

Iteration 0:   log likelihood = -234.39434  (not concave)
Iteration 1:   log likelihood = -221.82805  (not concave)
Iteration 2:   log likelihood = -218.56082
Iteration 3:   log likelihood = -195.60215
Iteration 4:   log likelihood = -194.16762
Iteration 5:   log likelihood = -194.16368
Iteration 6:   log likelihood = -194.16368
```

				Number of obs	=	74
				LR chi2(3)	=	80.46
Log likelihood = -194.16368				Prob > chi2	=	0.0000

mpg	Coef.	Std. Err.	z	P>|z|	[95% Conf. Interval]	
mu						
foreign	-1.600631	1.083131	-1.48	0.139	-3.72353	.5222673
weight	-.0067745	.0011346	-5.97	0.000	-.0089982	-.0045508
displacement	.0019286	.0097942	0.20	0.844	-.0172676	.0211249
_cons	41.84795	2.286289	18.30	0.000	37.3669	46.32899
lnsigma						
_cons	1.204895	.0821995	14.66	0.000	1.043787	1.366003

◁

13.5.7   Saving results in e()

`ml` automatically saves the standard things in `e()` for you. You need to add `e(cmd)` so that the rest of Stata will know that these results were produced by a full-fledged estimation command:

```
ereturn local cmd "newcmd"
```

You may want to store other things in `e()` for use by your `Replay` subroutine. Store them before setting `e(cmd)` so that, should the user press *Break* while you are still setting values in `e()`, Stata will know that the estimation is not complete.

Do not store anything in `e()` until after you have issued the last `ml model` command. `ml model` wipes out whatever is stored in `e()` every time it is run.

13.5.8   Displaying ancillary parameters

Ancillary parameters are often estimated in a transformed metric; for example, rather than estimating σ, we estimate $\ln \sigma$. The output would look much better if the estimation output table labeled the parameter with `/lnsigma` rather than `lnsigma:_cons`. It would also be an improvement if it displayed the respective point and interval estimates of σ.

`ml display` will label $\ln \hat{\sigma}$ with `/lnsigma` for us if we identify it as an ancillary parameter, which we do by filling in `e(k_aux)`. `ml display` assumes that `e(k_aux)` contains the number of ancillary (auxiliary) parameters in the model and that those parameters make up the last elements of the vector of point estimates. `e(k_aux)` will be effective only if it contains an integer smaller than the number of equations (`ml` stores the number of equations in `e(k_eq)`).

To take advantage of `e(k_aux)` in `ml display`, you must code your likelihood evaluator so that the ancillary parameters make up the last elements of the vector of coefficients, meaning that they are specified last in the `ml model` statement:

```
ml model lf mynormal_lf (y = x variables) /lnsigma , ...
```

Then somewhere after the `ml model` statement but before saving the name of the estimation command in `e(cmd)`, we must save the number of ancillary parameters in `e(k_aux)` by coding

```
ereturn scalar k_aux = 1
```

When `ml display` is used to display the output table, it looks at `e(k_aux)`, assumes that the last equation contains an ancillary parameter, and labels it appropriately in the output table.

Now that we know how to label our ancillary parameters, we might also want to display these parameters in another metric. To do this, use the `diparm()` option of `ml model`. `ml model` accepts multiple `diparm(`*diparm_args*`)` options, and a simplified syntax diagram for *diparm_args* is

$$eqname \; \Big[\, , \; \big[\texttt{exp} \, | \, \texttt{tanh} \, | \, \texttt{invlogit} \big] \; \underline{\texttt{lab}}\texttt{el}(string) \; \underline{\texttt{prob}} \; \texttt{dof}(\#) \; \underline{\texttt{level}}(\#) \Big]$$

where *eqname* is the equation name for the ancillary parameter. Basically, you specify the transformation (`exp`, `tanh`, or `invlogit`) for the ancillary parameter and use the `label()` option to give the transformation its own title in the output table.

▷ **Example**

When performing regression with homoskedastic errors, `mynormal8` will use `e(k_aux)` to display $\ln \hat{\sigma}$ with label `/lnsigma` and the `diparm()` option to display $\hat{\sigma}$ with label `sigma`.

Here are the differences between `mynormal7` and `mynormal8`:

```
──────────────────────────────────── fragment mynormal8.ado ───────────
program mynormal8
        version 11
        if replay() {
                if ('"'e(cmd)'"' != "mynormal8") error 301
                Replay '0'
        }
        else    Estimate '0'
end
program Estimate, eclass sortpreserve
        ...
        if "'log'" != "" {
                local qui quietly
        }
        if '"'hetero'"' == "" {                        // NEW block
                local diparm diparm(lnsigma, exp label("sigma"))
        }
        // mark the estimation sample
        marksample touse
        markout 'touse' 'hetero'
        ...
        if "'hetero'" == "" {                          // NEW block
                ereturn scalar k_aux = 1
        }
        else    ereturn scalar k_aux = 0
        ereturn local cmd mynormal8
        Replay , level('level')
end
program Replay
        ...
end
──────────────────────────── end mynormal8.ado ───────────
```

The output table looks much nicer with these new features:

```
. sysuse auto
(1978 Automobile Data)

. mynormal8 mpg foreign weight displacement, nolog
                                            Number of obs   =          74
                                            LR chi2(3)      =       80.46
Log likelihood = -194.16368                 Prob > chi2     =      0.0000
```

mpg	Coef.	Std. Err.	z	P>|z|	[95% Conf. Interval]	
foreign	-1.600631	1.083131	-1.48	0.139	-3.72353	.5222673
weight	-.0067745	.0011346	-5.97	0.000	-.0089982	-.0045508
displacement	.0019286	.0097942	0.20	0.844	-.0172676	.0211249
_cons	41.84795	2.286289	18.30	0.000	37.3669	46.32899
/lnsigma	1.204895	.0821995	14.66	0.000	1.043787	1.366003
sigma	3.336409	.2742511			2.839951	3.919653

◁

❏ **Technical note**

The `diparm()` option has many more features and correspondingly a much more complicated syntax than we have shown above. Most importantly, any transformation is available, including user-defined transformations. To learn about them, type

```
. help _diparm
```

Note the underscore in front of `diparm`. The contents of the `diparm()` option are in fact processed by the `_diparm` command.

❏

13.5.9   Exponentiated coefficients

For some models, exponentiating the coefficients from the fit assists interpretation. The `eform(`*string*`)` option of `ml display` will exponentiate the coefficients, use the delta method to compute the respective standard errors, transform the respective confidence intervals, and label the column of transformed coefficients with *string*.

In logistic regression, exponentiating the coefficients provides what are commonly known as odds ratios. Thus if we were working on an estimation command for logistic regression, we might add an option that will cause `eform("Odds Ratio")` to be passed to `ml display`. In fact, this is such a common occurrence that `ml display` already has an option that does this, namely, `or`. Table 13.1 lists all the shortcut options such as `or`.

Table 13.1. Common shortcuts for the eform(*string*) option

Shortcut	Equivalent
eform	eform("exp(b)")
hr	eform("Haz. Ratio")
shr	eform("SHR")
irr	eform("IRR")
or	eform("Odds Ratio")
rrr	eform("RRR")

▷ **Example**

No one would want a linear-regression routine that reported exponentiated coefficients, but it is probably better to stay with a familiar example rather than introduce a new one. mynormal9 has a new eform option.

Here are the differences between mynormal8 and mynormal9:

```
──────────────────────────────────── fragment mynormal9.ado ───────────
program mynormal9
        version 11
        if replay() {
                if (`"`e(cmd)'"' != "mynormal9") error 301
                Replay `0'
        }
        else    Estimate `0'
end
program Estimate, eclass sortpreserve
        syntax varlist(fv) [if] [in]          ///
                ...                           ///
                Level(cilevel)                /// -Replay- option
                EForm                         /// <- NEW
                *                             /// -mlopts- options
        ]

        ...
        ereturn local cmd mynormal9
        Replay , level(`level') `eform'                   // <- CHANGED
end
program Replay
        syntax [, Level(cilevel) EForm ]                  // <- CHANGED
        ml display, level(`level') `eform'                // <- CHANGED
end
─────────────────────────────────────── end mynormal9.ado ───────────
```

Here we see the new `eform` option in action.

```
. sysuse auto
(1978 Automobile Data)

. mynormal9 mpg foreign weight displacement, nolog eform
                                            Number of obs   =         74
                                            LR chi2(3)      =      80.46
Log likelihood = -194.16368                 Prob > chi2     =     0.0000
```

mpg	exp(b)	Std. Err.	z	P>|z|	[95% Conf. Interval]
foreign	.2017691	.2185425	-1.48	0.139	.0241486	1.685846
weight	.9932484	.0011269	-5.97	0.000	.9910421	.9954596
displacement	1.001931	.0098131	0.20	0.844	.9828806	1.02135
/lnsigma	1.204895	.0821995	14.66	0.000	1.043787	1.366003
sigma	3.336409	.2742511			2.839951	3.919653

◁

13.5.10   Offsetting linear equations

Suppose that you need to include a variable in an equation, and the coefficient on this variable must be a fixed value of your choosing. One way of doing this is by using constraints: set the fixed value of the coefficient using the `constraint` command, and specify the constraint in the `constraints()` option.

Another way is to specify the variable as an offset for the equation. In section 3.2 (page 43), we discussed the syntax for an equation in the `ml model` statement. We purposely omitted two options to keep the discussion simple. The full syntax for an equation in the `ml model` statement is

$$\left(\left[\,equation\_name:\,\right]\left[\,varlist_y\;=\right]\left[\,varlist_x\,\right]\left[\,,\;\underline{\text{noconst}}\text{ant}\right.\right.$$
$$\left.\underline{\text{off}}\text{set}(varname)\;\underline{\text{expo}}\text{sure}(varname)\,\right]\right)$$

The two new options are as follows.

`offset(`*varname*`)` specifies that the equation is to be $\mathbf{x}\boldsymbol{\beta} + varname$; in other words, the equation is to include *varname* with coefficient constrained to be 1.

`exposure(`*varname*`)` is an alternative to `offset(`*varname*`)`; it specifies that the equation is to be $\mathbf{x}\boldsymbol{\beta} + \ln(varname)$. The equation is to include $\ln(varname)$ with coefficient constrained to be 1.

See [R] **poisson** for examples using the `exposure()` option.

▷ **Example**

`mynormal10` has the new `offset(`*varname*`)` and `exposure(`*varname*`)` options. The options affect only the regression equation, that is, $\theta_{1j} = \mu_j = \mathbf{x}_{1j}\boldsymbol{\beta}_1$.

Here are the differences between `mynormal9` and `mynormal10`:

```
——————————————————————————— fragment mynormal10.ado ———————
program mynormal10
        version 11
        if replay() {
                if ('"'e(cmd)'"' != "mynormal10") error 301
                Replay '0'
        }
        else    Estimate '0'
end

program Estimate, eclass sortpreserve
        syntax varlist(fv) [if] [in]            ///
                [fweight pweight] [,            ///
                ...                             ///
                noLRTEST                        ///
                OFFset(varname numeric)         /// <- NEW
                EXPosure(varname numeric)       /// <- NEW
                Level(cilevel)                  /// -Replay- option
                ...
        ]

        ...
        if "'offset'" != "" {                          // NEW block
                local offopt "offset('offset')"
        }
        if "'exposure'" != "" {                        // NEW block
                local expopt "exposure('exposure')"
        }
        // mark the estimation sample
        marksample touse
        markout 'touse' 'hetero' 'offset' 'exposure'           // <- CHANGED
        ...
        if "'constant'" == "" {
                ...
                'qui' di as txt _n "Fitting constant-only model:"
                ml model lf mynormal_lf                 ///
                        (mu: 'lhs' =, 'offopt' 'expopt' )   /// <- CHANGED
                        (lnsigma: 'hetero')             ///
                        'wgt' if 'touse',               ///
                        ...
        }
        // fit the full model
        ml model lf2 mynormal_lf2                       ///
                (mu: 'lhs' = 'rhs',                     /// <- CHANGED
                        'constant' 'offopt' 'expopt'    /// <- CHANGED
                )                                       ///
                (lnsigma: 'hetero')                     ///
                'wgt' if 'touse',                       ///
                ...
        ...
end

program Replay
        ...
end
——————————————————————————— end mynormal10.ado ———————
```

Here we see the new offset() option in action.

```
. sysuse auto
(1978 Automobile Data)

. gen one = 1

. mynormal10 mpg foreign weight displacement, nolog offset(one)
                                              Number of obs   =         74
                                              LR chi2(3)      =      80.46
Log likelihood = -194.16368                   Prob > chi2     =     0.0000
```

mpg	Coef.	Std. Err.	z	P>|z|	[95% Conf. Interval]
foreign	-1.600631	1.083131	-1.48	0.139	-3.72353	.5222673
weight	-.0067745	.0011346	-5.97	0.000	-.0089982	-.0045508
displacement	.0019286	.0097942	0.20	0.844	-.0172676	.0211249
_cons	40.84795	2.286289	17.87	0.000	36.3669	45.32899
one	(offset)					
/lnsigma	1.204895	.0821995	14.66	0.000	1.043787	1.366003
sigma	3.336409	.2742511			2.839951	3.919653

We can see that it works by comparing the above output with

```
. mynormal10 mpg foreign weight displacement, nolog
  (output omitted)
```

The intercept (_cons) changed by 1.

◁

13.5.11 Program properties

Some commands use program properties to check for supported features in other commands; see [P] **program properties**. Program properties are associated with a program by using the properties() option of program. Associating your estimation command with program property *prop* advertises that your command meets the requirements that *prop* implies.

We discuss program properties for the svy prefix in section 14.1. Here we discuss the program properties that allow your estimation command to work with the nestreg prefix.

If command *mycmd* has the swml program property, nestreg assumes that the following items are all true:

- *mycmd* is an estimation command that follows standard Stata syntax and allows the if qualifier; see [U] **11 Language syntax** and [U] **11.1.3 if exp**.

In this case, the minimum requirement for standard Stata syntax is

$$mycmd \ \ varlist \ \ \big[\ if \ \big] \ \ \big[\ , \ \ options \ \big]$$

- *mycmd* saves the model coefficients and ancillary parameters in `e(b)`.

- `e(N)` contains the estimation sample size, and `e(sample)` identifies the estimation sample after *mycmd*.

- For the likelihood-ratio test, the log-likelihood value is saved in `e(ll)` and the model degrees of freedom is saved in `e(df_m)`.

- For the Wald test, the estimated covariance matrix for the coefficient estimates is saved in `e(V)`.

These requirements are easily met if you follow the outline specified in section 13.3. Notice that `ml` satisfies all but the first bulleted item; it is up to you to make sure that your command satisfies bullet 1.

▷ **Example**

mynormal11 has the new `swml` program property.

Here are the differences between `mynormal10` and `mynormal11`:

```
────────────────────────────── fragment mynormal11.ado ──────────
program mynormal11, properties(swml)       // <- NEW property
        version 11
        if replay() {
                if ('"`e(cmd)'"' != "mynormal11") error 301
                Replay `0'
        }
        else    Estimate `0'
end
program Estimate, eclass sortpreserve
        ...
        ereturn local cmd mynormal11
        Replay , level(`level') `eform'
end
program Replay
        ...
end
──────────────────────────────────── end mynormal11.ado ──────────
```

We revisit the regression example using `auto.dta` to see that the `nestreg` prefix is allowed with `mynormal11`.

```
. sysuse auto, clear
(1978 Automobile Data)

. nestreg, quietly lrtable: mynormal11 mpg (foreign weight) (displacement)

Block  1: foreign weight
Block  2: displacement
```

Block	LL	LR	df	Pr > LR	AIC	BIC
1	-194.1831	80.42	2	0.0000	396.3661	405.5824
2	-194.1637	0.04	1	0.8439	398.3274	409.8477

We conclude from the likelihood-ratio statistic that the coefficient on `displacement` (that is, Block 2) is not significant at the 5% level.

`mynormal11` contains every conceivable (non-`svy`) option and is bulletproof. The only features left to add are for the `svy` prefix command, discussed in chapter 14.

◁

14 Writing ado-files for survey data analysis

The way to write a survey estimation command is 1) write a standard (nonsurvey) estimation command and then 2) make a few changes so that your new estimation command will work with the svy prefix; see [SVY] **svy**. This chapter details how to do that. We will assume that the first step is already complete—we will start with mynormal11 from the end of chapter 13. Although mynormal11 is a highly developed estimator, we could just as well start with mynormal1. The extra features mentioned in chapter 13 are not required for making your estimator work with the svy prefix.

Stata's svy prefix command provides the following variance estimation methods:

- svy brr: balanced repeated replication (BRR)
- svy jackknife: jackknife for complex survey data
- svy linearized: linearization/Taylor expansion

The programming changes required by the BRR and jackknife variance estimation methods are minimal. Linearization requires that your evaluator support the robust variance estimator; thus your likelihood evaluator must either be method lf or a member of the lf family.

14.1 Program properties

We introduced program properties in section 13.5.11; also see [P] **program properties**. The svy prefix command uses program properties to identify which of the variance estimation methods are supported by the prefixed estimation command. The program properties are svyb for BRR, svyj for the complex survey jackknife, and svyr for linearization.

The syntax for the svy prefix is

svy *vcetype* [, *svy_options*] : *command varlist* [*if*] [*in*] [, *options*]

Here are the extra requirements for programs (ado-files) to support the `svy` prefix:

1. *command* must be `eclass`, allow `iweight`s, and accept the standard estimation syntax; see [P] **program**, [P] **syntax**, and [P] **mark**.

   command varlist $\begin{bmatrix} if \end{bmatrix}$ $\begin{bmatrix} in \end{bmatrix}$ $\begin{bmatrix} weight \end{bmatrix}$ $\begin{bmatrix} , & options \end{bmatrix}$

   varlist may contain multiple-equation specifications as in the `ml model` statement.

2. *command* must save the model coefficients and ancillary parameters in `e(b)` and the estimation sample size in `e(N)`, and it must identify the estimation subsample in `e(sample)`; see [P] **ereturn**.

3. `svy brr` requires that *command* have `svyb` as a program property. For example, the program definition for `regress` appears as

   ```
   program regress, properties(...  svyb ...)  ...
   ```

4. `svy bootstrap` requires that *command* have `svyb` as a program property.

5. `svy sdr` requires that *command* have `svyb` as a program property.

6. `svy jackknife` requires that *command* have `svyj` as a program property.

7. `svy linearized` has the following requirements:

   a. *command* must have `svyr` as a program property.

   b. `predict` after *command* must be able to generate scores using the following syntax:

      `predict` $\begin{bmatrix} type \end{bmatrix}$ *stub** $\begin{bmatrix} if \end{bmatrix}$ $\begin{bmatrix} in \end{bmatrix}$, s̲cores

   This syntax implies that estimation results with k equations will cause `predict` to generate k new equation-level score variables. These new equation-level score variables are *stub*1 for the first equation, *stub*2 for the second equation, ..., and *stub*k for the last equation. Actually, `svy` does not strictly require that these new variables be named this way, but this is a good convention to follow.

8. *command* must save the model-based variance estimator for the coefficients and ancillary parameters in `e(V)`; see [SVY] **variance estimation**.

Item 1 is satisfied by following the outline in section 13.3, and items 2 and 6 are handled by the `ml` command and by setting the appropriate program properties.

▷ **Example**

mynormal12 has the new program properties required by `svy`.

Here are the differences between `mynormal11` and `mynormal12`:

```
─────────────────────────────────── fragment mynormal12.ado ───────
program mynormal12, properties(svyb svyj svyr swml)
        // NEW properties      ^^^^^^^^^^^^^^^^^^
        version 11
        if replay() {
                if (`"`e(cmd)'"' != "mynormal12") error 301
                Replay `0'
        }
        else    Estimate `0'
end

program Estimate, eclass sortpreserve
        syntax varlist(fv) [if] [in]          ///
                [fweight pweight iweight] [,   /// <- ADDED iweight
                ...                            ///
        ]

        ...

        ereturn local cmd mynormal12

        Replay , level(`level') `eform'
end

program Replay
        ...
end
─────────────────────────────────── end mynormal12.ado ───────
```

Here we use `svy: mynormal12` with some survey data:

```
. use http://www.stata-press.com/data/r11/nhanes2a

. svyset [pweight=leadwt], strata(stratid) psu(psuid) vce(jackknife)

      pweight: leadwt
          VCE: jackknife
          MSE: off
  Single unit: missing
     Strata 1: stratid
         SU 1: psuid
        FPC 1: <zero>

. svy, subpop(rural): mynormal12 loglead age female
(running mynormal12 on estimation sample)

Jackknife replications (62)
──────┼─── 1 ───┼─── 2 ───┼─── 3 ───┼─── 4 ───┼─── 5
ss....................................ss..........   50
...........

Survey data analysis

Number of strata   =        29        Number of obs   =        4678
Number of PSUs     =        58        Population size =   106180946
                                      Subpop. no. of obs =      1837
                                      Subpop. size    =    36182921
                                      Replications    =          58
                                      Design df       =          29
                                      F(   2,     28) =      128.08
                                      Prob > F        =      0.0000
```

loglead	Coef.	Jackknife Std. Err.	t	P>|t|	[95% Conf. Interval]	
age	.0023002	.0008534	2.70	0.012	.0005549	.0040455
female	-.4111291	.0253709	-16.20	0.000	-.4630183	-.3592399
_cons	2.60715	.051626	50.50	0.000	2.501563	2.712737
/lnsigma	-.961362	.045809	-20.99	0.000	-1.055052	-.8676721

Note: 2 strata omitted because they contain no subpopulation members.
. estat effects, deff deft meff meft

loglead	Coef.	Jackknife Std. Err.	DEFF	DEFT	MEFF	MEFT
age	.0023002	.0008534	1.75361	1.32424	2.5775	1.60546
female	-.4111291	.0253709	1.73034	1.31543	2.08058	1.44242
_cons	2.60715	.051626	3.04471	1.74491	3.27341	1.80926
/lnsigma	-.961362	.045809	4.60444	2.1458	7.70975	2.77664

◁

14.2 Writing your own predict command

Each estimation command could have its own `predict` subroutine. Stata's `predict` command will pass control to the subroutine identified in `e(predict)`. ml's `predict` subroutine is `ml_p`, and it accepts the standard options documented in [R] **predict**. Specifically, `ml_p` accepts the `scores` option and uses `ml score` to compute the equation-level scores, which is a significant part of the requirements for `svy linearized`.

`ml score` automatically computes equation-level scores for method-lf and lf-family likelihood evaluators; no extra work is required. `ml score` does not work with d-family evaluators, because observation-level log-likelihood values (and their derivatives) are not available.

Write your own `predict` subroutine when you need `predict` to generate variables that are functions of `e(b)` and your data. Keep in mind that your `predict` subroutine could call `ml_p` for computing linear predictions (`xb`), standard error of the linear predictions (`stdp`), equation-level scores, or any other standard prediction documented in [R] **predict**.

▷ **Example**

Suppose we wanted `predict` to generate a variable with the predicted values of $\ln \sigma$ or σ. We can use `_predict` to get the predicted values of $\ln \sigma$ (which is a linear prediction of the `lnsigma` equation). Then we just exponentiate those values to get the predicted value of σ. Here is a listing of `mynormal_p.ado`, a `predict` subroutine for our maximum likelihood regression command. `mynormal_p` calls `ml_p` to handle the situations when the `lnsigma` and `sigma` options are not specified.

─────────────────────────────────── begin `mynormal_p.ado` ───────────

```
program mynormal_p
        syntax anything(id="newvarname") [if] [in] [, LNSigma SIGma * ]
        if "`lnsigma'" != "" {
                syntax newvarname [if] [in] [, LNSigma ]
                _predict `typlist' `varlist' `if' `in', equation(lnsigma)
                label variable `varlist' "predicted ln(sigma)"
                exit
        }
        if "`sigma'" != "" {
                syntax newvarname [if] [in] [, SIGma ]
                _predict `typlist' `varlist' `if' `in', equation(lnsigma)
                quietly replace `varlist' = exp(`varlist')
                label variable `varlist' "predicted sigma"
                exit
        }
        // let -ml_p- handle all other cases
        ml_p `0'
end
```

──────────────────────────────────── end `mynormal_p.ado` ───────────

`mynormal13` uses `mynormal_p` as its `predict` subroutine. Here are the differences between `mynormal12` and `mynormal13`:

─────────────────────────────────── fragment `mynormal13.ado` ───────────

```
program mynormal13, properties(svyb svyj svyr swml)
        version 11
        if replay() {
                if (`"`e(cmd)'"' != "mynormal13") error 301
                Replay `0'
        }
        else    Estimate `0'
end

program Estimate, eclass sortpreserve
        syntax varlist(fv) [if] [in]                ///
                [fweight pweight iweight] [,         ///
                ...                                  ///
        ]
        ...
        ereturn local predict mynormal_p           // <- NEW e(predict)
        ereturn local cmd mynormal13
        Replay , level(`level') `eform'
end

program Replay
        ...
end
```

──────────────────────────────────── end `mynormal13.ado` ───────────

Here we use `svy: mynormal13` with some survey data and `predict` to generate a variable containing the estimated value of σ:

```
. use http://www.stata-press.com/data/r11/nhanes2a

. svyset [pweight=leadwt], strata(stratid) psu(psuid)
      pweight: leadwt
          VCE: linearized
  Single unit: missing
     Strata 1: stratid
        SU 1: psuid
       FPC 1: <zero>

. svy: mynormal13 loglead age female
(running mynormal13 on estimation sample)

Survey data analysis

Number of strata   =        31        Number of obs      =        4948
Number of PSUs     =        62        Population size    =   112915016
                                      Design df          =          31
                                      F(   2,     30)    =      625.61
                                      Prob > F           =      0.0000
```

loglead	Coef.	Linearized Std. Err.	t	P>|t|	[95% Conf. Interval]	
age	.0024985	.0004326	5.78	0.000	.0016162	.0033807
female	-.3648611	.0101531	-35.94	0.000	-.3855684	-.3441537
_cons	2.663709	.0295542	90.13	0.000	2.603432	2.723985
/lnsigma	-.9892149	.0294101	-33.64	0.000	-1.049197	-.9292325
sigma	.3718685	.0109367			.3502188	.3948566

```
. predict shat, sigma

. describe shat
```

variable name	storage type	display format	value label	variable label
shat	float	%9.0g		predicted sigma

```
. summarize shat
```

Variable	Obs	Mean	Std. Dev.	Min	Max
shat	4948	.3718686	0	.3718686	.3718686

◁

15 Other examples

This chapter contains example estimation commands. In each example, we derive the log-likelihood function, show the derivatives that make up the gradient and Hessian functions (where applicable), list one or more likelihood evaluators, provide a fully functional estimation command, and use the estimation command to fit the model to a dataset. Appendix C contains the ado-files for each estimation command discussed in this chapter.

15.1 The logit model

In the logit model, the dependent variable Y is an indicator variable, meaning that it takes on only the values 0 or 1. The probability mass function from the Bernoulli distribution (basically, the only distribution for random indicator variables) is

$$f(y_j; \pi_j) = \begin{cases} \pi_j & \text{if } y_j = 1 \\ 1 - \pi_j & \text{if } y_j = 0 \end{cases}$$

where $0 \leq \pi_j \leq 1$ and we identify π_j as the probability for a success (we arbitrarily call $y_j = 1$ a success). The logit model then uses the log of the odds ratio (the logit function) to describe the relationship between the independent variables $\mathbf{x}_j$ and π_j by

$$h^{-1}(\pi_j) = \ln \frac{\pi_j}{1 - \pi_j} = \mathbf{x}_j \boldsymbol{\beta}$$

which can be rewritten as

$$\pi_j = h(\mathbf{x}_j \boldsymbol{\beta}) = \frac{e^{\mathbf{x}_j \boldsymbol{\beta}}}{1 + e^{\mathbf{x}_j \boldsymbol{\beta}}} = \frac{1}{1 + e^{-\mathbf{x}_j \boldsymbol{\beta}}}$$

where $h()$ is generically called the *link function*. This particular function is commonly referred to as the *logistic function* or *inverse logit function*. The log-likelihood function for the jth observation is

$$\ln \ell_j = \begin{cases} \ln h(\mathbf{x}_j \boldsymbol{\beta}) & \text{if } y_j = 1 \\ \ln h(-\mathbf{x}_j \boldsymbol{\beta}) & \text{if } y_j = 0 \end{cases}$$

Thus the logit model meets the linear-form restrictions. Our method-`lf` likelihood evaluator is

```
                                                          begin mylogit_lf.ado
     program mylogit_lf
             version 11
             args lnf xb
             quietly replace 'lnf' = ln(invlogit( 'xb')) if $ML_y1 == 1
             quietly replace 'lnf' = ln(invlogit(-'xb')) if $ML_y1 == 0
     end
                                                            end mylogit_lf.ado
```

Note that

$$\frac{\partial h(\mathbf{x}_j\boldsymbol{\beta})}{\partial \mathbf{x}_j\boldsymbol{\beta}} = h(\mathbf{x}_j\boldsymbol{\beta})h(-\mathbf{x}_j\boldsymbol{\beta})$$

Thus the derivatives of the log-likelihood function are

$$g_j = \frac{\partial \ln \ell_j}{\partial \mathbf{x}_j\boldsymbol{\beta}} = \begin{cases} h(-\mathbf{x}_j\boldsymbol{\beta}) & \text{if } y_j = 1 \\ -h(\mathbf{x}_j\boldsymbol{\beta}) & \text{if } y_j = 0 \end{cases}$$

$$H_j = \frac{\partial^2 \ln \ell_j}{\partial^2 \mathbf{x}_j\boldsymbol{\beta}} = -h(\mathbf{x}_j\boldsymbol{\beta})h(-\mathbf{x}_j\boldsymbol{\beta})$$

and our method-lf2 likelihood evaluator is

```
                                                         begin mylogit_lf2.ado
     program mylogit_lf2
             version 11
             args todo b lnfj g1 H
             tempvar xb lj
             mleval 'xb' = 'b'
             quietly {
                     gen double 'lj' = invlogit( 'xb')  if $ML_y1 == 1
                     replace     'lj' = invlogit(-'xb')  if $ML_y1 == 0
                     replace 'lnfj' = ln('lj')
                     if ('todo'==0) exit

                     replace 'g1' =  invlogit(-'xb')  if $ML_y1 == 1
                     replace 'g1' = -invlogit( 'xb')  if $ML_y1 == 0
                     if ('todo'==1) exit

                     mlmatsum 'lnfj' 'H' = -1*abs('g1')*'lj', eq(1,1)
             }
     end
                                                           end mylogit_lf2.ado
```

Our estimation command for this model is called mylogit and is listed in appendix C.1. mylogit uses mylogit_lf2 as its likelihood evaluator. Because mylogit_lf2 computes the equation-level scores, mylogit allows all optimization algorithms, robust and outer product of the gradients (OPG) variance estimates, and the svy option. The maximum likelihood estimate (MLE) for π in the constant-only model is $\bar{p}$, the average number of positive outcomes observed in the estimation sample. Thus the MLE for β_0 is $\widehat{\beta}_0 = h^{-1}(\bar{p})$.

Here is `mylogit` in use:

```
. sysuse cancer
(Patient Survival in Drug Trial)

. mylogit died i.drug age

Fitting full model:

Iteration 0:    log likelihood = -31.199418
Iteration 1:    log likelihood = -22.259892
Iteration 2:    log likelihood = -21.539075
Iteration 3:    log likelihood = -21.525583
Iteration 4:    log likelihood = -21.525537
Iteration 5:    log likelihood = -21.525537

My logit estimates                      Number of obs   =         48
                                        LR chi2(3)      =      19.35
Log likelihood = -21.525537             Prob > chi2     =     0.0002
```

died	Coef.	Std. Err.	z	P>|z|	[95% Conf. Interval]	
drug						
2	-3.550783	1.222919	-2.90	0.004	-5.94766	-1.153905
3	-3.249504	1.180797	-2.75	0.006	-5.563824	-.935184
age	.1203223	.0718246	1.68	0.094	-.0204513	.2610959
_cons	-3.625338	3.966999	-0.91	0.361	-11.40051	4.149838

```
. mylogit, or

My logit estimates                      Number of obs   =         48
                                        LR chi2(3)      =      19.35
Log likelihood = -21.525537             Prob > chi2     =     0.0002
```

died	Odds Ratio	Std. Err.	z	P>|z|	[95% Conf. Interval]	
drug						
2	.0287022	.0351004	-2.90	0.004	.0026119	.3154027
3	.0387934	.0458072	-2.75	0.006	.0038341	.3925136
age	1.12786	.0810081	1.68	0.094	.9797564	1.298352

15.2   The probit model

Similarly to the logit model in the previous section, the dependent variable Y is an indicator variable; however, the probit model uses the standard normal cumulative distribution function $\Phi()$ to link the independent variables $\mathbf{x}_j$ and the probability of success, π_j, by

$$\pi_j = \Phi(\mathbf{x}_j\boldsymbol{\beta})$$

The log-likelihood function for the jth observation is

$$\ln \ell_j = \begin{cases} \ln \Phi(\mathbf{x}_j\boldsymbol{\beta}) & \text{if } y_j = 1 \\ \ln \Phi(-\mathbf{x}_j\boldsymbol{\beta}) & \text{if } y_j = 0 \end{cases}$$

Thus the probit model meets the linear-form restrictions. Our method-1f likelihood evaluator is

```
                                         begin myprobit_lf.ado
   program myprobit_lf
        version 11
        args lnfj xb
        quietly replace 'lnfj' = ln(normal( 'xb')) if $ML_y1 == 1
        quietly replace 'lnfj' = ln(normal(-'xb')) if $ML_y1 == 0
   end
                                         end myprobit_lf.ado
```

Taking the derivatives of the log-likelihood function yields

$$g_j = \frac{\partial \ln \ell_j}{\partial \mathbf{x}_j \boldsymbol{\beta}} = \begin{cases} \phi(\mathbf{x}_j \boldsymbol{\beta})/\Phi(\mathbf{x}_j \boldsymbol{\beta}) & \text{if } y_j = 1 \\ -\phi(\mathbf{x}_j \boldsymbol{\beta})/\Phi(-\mathbf{x}_j \boldsymbol{\beta}) & \text{if } y_j = 0 \end{cases}$$

$$H_j = \frac{\partial^2 \ln \ell_j}{\partial^2 \mathbf{x}_j \boldsymbol{\beta}} = -g_j(g_j + \mathbf{x}_j \boldsymbol{\beta})$$

and our method-1f2 likelihood evaluator is

```
                                         begin myprobit_lf2.ado
   program myprobit_lf2
        version 11
        args todo b lnfj g1 H
        tempvar xb lj
        mleval 'xb' = 'b'
        quietly {
                // Create temporary variable used in both likelihood
                // and scores
                gen double 'lj'  = normal( 'xb')  if $ML_y1 == 1
                replace     'lj'  = normal(-'xb')  if $ML_y1 == 0
                replace     'lnfj' = log('lj')
                if ('todo'==0) exit

                replace 'g1' =  normalden('xb')/'lj'  if $ML_y1 == 1
                replace 'g1' = -normalden('xb')/'lj'  if $ML_y1 == 0
                if ('todo'==1) exit

                mlmatsum 'lnfj' 'H' = -'g1'*('g1'+'xb'), eq(1,1)
        }
   end
                                         end myprobit_lf2.ado
```

Our estimation command for this model is called myprobit and is listed in appendix C.2. myprobit uses myprobit_d2 as its likelihood evaluator. Because myprobit_d2 computes the equation-level scores, myprobit allows all optimization algorithms, robust and OPG variance estimates, and the svy option. As with the logit model, the MLE for π in the constant-only model is $\bar{p}$, so the MLE for β_0 is $\widehat{\beta}_0 = \Phi^{-1}(\bar{p})$.

Here is `myprobit` in use:

```
. use http://www.stata-press.com/data/r11/nhanes2a
. svyset [pweight=leadwt], strata(stratid) psu(psuid)

      pweight: leadwt
          VCE: linearized
  Single unit: missing
     Strata 1: stratid
         SU 1: psuid
        FPC 1: <zero>
. svy: myprobit heartatk diabetes female age
(running myprobit on estimation sample)

Survey: My probit estimates

Number of strata   =        31          Number of obs      =       4946
Number of PSUs     =        62          Population size    =  112861252
                                        Design df          =         31
                                        F(   3,     29)    =      57.62
                                        Prob > F           =     0.0000
```

heartatk	Coef.	Linearized Std. Err.	t	P>|t|	[95% Conf. Interval]	
diabetes	.1375559	.1223278	1.12	0.269	-.1119333	.3870451
female	-.2875907	.0928203	-3.10	0.004	-.476899	-.0982825
age	.0383615	.0030946	12.40	0.000	.0320501	.044673
_cons	-3.741134	.1901302	-19.68	0.000	-4.128907	-3.353361

```
. estat effects, deft
```

heartatk	Coef.	Linearized Std. Err.	DEFT
diabetes	.1375559	.1223278	.737652
female	-.2875907	.0928203	1.09954
age	.0383615	.0030946	1.16025
_cons	-3.741134	.1901302	1.28455

15.3 Normal linear regression

In classical normal linear regression, the dependent variable Y is assumed to be normally distributed. We denote the mean and standard deviation of the jth observation with $\mu_j = \mathbf{x}_j \boldsymbol{\beta}$ and $\sigma_j > 0$, respectively. The probability density function is

$$f(y; \mu_j, \sigma_j) = \frac{1}{\sigma_j \sqrt{2\pi}} \exp\left\{ -\frac{1}{2} \left(\frac{y - \mu_j}{\sigma_j} \right)^2 \right\} = \frac{1}{\sigma_j} \phi\left(\frac{y - \mu_j}{\sigma_j} \right)$$

where $\phi()$ is the probability density function of the standard normal distribution. The log likelihood for the jth observation is

$$\ln \ell_j = \ln \phi \left(\frac{y_j - \mu_j}{\sigma_j} \right) - \ln \sigma_j$$

Thus the linear regression model meets the linear-form restrictions. We will model the standard deviation in the log space,

$$\theta_{2j} = \ln \sigma_j$$

so our method-`lf` likelihood evaluator is

————————————————————————— begin `mynormal_lf.ado` —————————

```
program mynormal_lf
        version 11
        args lnf mu lnsigma
        quietly replace 'lnf' = ln(normalden($ML_y1,'mu',exp('lnsigma')))
end
```
————————————————————————— end `mynormal_lf.ado` —————————

The derivative of $\phi()$ is

$$\phi'(z) = -z\phi(z)$$

so taking derivatives of the log-likelihood function yields

$$g_{1j} = \frac{\partial \ln \ell_j}{\partial \mu_j} = \frac{y_j - \mu_j}{\sigma_j^2} \qquad\qquad H_{11j} = \frac{\partial^2 \ell_j}{\partial^2 \mu_j} = -\frac{1}{\sigma_j^2}$$

$$g_{2j} = \frac{\partial \ln \ell_j}{\partial \ln \sigma_j} = \left(\frac{y_j - \mu_j}{\sigma_j} \right)^2 - 1 \qquad\qquad H_{22j} = \frac{\partial^2 \ell_j}{\partial^2 \ln \sigma_j} = -2 \left(\frac{y_j - \mu_j}{\sigma_j} \right)^2$$

$$H_{12j} = \frac{\partial^2 \ell_j}{\partial \mu_j \, \partial \ln \sigma_j} = -2 \frac{y_j - \mu_j}{\sigma_j^2}$$

and our method-`lf2` likelihood evaluator is

————————————————————————— begin `mynormal_lf2.ado` —————————

```
program mynormal_lf2
        version 11
        args todo b lnfj g1 g2 H
        tempvar mu lnsigma sigma
        mleval 'mu' = 'b', eq(1)
        mleval 'lnsigma' = 'b', eq(2)
        quietly {
                gen double 'sigma' = exp('lnsigma')
                replace 'lnfj' = ln( normalden($ML_y1,'mu','sigma') )
                if ('todo'==0) exit

                tempvar z
                tempname dmu dlnsigma
                gen double 'z' = ($ML_y1-'mu')/'sigma'
                replace 'g1' = 'z'/'sigma'
                replace 'g2' = 'z'*'z'-1
                if ('todo'==1) exit
```

```
                   tempname d11 d12 d22
                   mlmatsum 'lnfj' 'd11' = -1/'sigma'^2      , eq(1)
                   mlmatsum 'lnfj' 'd12' = -2*'z'/'sigma'    , eq(1,2)
                   mlmatsum 'lnfj' 'd22' = -2*'z'*'z'        , eq(2)
                   matrix 'H' = ('d11', 'd12' \ 'd12'', 'd22')
           }
      end
```
--- end mynormal_lf2.ado ---------

Our estimation command for this model is called mynormal and is listed in appendix C.3. mynormal uses mynormal_d2 as its likelihood evaluator. Because mynormal_d2 computes the equation-level scores, mynormal allows all optimization algorithms, robust and OPG variance estimates, and the svy option. The MLEs for the constant-only model are

$$\widehat{\mu} = \overline{y} = \frac{1}{N} \sum_{j=1}^{N} y_j \quad \text{and} \quad \widehat{\sigma} = \left\{ \frac{1}{N} \sum_{j=1}^{N} (y_j - \overline{y})^2 \right\}^{1/2}$$

Here is mynormal in use:

```
. use http://www.stata-press.com/data/r11/nhanes2a
. svyset [pweight=leadwt], strata(stratid) psu(psuid)
       pweight: leadwt
           VCE: linearized
  Single unit: missing
     Strata 1: stratid
        SU 1: psuid
       FPC 1: <zero>
. svy: mynormal loglead age female
(running mynormal on estimation sample)
Survey: My normal estimates
Number of strata   =        31          Number of obs     =        4948
Number of PSUs     =        62          Population size   =   112915016
                                        Design df         =          31
                                        F(   2,     30)   =      625.61
                                        Prob > F          =      0.0000
```

loglead	Coef.	Linearized Std. Err.	t	P>|t|	[95% Conf. Interval]	
age	.0024985	.0004326	5.78	0.000	.0016162	.0033807
female	-.3648611	.0101531	-35.94	0.000	-.3855684	-.3441537
_cons	2.663709	.0295542	90.13	0.000	2.603432	2.723985
/lnsigma	-.9892149	.0294101	-33.64	0.000	-1.049197	-.9292325

```
. estat effects, deff meff
```

loglead	Coef.	Linearized Std. Err.	DEFF	MEFF
age	.0024985	.0004326	1.63378	1.96588
female	-.3648611	.0101531	.921835	.920059
_cons	2.663709	.0295542	3.52743	3.25028
/lnsigma	-.9892149	.0294101	6.40399	8.5596

15.4 The Weibull model

In the Weibull model, the dependent variable measures time, $T > 0$, until failure (such as death or some other specific event). The Weibull density function is

$$f(t_j; \eta_j, \gamma_j) = \frac{\gamma_j}{\eta_j} \left(\frac{t_j}{\eta_j} \right)^{\gamma_j - 1} \exp \left\{ - \left(\frac{t_j}{\eta_j} \right)^{\gamma_j} \right\}$$

with $\eta_j > 0$ and $\gamma_j > 0$. Because the Weibull model is often used in survival analysis, we should also account for censoring, for example, the last time the subject came under observation when failure was not observed during the study.

We need a variable to distinguish between observed failures and censored observations, so let $d_j = 1$ indicate that t_j is the time of failure and $d_j = 0$ indicate that t_j is the time of censoring. If an observation is censored, we know that the time of failure is greater than the time of censoring.

$$\Pr(T > t_j) = \int_{t_j}^{\infty} \frac{\gamma_j}{\eta_j} \left(\frac{s}{\eta_j} \right)^{\gamma_j - 1} \exp \left\{ - \left(\frac{s}{\eta_j} \right)^{\gamma_j} \right\} ds$$

$$= \exp \left\{ - \left(\frac{t_j}{\eta_j} \right)^{\gamma_j} \right\}$$

Thus the log likelihood for the jth observation is

$$\ln \ell_j = \begin{cases} -(t_j/\eta_j)^{\gamma_j} + \ln \gamma_j - \ln \eta_j + (\gamma_j - 1)(\ln t_j - \ln \eta_j) & \text{if } d_j = 1 \\ -(t_j/\eta_j)^{\gamma_j} & \text{if } d_j = 0 \end{cases}$$

with $\eta_j > 0$ and $\gamma_j > 0$, and this Weibull model meets the linear-form restrictions.

We will model both η_j and γ_j in the log space. To emphasize this, we rewrite the log likelihood as

$$\ln \ell_j = \begin{cases} -(t_j e^{-\theta_{1j}})^{e^{\theta_{2j}}} + \theta_{2j} - \theta_{1j} + (e^{\theta_{2j}} - 1)(\ln t_j - \theta_{1j}) & \text{if } d_j = 1 \\ -(t_j e^{-\theta_{1j}})^{e^{\theta_{2j}}} & \text{if } d_j = 0 \end{cases} \tag{15.1}$$

$$\theta_{1j} = \ln \eta_j = \mathbf{x}_{1j} \boldsymbol{\beta}_1$$
$$\theta_{2j} = \ln \gamma$$

and our method-`lf` likelihood evaluator is

```
——————————————————————————————— begin myweibull_lf.ado ———————
program myweibull_lf
        version 11
        args lnfj leta lgam
        tempvar p M R
        quietly {
                gen double `p' = exp(`lgam')
                gen double `M' = ($ML_y1*exp(-`leta'))^`p'
                gen double `R' = ln($ML_y1)-`leta'
                replace `lnfj' = -`M' + $ML_y2*(`lgam'-`leta'+(`p'-1)*`R')
        }
end
——————————————————————————————— end myweibull_lf.ado ———————
```

where `$ML_y1` is the time variable t_j and `$ML_y2` is the censor variable d_j. We use temporary variables to evaluate some of the terms in the log likelihood. This is not necessary, but it makes the program slightly more readable and keeps us from having to evaluate `exp(`lgam')` more than once. Remember, those temporary variables were generated as `doubles`.

Before we take derivatives of the log likelihood, let's get better acquainted with the above intermediate results to help us with notation. Using the notation in (15.1), we have

$$p_j = e^{\theta_{2j}}$$
$$R_j = \ln t_j - \theta_{1j}$$
$$M_j = (t_j e^{-\theta_{1j}})^{e^{\theta_{2j}}}$$

so our log likelihood is

$$\ln \ell_j = d_j\{\theta_{2j} - \theta_{1j} + (p_j - 1)R_j\} - M_j$$

The derivatives of p_j, R_j, and M_j are

$$\frac{\partial p_j}{\partial \theta_{1j}} = 0 \qquad\qquad\qquad \frac{\partial p_j}{\partial \theta_{2j}} = p_j$$

$$\frac{\partial R_j}{\partial \theta_{1j}} = -1 \qquad\qquad\qquad \frac{\partial R_j}{\partial \theta_{2j}} = 0$$

$$\frac{\partial M_j}{\partial \theta_{1j}} = -p_j M_j \qquad\qquad\qquad \frac{\partial M_j}{\partial \theta_{2j}} = R_j p_j M_j$$

Thus the derivatives of the log-likelihood function are

$$g_{1j} = \frac{\partial \ln \ell_j}{\partial \theta_{1j}} = p_j(M_j - d_j) \qquad\qquad H_{11j} = \frac{\partial^2 \ln \ell_j}{\partial^2 \theta_{1j}} = -p_j^2 M_j$$

$$g_{2j} = \frac{\partial \ln \ell_j}{\partial \theta_{2j}} = d_j - R_j p_j(M_j - d_j) \quad H_{22j} = \frac{\partial^2 \ln \ell_j}{\partial^2 \theta_{2j}} = -p_j R_j(R_j p_j M_j + M_j - d_j)$$

$$H_{12j} = \frac{\partial^2 \ln \ell_j}{\partial \theta_{1j} \partial \theta_{2j}} = p_j(M_j - d_j + R_j p_j M_j)$$

and our method-`lf2` likelihood evaluator is

──────────────────────────────────── begin `myweibull_lf2.ado` ────────

```
program myweibull_lf2
        version 11
        args todo b lnf g1 g2 H
        tempvar leta lgam p M R
        mleval 'leta' = 'b', eq(1)
        mleval 'lgam' = 'b', eq(2)
        local t "$ML_y1"
        local d "$ML_y2"
        quietly {
                gen double 'p' = exp('lgam')
                gen double 'M' = ('t'*exp(-'leta'))^'p'
                gen double 'R' = ln('t')-'leta'
                replace 'lnf' = -'M' + 'd'*('lgam'-'leta' + ('p'-1)*'R')
                if ('todo'==0) exit

                replace 'g1' = 'p'*('M'-'d')
                replace 'g2' = 'd' - 'R'*'p'*('M'-'d')
                if ('todo'==1) exit

                tempname d11 d12 d22
                mlmatsum 'lnf' 'd11' = -'p'^2 * 'M'                  , eq(1)
                mlmatsum 'lnf' 'd12' = 'p'*('M'-'d'+'R'*'p'*'M')    , eq(1,2)
                mlmatsum 'lnf' 'd22' = -'p'*'R'*('R'*'p'*'M'+'M'-'d') , eq(2)
                matrix 'H' = ('d11','d12' \ 'd12'','d22')
        }
end
```

──────────────────────────────────── end `myweibull_lf2.ado` ────────

Our estimation command for this model is called `myweibull` and is listed in appendix C.4. `myweibull` uses `myweibull_lf2` as its likelihood evaluator. Because `myweibull_lf2` computes the equation-level scores, `myweibull` allows all optimization algorithms, robust and OPG variance estimates, and the `svy` option. When $\gamma_j = 1$, we have the exponential model (with censoring). The MLE for η in the constant-only exponential model is

$$\widehat{\eta} = \frac{\sum_{j=1}^N t_j}{\sum_{j=1}^N d_j}$$

Thus `myweibull` uses $\widehat{\beta}_0 = \ln \widehat{\eta}$ as the initial value.

Here is `myweibull` in use:

```
. sysuse cancer
(Patient Survival in Drug Trial)

. constraint 1 [lngamma]_cons = 0

. myweibull studytime i.drug age, fail(died) constraint(1) hr

Fitting constant-only model:

Iteration 0:   log likelihood = -129.51967
Iteration 1:   log likelihood = -129.51967

Fitting full model:

Iteration 0:   log likelihood = -129.51967
Iteration 1:   log likelihood = -122.88547
Iteration 2:   log likelihood = -116.36177
Iteration 3:   log likelihood = -116.34506
Iteration 4:   log likelihood = -116.34505
Iteration 5:   log likelihood = -116.34505
```

```
My Weibull estimates                        Number of obs  =        48
                                            Wald chi2(3)   =     25.16
Log likelihood = -116.34505                 Prob > chi2    =    0.0000
  ( 1)  [lngamma]_cons = 0
```

	Haz. Ratio	Std. Err.	z	P>|z|	[95% Conf. Interval]	
drug						
2	4.313332	2.053639	3.07	0.002	1.69646	10.96686
3	6.407428	3.000648	3.97	0.000	2.558936	16.04383
age	.9181277	.0307545	-2.55	0.011	.8597862	.9804281
/lngamma	(omitted)					
gamma	1	.			.	.
1/gamma	1	.			.	.

15.5   The Cox proportional hazards model

In the Cox proportional hazards model, the dependent variable measures time, $T > 0$, until an individual fails or is censored. Given the vector of covariates $\mathbf{x}_j$ for the jth individual, the log-likelihood function for this model is

$$\ln L = \sum_{i=1}^{N} \left[\sum_{k \in D_i} \mathbf{x}_k \boldsymbol{\beta} - d_i \ln \left\{ \sum_{j \in R_i} \exp(\mathbf{x}_j \boldsymbol{\beta}) \right\} \right]$$

where i indexes the ordered failure times $t_{(i)}$ $(i = 1, \ldots, N)$, d_i is the number who fail at $t_{(i)}$, D_i is the set of observations that fail at $t_{(i)}$, and R_i (the risk pool at time $t_{(i)}$) is the set of observations that are at risk at time $t_{(i)}$. A more complete discussion of this model, including other methods for handling tied failure times, is in Cleves et al. (2010).

This model fails to meet the linear-form restrictions, so we will have to use method d0 rather than method lf to maximize the log-likelihood function. Without equation-level scores, the BHHH algorithm, robust and OPG variance estimates, and the svy option are not allowed.

We can sort the data from within our estimation command, so in accordance with the technical notes in section 6.5 (pages 149–150), our method-d0 likelihood evaluator is

```
                                                            begin mycox2_d0.ado
    program mycox2_d0
            version 11
            args todo b lnf
            local negt "$ML_y1"              // $ML_y1 is -t
            local d "$ML_y2"                 // $ML_y2 is d
            tempvar xb B A sumd last L
            mleval 'xb' = 'b'
            // data assumed already sorted by 'negt' and 'd'
            local byby "by 'negt' 'd'"
            local wxb "$ML_w*'xb'"
            quietly {
                    gen double 'B' = sum($ML_w*exp('xb'))
                    'byby': gen double 'A' = cond(_n==_N, sum('wxb'), .) if 'd'==1
                    'byby': gen 'sumd' = cond(_n==_N, sum($ML_w*'d'), .)
                    'byby': gen byte 'last' = (_n==_N & 'd' == 1)
                    gen double 'L' = 'A' - 'sumd'*ln('B') if 'last'
                    mlsum 'lnf' = 'L' if 'last', noweight
            }
    end
                                                            end mycox2_d0.ado
```

Our estimation command for this model is called mycox and is listed in appendix C.5. We use the default initial values for this estimation command; however, fitting a constant-only model is a little tricky because of the lack of a constant. To get around this, we use the iterate(0) option to compute the value of the likelihood ll_0 at the default initial values (vector of zeros) and then pass lf0(0 ll_0) to the ml model statement in the second step. The relevant code fragment is

——————————————————————————— fragment `mycox.ado` ———————

```
      ...
      gettoken time rhs : varlist
      tempvar negt
      quietly gen double `negt' = -`time'
      sort `negt' `failure'
      ...
      // fit constant-only model
      if "`lrtest'" == "" {
              quietly ml model d0 mycox2_d0                      ///
                      (xb: `negt' `failure' = `rhs',            ///
                              noconstant                         ///
                      )                                          ///
                      `wgt' if `touse',                          ///
                      iterate(0)                                 ///
                      nocnsnotes missing maximize
              local initopt lf0(0 `e(ll)')
      }
      // fit the full model
      `qui' di as txt _n "Fitting full model:"
      ml model d0 mycox2_d0                                       ///
              (xb: `negt' `failure' = `rhs',                    ///
                      noconstant                                 ///
              )                                                  ///
              `wgt' if `touse',                                  ///
              `log' `mlopts' `vce' `initopt'                     ///
              missing maximize
      ...
```

——————————————————————————————— end `mycox.ado` ———————

Here is `mycox` in use:

```
. sysuse cancer
(Patient Survival in Drug Trial)

. mycox studytime i.drug age, fail(died)

Fitting full model:

initial:      log likelihood = -99.911448
alternative:  log likelihood = -99.606964
rescale:      log likelihood = -99.101334
Iteration 0:  log likelihood = -99.101334
Iteration 1:  log likelihood = -82.649192
Iteration 2:  log likelihood = -81.729635
Iteration 3:  log likelihood = -81.652727
Iteration 4:  log likelihood = -81.652567
Iteration 5:  log likelihood = -81.652567

My cox estimates                                Number of obs   =         48
                                                LR chi2(3)      =      36.52
Log likelihood = -81.652567                     Prob > chi2     =     0.0000
```

	Coef.	Std. Err.	z	P>|z|	[95% Conf.	Interval]
drug						
2	-1.71156	.4943639	-3.46	0.001	-2.680495	-.7426242
3	-2.956384	.6557432	-4.51	0.000	-4.241617	-1.671151
age	.11184	.0365789	3.06	0.002	.0401467	.1835333

15.6   The random-effects regression model

The random-effects regression model is typically written

$$y_{it} = \mathbf{x}_{it}\boldsymbol{\beta} + u_i + e_{it}$$

where u_i (the random effects) are assumed independent and identically distributed $N(0, \sigma_u^2)$, e_{it} are assumed independent and identically distributed $N(0, \sigma_e^2)$, and u_i and e_{it} are assumed uncorrelated (independent). The joint likelihood function for the ith group or panel is given by

$$f(y_{i1}, \ldots, y_{iT_i}) = \int_{-\infty}^{\infty} f(u_i) \prod_{t=1}^{T_i} f(y_{it}|u_i)\, du_i$$

where

$$f(u_i) = \frac{1}{\sigma_u}\phi\left(\frac{u_i}{\sigma_u}\right) \quad \text{and} \quad f(y_{it}|u_i) = \frac{1}{\sigma_e}\phi\left(\frac{y_{it} - \mathbf{x}_{it}\boldsymbol{\beta} - u_i}{\sigma_e}\right)$$

Thus the log likelihood for the ith group of observations is

$$\ln L_i = -\frac{1}{2}\left\{ \frac{\sum_t z_{it}^2 - a_i(\sum_t z_{it})^2}{\sigma_e^2} + \ln(T_i\sigma_u^2/\sigma_e^2 + 1) + T_i\ln(2\pi\sigma_e^2)\right\}$$

$$T_i = \text{number of observations in the } i\text{th group}$$
$$z_{it} = y_{it} - \mathbf{x}_{it}\boldsymbol{\beta}$$
$$a_i = \sigma_u^2/(T_i\sigma_u^2 + \sigma_e^2)$$

and this model does not meet the linear-form restrictions. Without equation-level scores, the BHHH algorithm, robust and OPG variance estimates, and the svy option are not allowed.

We wish to model the variances in the log space, so the equations for our evaluator are

$$\theta_{1it} = \mathbf{x}_{it}\boldsymbol{\beta}$$
$$\theta_{2it} = \ln\sigma_u$$
$$\theta_{3it} = \ln\sigma_e$$

Taking derivatives with respect to $\boldsymbol{\beta}$, $\ln \sigma_u$, and $\ln \sigma_e$ yields

$$\frac{\partial \ln L_i}{\partial \boldsymbol{\beta}} = \sum_{t=1}^{T_i} \frac{\partial \ln L_i}{\partial \mathbf{x}_{it}\boldsymbol{\beta}} \mathbf{x}_{it}$$

$$\frac{\partial \ln L_i}{\partial \mathbf{x}_{it}\boldsymbol{\beta}} = \frac{1}{\sigma_e^2} \left(z_{it} - a_i \sum_{s=1}^{T_i} z_{is} \right)$$

$$\frac{\partial \ln L_i}{\partial \ln \sigma_u} = \frac{a_i^2}{\sigma_u^2} \left(\sum_{t=1}^{T_i} z_{it} \right)^2 - T_i a_i$$

$$\frac{\partial \ln L_i}{\partial \ln \sigma_e} = \frac{1}{\sigma_e^2} \left\{ \sum_{t=1}^{T_i} z_{it}^2 - a_i \left(\sum_{t=1}^{T_i} z_{it} \right)^2 \right\} - (T_i - 1) - \frac{a_i^2}{\sigma_u^2} \left(\sum_{t=1}^{T_i} z_{it} \right)^2 - \frac{\sigma_e^2}{\sigma_u^2} a_i$$

The second (and cross) derivatives, with respect to $\ln \sigma_u$ and $\ln \sigma_e$, are

$$\frac{\partial^2 \ln L_i}{\partial^2 \ln \sigma_u} = - \left\{ \frac{2a_i^2}{\sigma_u^2} \left(\sum_{t=1}^{T_i} z_{it} \right)^2 - \frac{4a_i^3 \sigma_e^2}{\sigma_u^4} \left(\sum_{t=1}^{T_i} z_{it} \right)^2 + \frac{2T_i a_i^2 \sigma_e^2}{\sigma_u^2} \right\}$$

$$\frac{\partial^2 \ln L_i}{\partial^2 \ln \sigma_e} = - \left[\frac{2}{\sigma_e^2} \left\{ \sum_{t=1}^{T_i} z_{it}^2 - a_i \left(\sum_{t=1}^{T_i} z_{it} \right)^2 \right\} \right.$$
$$\left. - \left(\frac{2a_i^2}{\sigma_u^2} + \frac{4a_i^3 \sigma_e^2}{\sigma_u^4} \right) \left(\sum_{t=1}^{T_i} z_{it} \right)^2 + \frac{2a_i \sigma_e^2}{\sigma_u^2} - \frac{2a_i^2 \sigma_e^4}{\sigma_u^4} \right]$$

$$\frac{\partial^2 \ln L_i}{\partial \ln \sigma_e \, \partial \ln \sigma_u} = - \left\{ \frac{4a_i^3 \sigma_e^2}{\sigma_u^4} \left(\sum_{t=1}^{T_i} z_{it} \right)^2 - \frac{2T_i a_i^2 \sigma_e^2}{\sigma_u^2} \right\}$$

(*Continued on next page*)

The rest of the second derivatives are

$$\frac{\partial^2 \ln L_i}{\partial \boldsymbol{\beta} \partial \boldsymbol{\beta}'} = -\frac{1}{\sigma_e^2} \left\{ \sum_{t=1}^{T_i} \mathbf{x}'_{it}\mathbf{x}_{it} - a_i \left(\sum_{s=1}^{T_i} \mathbf{x}'_{is} \right) \left(\sum_{t=1}^{T_i} \mathbf{x}_{it} \right) \right\}$$

$$\frac{\partial^2 \ln L_i}{\partial \boldsymbol{\beta} \, \partial \ln \sigma_u} = \sum_{t=1}^{T_i} \frac{\partial^2 \ln L_i}{\partial \mathbf{x}_{it}\boldsymbol{\beta} \, \partial \ln \sigma_u} \mathbf{x}_{it}$$

$$\frac{\partial^2 \ln L_i}{\partial \mathbf{x}_{it}\boldsymbol{\beta} \, \partial \ln \sigma_u} = -\frac{2a_i^2}{\sigma_u^2} \sum_{t=1}^{T_i} z_{it}$$

$$\frac{\partial^2 \ln L_i}{\partial \boldsymbol{\beta} \, \partial \ln \sigma_e} = \sum_{t=1}^{T_i} \frac{\partial^2 \ln L_i}{\partial \mathbf{x}_{it}\boldsymbol{\beta} \, \partial \ln \sigma_e} \mathbf{x}_{it}$$

$$\frac{\partial^2 \ln L_i}{\partial \mathbf{x}_{it}\boldsymbol{\beta} \, \partial \ln \sigma_e} = -\left\{ \frac{2}{\sigma_e^2} \left(z_{it} - a_i \sum_{t=1}^{T_i} z_{it} \right) - \frac{2a_i^2}{\sigma_u^2} \sum_{t=1}^{T_i} z_{it} \right\}$$

We cannot use `mlmatsum` to calculate $\partial^2 \ln L_i / \partial \boldsymbol{\beta} \partial \boldsymbol{\beta}'$ directly, but we can use it to calculate

$$\sum_{t=1}^{T_i} \mathbf{x}'_{it}\mathbf{x}_{it}$$

and we can use `mlmatbysum` to calculate

$$a_i \left(\sum_{s=1}^{T_i} \mathbf{x}'_{is} \right) \left(\sum_{t=1}^{T_i} \mathbf{x}_{it} \right)$$

Our method-`d2` evaluator for this model is called `myrereg2_d2` and is listed in appendix C.6. If you read closely, you will notice that we took the `sort` command from the evaluator (`myrereg_d2` from section 6.4.3) and placed it in the estimation command; this prevents the evaluator from performing unnecessary tasks.

Our estimation command for this model is called `myrereg` and is listed in appendix C.6. The initial value for the intercept in the constant-only model is the sample mean of the response variable. The initial values for the variances are taken from the variance components from a one-way analysis of variance of the constant-only model with the groups treated as random factors:

$$\ln s_e = \frac{1}{2} \ln(\text{MS}_e)$$

$$\ln s_u = \frac{1}{2} \ln \left(\frac{\text{MS}_b - \text{MS}_e}{\overline{\overline{T}}} \right)$$

where MS_b is the mean squares between groups, MS_e is the within-group (or error) mean squares, and $\overline{T}$ is the average number of observations per group.

Here is `myrereg` in use:

```
. use http://www.stata-press.com/data/ml4/tablef7-1
. myrereg lnC lnQ lnPF lf, panel(i) nolog
My rereg estimates                              Number of obs   =         90
                                                LR chi2(3)      =     436.32
Log likelihood =  114.72904                     Prob > chi2     =     0.0000
```

lnC	Coef.	Std. Err.	z	P>|z|	[95% Conf. Interval]	
lnQ	.90531	.0253759	35.68	0.000	.8555742	.9550459
lnPF	.4233757	.013888	30.48	0.000	.3961557	.4505957
lf	-1.064456	.1962308	-5.42	0.000	-1.449061	-.6798508
_cons	9.618648	.2066219	46.55	0.000	9.213677	10.02362
/lns_u	-2.170817	.3026644	-7.17	0.000	-2.764028	-1.577605
/lns_e	-2.828404	.077319	-36.58	0.000	-2.979947	-2.676862
sigma_u	.1140844	.0345293			.0630373	.2064689
sigma_e	.0591071	.0045701			.0507955	.0687786

15.7   The seemingly unrelated regression model

Models for systems of equations can also be fit using `ml`. To illustrate, we will implement seemingly unrelated regression. An early challenge we will face is determining a strategy to ensure convergence when fitting this model; all the other models we have looked at thus far rarely have this problem.

The seemingly unrelated regression model is a system of regression equations where the errors are assumed correlated between equations. A system of p equations may be written

$$y_{1j} = \beta_{1,0} + \beta_{1,1}x_{1j} + \cdots + \beta_{1,k}x_{kj} + e_{1j}$$
$$y_{2j} = \beta_{2,0} + \beta_{2,1}x_{1j} + \cdots + \beta_{2,k}x_{kj} + e_{2j}$$
$$\vdots$$
$$y_{pj} = \beta_{p,0} + \beta_{p,1}x_{1j} + \cdots + \beta_{p,k}x_{kj} + e_{pj}$$

Using vector notation, we have

$$\mathbf{y}_j = \boldsymbol{\beta}'\mathbf{x}_j + \mathbf{e}_j = \boldsymbol{\mu}_j + \mathbf{e}_j$$

where

$$\mathbf{y}_j = (y_{1j}, \ldots, y_{pj})'$$
$$\boldsymbol{\beta} = (\boldsymbol{\beta}_1, \ldots, \boldsymbol{\beta}_p)$$
$$\mathbf{x}_j = (x_{1j}, \ldots, x_{kj})'$$
$$\mathbf{e}_j = (e_{1j}, \ldots, e_{pj})'$$

Here $\boldsymbol{\beta}$ is a matrix with p columns and k rows, and the error vectors $\mathbf{e}_j$ are assumed to be independent across observations and distributed multivariate normal with a zero-mean vector and variance matrix Σ.

Even though $\mathbf{x}_j$ is the same for all p equations, we can impose exclusion restrictions on the elements of the coefficient matrix to remove a variable from an equation. Given ml's flexible equation notation, we can just specify which independent variables go with a given dependent variable. For example, if our system was

$$y_{1j} = \beta_{1,0} + \beta_{1,1}x_{1j} + \beta_{1,2}x_{2j} + \beta_{1,3}x_{3j} + e_{1j}$$
$$y_{2j} = \beta_{2,0} + \beta_{2,1}x_{1j} + \beta_{2,2}x_{2j} + \beta_{2,3}x_{3j} + e_{2j}$$

we would type something like

```
. ml model lf1 mysureg_lf1 (y1 = x1 x2 x3) (y2 = x1 x2 x3)
```

But if our system was

$$y_{1j} = \beta_{1,0} + \beta_{1,1}x_{1j} + e_{1j}$$
$$y_{2j} = \beta_{2,0} + \beta_{2,1}x_{2j} + e_{2j}$$
$$y_{3j} = \beta_{3,0} + \beta_{3,1}x_{3j} + e_{3j}$$

we would type

```
. ml model lf1 mysureg_lf1 (y1 = x1) (y2 = x2) (y3 = x3)
```

The log likelihood for the jth observation is

$$\ln L_j = -\frac{1}{2} \left\{ p \ln(2\pi) + \ln |\Sigma| + (\mathbf{y}_j - \boldsymbol{\beta}'\mathbf{x}_j)'\Sigma^{-1}(\mathbf{y}_j - \boldsymbol{\beta}'\mathbf{x}_j) \right\}$$

where $|\Sigma|$ is the determinant of the error variance matrix. Although this model meets the linear-form restrictions, we will need to write an evaluator for the *concentrated likelihood* (to be explained shortly), so we will develop a method-d1 evaluator instead (we started with a d0 evaluator and then added derivatives for the d1 evaluator). Later we will develop a method-lf0 evaluator for the unconcentrated likelihood so that we can obtain scores and hence compute robust standard errors.

As in the previous examples, we will break this rather large formula into smaller, more manageable pieces. First, we will denote the jth residual vector by

$$\mathbf{r}_j = \mathbf{y}_j - \boldsymbol{\beta}'\mathbf{x}_j = (r_{1j}, \ldots, r_{pj})'$$

and the jth quadratic form by

$$Q_j = (\mathbf{y}_j - \boldsymbol{\beta}'\mathbf{x}_j)'\Sigma^{-1}(\mathbf{y}_j - \boldsymbol{\beta}'\mathbf{x}_j) = \mathbf{r}_j'\Sigma^{-1}\mathbf{r}_j$$

By σ_{ab}, we mean the element of Σ from row a and column b and, similarly, for σ^{ab} from Σ^{-1}. I_{ab} is the matrix with one in row a and column b and zero everywhere else, and

$$\delta_{ab} = \left\{ \begin{array}{ll} 1 & \text{if } a = b \\ 0 & \text{if } a \neq b \end{array} \right.$$

The relevant derivatives, using Lütkepohl (1996) as a reference, are then

$$\frac{\partial Q_j}{\partial \boldsymbol{\beta}'\mathbf{x}_j} = -2\Sigma^{-1}\mathbf{r}_j$$

$$\frac{\partial \ln |\Sigma|}{\partial \sigma_{ab}} = (2 - \delta_{ab})\sigma^{ab}$$

$$\frac{\partial Q_j}{\partial \sigma_{ab}} = -(2 - \delta_{ab})\mathbf{r}_j'\Sigma^{-1}I_{ab}\Sigma^{-1}\mathbf{r}_j$$

Thus the derivatives we will code in our likelihood evaluator are

$$\frac{\partial L_j}{\partial \boldsymbol{\beta}'\mathbf{x}_j} = \Sigma^{-1}\mathbf{r}_j$$

$$\frac{\partial L_j}{\partial \sigma_{ab}} = \frac{1}{2}(2 - \delta_{ab})\left(\mathbf{r}_j'\Sigma^{-1}I_{ab}\Sigma^{-1}\mathbf{r}_j - \sigma^{ab}\right)$$

The analytical derivatives necessary for implementing a method-d2 evaluator are not difficult; however, the investment adds little to this discussion, so we will let `ml` compute the Hessian numerically. Our `d1` evaluator for this model is

——————————————————————————————— begin `mysureg_d1.ado` ———————

```
program mysureg_d1
        version 11
        args todo b lnf g H
        local p : word count $ML_y
        local k = `p'*(`p'+1) / 2 + `p'

        tempname S iS sij isij isi ip
        matrix `S' = J(`p',`p',0)
quietly {
        // get residuals and build variance matrix
        local e 1
        forval i = 1/`p' {
                tempvar r`i'
                mleval `r`i'' = `b', eq(`i')
                replace `r`i'' = ${ML_y`i'} - `r`i''
                local resids `resids' `r`i''
                mleval `sij' = `b', eq(`=`p'+`e'') scalar
                matrix `S'[`i',`i'] = `sij'
                local ++e
                forval j = `=`i'+1'/`p' {
                        mleval `sij' = `b', eq(`=`p'+`e'') scalar
                        matrix `S'[`i',`j'] = `sij'
                        matrix `S'[`j',`i'] = `sij'
                        local ++e
                }
        }
        matrix `iS' = invsym(`S')
        // get score variables
        forval i = 1/`k' {
                tempvar g`i'
                quietly gen double `g`i'' = .
        }
        gen double `ip' = 0
        forval i = 1/`p' {
                matrix `isi' = `iS'[`i',1...]
                matrix colnames `isi' = `resids'
                matrix score `g`i'' = `isi', replace
                replace `ip' = `ip' + `r`i''*`g`i''
        }
        mlsum `lnf' = -0.5*(`p'*ln(2*c(pi))+ln(det(`S'))+`ip')

} // quietly
        if (`todo' == 0 | missing(`lnf')) exit

        // compute the scores and gradient
        tempname gi gs
        capture matrix drop `g'
quietly {
        local e = `p'+1
        forval i = 1/`p' {
                mlvecsum `lnf' `gi' = `g`i'', eq(`i')
                matrix `g' = nullmat(`g'), `gi'
                replace `g`e'' = 0.5*(`g`i''*`g`i''-`iS'[`i',`i'])
                mlvecsum `lnf' `gi' = `g`e'' , eq(`e')
                matrix `gs' = nullmat(`gs'), `gi'
                local ++e
                forval j = `=`i'+1'/`p' {
                        replace `g`e'' = `g`i''*`g`j'' - `iS'[`i',`j']
```

```
                           mlvecsum 'lnf' 'gi' = 'g'e'' , eq('e')
                           matrix 'gs' = nullmat('gs'), 'gi'
                           local ++e
                    }
            }
            matrix 'g' = 'g' , 'gs'
       } // quietly
   end
```

――――――――――――――――――――――――――――――――――――――― end `mysureg_d1.ado` ―――――――――

Here we try to fit this model using a subset of the variables in the dataset from
Grunfeld and Griliches (1960); the data were taken from the website for Greene (2008).
Unfortunately, optimizing this likelihood is difficult.

```
. set seed 123456789

. use http://www.stata-press.com/data/ml4/tablef13-1wide

. ml model d1 mysureg_d1
>        (iGM: iGM = fGM cGM)
>        (iCH: iCH = fCH cCH)
>        (iGE: iGE = fGE cGE)
>        /sigma1_1 /sigma1_2 /sigma1_3
>        /sigma2_2 /sigma2_3
>        /sigma3_3
. ml max, difficult iterate(300)
initial:       log likelihood =    -<inf>  (could not be evaluated)
feasible:      log likelihood = -1111524.6
rescale:       log likelihood = -106428.32
rescale eq:    log likelihood = -391.02719
Iteration 0:   log likelihood = -391.02719  (not concave)
Iteration 1:   log likelihood = -376.74478  (not concave)
Iteration 2:   log likelihood = -376.04951  (not concave)
(output omitted )
Iteration 300: log likelihood = -364.62605  (not concave)
convergence not achieved

(output omitted )
```

One clear problem with optimizing this likelihood is that the current guess for Σ must
always be positive definite, something that cannot be guaranteed until `ml` is already close
to the solution. We might be able to get around this problem by using transformations
of the elements of Σ, but there is a less complicated solution.

The log likelihood for this model can be written as

$$\ln L = -\frac{N}{2}\left[p\ln(2\pi) + \ln|\Sigma| + \text{trace}\{\Sigma^{-1}\mathbf{W}(\beta)\}\right]$$

where

$$\mathbf{W}(\beta) = \frac{1}{N}\sum_{j=1}^{N}(\mathbf{y}_j - \beta'\mathbf{x}_j)(\mathbf{y}_j - \beta'\mathbf{x}_j)'$$

is the average of the outer product of the residual vectors. It turns out that the MLE of Σ is

$$\widehat{\Sigma} = \mathbf{W}(\widehat{\boldsymbol{\beta}}) \tag{15.2}$$

If we were to substitute $\mathbf{W}(\beta)$ for Σ in our log-likelihood function, we get what is called a *concentrated* log-likelihood function:

$$\ln L_{\beta} = -\frac{N}{2} \Big[p\{\ln(2\pi) + 1\} + \ln|\mathbf{W}(\beta)| \Big]$$

(see Greene 2008, sec. 16.9.3). Let w_{ab} be the element of $\mathbf{W}(\boldsymbol{\beta})$ from row a and column b, and then

$$w_{ab} = \frac{1}{N} \sum_{j=1}^{N} r_{aj} r_{bj}$$

where r_{aj} is the ath element of $\mathbf{r}_j$. Taking derivatives, we have

$$\frac{\partial w_{ab}}{\partial \beta_a' \mathbf{x}_j} = -\frac{1 + \delta_{ab}}{N} \sum_{j=1}^{N} r_{bj}$$

$$\frac{\partial \ln L_{\beta}}{\partial \beta_a' \mathbf{x}_j} = -\frac{N}{2} \sum_{b=1}^{p} \frac{\partial \ln|\mathbf{W}|}{\partial w_{ab}} \frac{\partial w_{ab}}{\partial \beta_a' \mathbf{x}_j} = \sum_{b=1}^{p} w^{ab} \sum_{j=1}^{N} r_{bj}$$

and

$$\frac{\partial^2 \ln L_{\beta}}{\partial \beta_a' \mathbf{x}_j \partial \beta_b' \mathbf{x}_j} = -w^{ab} + \frac{\partial w^{ab}}{\partial \beta_b' \mathbf{x}_j} \sum_{j=1}^{N} r_{bj}$$

$$\approx -w^{ab} \tag{15.3}$$

which we will use to approximate the Hessian. In fact, the second term in (15.3) will go to zero at the same rate as the elements of the gradient vector, so this approximation becomes the correct calculation close to the solution. Our d2 evaluator for the concentrated model is

─────────────────────────────── begin `mysuregc_d2.ado` ───────────

```
program mysuregc_d2
        // concentrated likelihood for the SUR model
        version 11
        args todo b lnf g H
        local p : word count $ML_y
        tempname W sumw

        // get residuals and build variance matrix
        forval i = 1/`p' {
                tempvar r`i'
                mleval `r`i'' = `b', eq(`i')
                quietly replace `r`i'' = ${ML_y`i'} - `r`i''
                local resids `resids' `r`i''
        }
```

```
        quietly matrix accum 'W' = 'resids' [iw=$ML_w] if $ML_samp, noconstant
        scalar 'sumw' = r(N)
        matrix 'W' = 'W'/'sumw'
        scalar 'lnf' = -0.5*'sumw'*('p'*(ln(2*c(pi))+1)+ln(det('W')))

        if ('todo' == 0 | missing('lnf')) exit

        // compute gradient
        tempname gi iW iWi
        tempvar  scorei
        quietly generate double 'scorei' = .
        capture matrix drop 'g'
        matrix 'iW' = invsym('W')
        forval i = 1/'p' {
                matrix 'iWi' = 'iW'['i',1...]
                matrix colnames 'iWi' = 'resids'
                matrix score 'scorei' = 'iWi', replace
                mlvecsum 'lnf' 'gi' = 'scorei', eq('i')
                matrix 'g' = nullmat('g'), 'gi'
        }

        if ('todo' == 1 | missing('lnf')) exit

        // compute the Hessian, as if we were near the solution
        tempname hij
        local k = colsof('b')
        matrix 'H' = J('k','k',0)
        local r 1
        forval i = 1/'p' {
                local c 'r'
                mlmatsum 'lnf' 'hij' = -1*'iW'['i','i'], eq('i')
                matrix 'H'['r','c'] = 'hij'
                local c = 'c' + colsof('hij')
                forval j = '='i'+1'/'p' {
                        mlmatsum 'lnf' 'hij' = -1*'iW'['i','j'], eq('i','j')
                        matrix 'H'['r','c'] = 'hij'
                        matrix 'H'['c','r'] = 'hij''
                        local c = 'c' + colsof('hij')
                }
                local r = 'r' + rowsof('hij')
        }
end
```

─── end `mysuregc_d2.ado` ─────────

We can now interactively compute the maximum likelihood estimates of β for the concentrated log likelihood using the data from Grunfeld and Griliches (1960).

```
. use http://www.stata-press.com/data/ml4/tablef13-1wide

. ml model d2 mysuregc_d2
>         (iGM: iGM = fGM cGM)
>         (iCH: iCH = fCH cCH)
>         (iGE: iGE = fGE cGE)
```

```
. ml max

initial:        log likelihood = -341.95727
alternative:    log likelihood = -341.86494
rescale:        log likelihood = -340.01308
rescale eq:     log likelihood = -334.19438
Iteration 0:    log likelihood = -334.19438
Iteration 1:    log likelihood =  -294.5609
Iteration 2:    log likelihood = -287.39098
Iteration 3:    log likelihood = -287.24995
Iteration 4:    log likelihood = -287.24908
Iteration 5:    log likelihood = -287.24907
```

				Number of obs	=		20
				Wald chi2(2)	=		234.83
Log likelihood = -287.24907				Prob > chi2	=		0.0000

	Coef.	Std. Err.	z	P>|z|	[95% Conf.	Interval]
iGM						
fGM	.1149511	.0229982	5.00	0.000	.0698755	.1600268
cGM	.3756614	.0336697	11.16	0.000	.3096699	.4416528
_cons	-133.7523	94.7783	-1.41	0.158	-319.5144	52.00973
iCH						
fCH	.0720636	.0181543	3.97	0.000	.0364818	.1076454
cCH	.3211742	.0264877	12.13	0.000	.2692593	.3730891
_cons	-2.772468	12.32262	-0.22	0.822	-26.92436	21.37942
iGE						
fGE	.0279556	.0140716	1.99	0.047	.0003758	.0555353
cGE	.1518845	.0236063	6.43	0.000	.105617	.1981521
_cons	-12.75894	28.52304	-0.45	0.655	-68.66307	43.14519

Although we can compute $\widehat{\Sigma}$ using (15.2), doing so would not provide us with its variance estimates. This obstacle is easily remedied by using the estimated parameters from the concentrated model as initial values for maximizing the unconcentrated likelihood function. Moreover, as we previously indicated, the unconcentrated model satisfies the linear-form restrictions, so we can obtain scores. Because we have excellent initial values based on the concentrated model, there is little advantage to providing analytic derivatives for the unconcentrated likelihood. Thus a method-lf0 evaluator suffices. In fact, we could write a method-lf evaluator, but it is slightly easier to convert a d-family evaluator into an lf-family one because both d- and lf-family evaluators receive the coefficient vector 'b'. Our method-lf0 evaluator is

———————————————————————————— begin `mysureg_1f0.ado` ————————

```
program mysureg_1f0
        version 11
        args todo b lnfj
        local p : word count $ML_y
        local k = `p'*(`p'+1) / 2 + `p'

        tempname S iS sij isij isi
        tempvar ip
        matrix `S' = J(`p',`p',0)
quietly {
        // get residuals and build variance matrix
        local e 1
        forval i = 1/`p' {
                tempvar r`i'
                mleval `r`i'' = `b', eq(`i')
                replace `r`i'' = ${ML_y`i'} - `r`i''
                local resids `resids' `r`i''
                mleval `sij' = `b', eq(`=`p'+`e'') scalar
                matrix `S'[`i',`i'] = `sij'
                local ++e
                forval j = `=`i'+1'/`p' {
                        mleval `sij' = `b', eq(`=`p'+`e'') scalar
                        matrix `S'[`i',`j'] = `sij'
                        matrix `S'[`j',`i'] = `sij'
                        local ++e
                }
        }
        matrix `iS' = invsym(`S')
        // get score variables
        tempvar scorei
        gen double `scorei' = .
        gen double `ip' = 0
        forval i = 1/`p' {
                matrix `isi' = `iS'[`i',1...]
                matrix colnames `isi' = `resids'
                matrix score `scorei' = `isi', replace
                replace `ip' = `ip' + `r`i''*`scorei'
        }
        replace `lnfj' = -0.5*(`p'*ln(2*c(pi))+ln(det(`S'))+`ip')
} // quietly
end
```

———————————————————————————————— end `mysureg_1f0.ado` ————————

In the following example, we compute the residuals from each equation, use them and `matrix accum` to compute $\widehat{\Sigma}$, and then supply $\widehat{\beta}$ and the upper-triangle elements of $\widehat{\Sigma}$ (with the appropriate equation and column names) as initial values using `ml init`.

```
. // get the residuals
. predict r1, eq(#1)

. replace r1 = iGM - r1
(20 real changes made)

. predict r2, eq(#2)

. replace r2 = iCH - r2
(20 real changes made)

. predict r3, eq(#3)

. replace r3 = iGE - r3
(20 real changes made)

. // compute the MLE of Sigma from the residuals
. matrix accum Sigma = r1 r2 r3, noconstant
(obs=20)

. matrix Sigma = Sigma/r(N)

. // initial values of the coefficients
. matrix b = e(b)

. // initial values of the upper triangle elements of Sigma
. matrix s = Sigma[1,1..3]

. matrix s = s, Sigma[2,2..3]

. matrix s = s, Sigma[3,3]

. matrix colnames s = _cons

. matrix coleq    s = sigma1_1 sigma1_2 sigma1_3 sigma2_2 sigma2_3 sigma3_3

. ml model lf0 mysureg_lf0
>          (iGM: iGM = fGM cGM)
>          (iCH: iCH = fCH cCH)
>          (iGE: iGE = fGE cGE)
>          /sigma1_1 /sigma1_2 /sigma1_3
>                    /sigma2_2 /sigma2_3
>                              /sigma3_3

. ml init b

. ml init s
```

```
. ml max
initial:        log likelihood = -287.24907
rescale:        log likelihood = -287.24907
rescale eq:     log likelihood = -287.24907
Iteration 0:    log likelihood = -287.24907
Iteration 1:    log likelihood = -287.24907
```

		Number of obs	=	20
		Wald chi2(2)	=	234.75
Log likelihood = -287.24907		Prob > chi2	=	0.0000

	Coef.	Std. Err.	z	P>|z|	[95% Conf. Interval]	
iGM						
fGM	.1149513	.0231193	4.97	0.000	.0696382	.1602643
cGM	.3756612	.0337378	11.13	0.000	.3095364	.441786
_cons	-133.7528	95.19374	-1.41	0.160	-320.3291	52.8235
iCH						
fCH	.0720611	.0190324	3.79	0.000	.0347583	.1093639
cCH	.3211765	.0269839	11.90	0.000	.2682891	.3740639
_cons	-2.770999	12.76672	-0.22	0.828	-27.79332	22.25132
iGE						
fGE	.0279566	.0146924	1.90	0.057	-.00084	.0567533
cGE	.1518841	.0236933	6.41	0.000	.105446	.1983221
_cons	-12.76079	29.48049	-0.43	0.665	-70.54148	45.0199
sigma1_1						
_cons	7173.009	2272.075	3.16	0.002	2719.824	11626.19
sigma1_2						
_cons	-307.9035	248.6466	-1.24	0.216	-795.2419	179.4348
sigma1_3						
_cons	615.6388	507.0188	1.21	0.225	-378.0998	1609.377
sigma2_2						
_cons	150.7304	47.95484	3.14	0.002	56.74066	244.7202
sigma2_3						
_cons	-16.29927	73.92042	-0.22	0.825	-161.1806	128.5821
sigma3_3						
_cons	661.1597	209.1859	3.16	0.002	251.1629	1071.156

The value of the log likelihood does not change between the concentrated and unconcentrated model fits; in fact, `ml` converges in one iteration (as it should). The variance estimates change because `ml` can compute the covariances between the MLEs of β and Σ. Note that

$$\mathrm{Var}\{\mathrm{vec}(\widehat{\beta})\} = \widehat{\Sigma} \otimes (\mathbf{X}'\mathbf{X})^{-1}$$

(with the implied exclusion restrictions) is true only asymptotically and for some special cases of the seemingly unrelated regression model. One special case is multivariate normal regression where there are no exclusion restrictions; that is, the same independent variables are present in all the equations.

Our estimation command for this model is called `mysureg` and is listed in appendix C.7. In the constant-only model, the MLE for β_0 is

$$\widehat{\beta}_0 = \overline{\mathbf{y}} = \frac{1}{N} \sum_{j=1}^{N} \mathbf{y}_j$$

the sample mean of the dependent variables. `mysureg` uses `matrix vecaccum` to compute $\widehat{\beta}_0$. The initial values for the full model fit are $\widehat{\beta}$ and $\widehat{\Sigma}$ from the concentrated likelihood fit.

Here is mysureg in use:

```
. use http://www.stata-press.com/data/ml4/tablef13-1wide, clear
. mysureg (iGM fGM cGM)
>        (iCH fCH cCH)
>        (iGE fGE cGE), vce(robust)
Fitting constant-only model:
Iteration 0:   log likelihood = -323.88722
Iteration 1:   log likelihood = -323.88722
Fitting concentrated model:
Iteration 0:   log likelihood = -323.88722
Iteration 1:   log likelihood = -291.22787
Iteration 2:   log likelihood = -287.27972
Iteration 3:   log likelihood = -287.24921
Iteration 4:   log likelihood = -287.24908
Iteration 5:   log likelihood = -287.24907
Fitting full model:
Iteration 0:   log pseudolikelihood = -287.24907
Iteration 1:   log pseudolikelihood = -287.24907
```

```
My sureg estimates                          Number of obs    =        20
                                            Wald chi2(6)     =   1396.39
Log pseudolikelihood = -287.24907           Prob > chi2      =    0.0000
```

	Coef.	Robust Std. Err.	z	P>|z|	[95% Conf. Interval]	
iGM						
fGM	.1149513	.0228856	5.02	0.000	.0700963	.1598062
cGM	.3756612	.0429121	8.75	0.000	.291555	.4597674
_cons	-133.7528	88.83564	-1.51	0.132	-307.8675	40.36185
iCH						
fCH	.0720611	.0144682	4.98	0.000	.0437039	.1004182
cCH	.3211765	.0244714	13.12	0.000	.2732135	.3691395
_cons	-2.770999	8.789279	-0.32	0.753	-19.99767	14.45567
iGE						
fGE	.0279566	.0122976	2.27	0.023	.0038537	.0520595
cGE	.1518841	.0161167	9.42	0.000	.120296	.1834722
_cons	-12.76079	23.70002	-0.54	0.590	-59.21198	33.6904
/sigma1_1	7173.009	1449.301	4.95	0.000	4332.43	10013.59
/sigma1_2	-307.9035	299.5417	-1.03	0.304	-894.9944	279.1873
/sigma1_3	615.6388	449.4178	1.37	0.171	-265.2039	1496.482
/sigma2_2	150.7304	77.33421	1.95	0.051	-.8418424	302.3027
/sigma2_3	-16.29927	59.50743	-0.27	0.784	-132.9317	100.3332
/sigma3_3	661.1597	196.7761	3.36	0.001	275.4856	1046.834

```
. mysureg, notable corr
Correlation matrix of residuals:
        iGM      iCH      iGE
iGM       1
iCH  -.2961        1
iGE   .2827   -.0516        1
```

A   Syntax of ml

Syntax

ml model in interactive mode

| ml <u>mod</u>el | *method progname eq* $\big[\, eq\, \dots\,\big]\ \big[\, if\,\big]\ \big[\, in\,\big]\ \big[\, weight\,\big]$ |
| | $\big[\ ,\ model\_options$ svy $diparm\_options\,\big]$ |

| ml <u>mod</u>el | *method funcname*() *eq* $\big[\, eq\, \dots\,\big]\ \big[\, if\,\big]\ \big[\, in\,\big]\ \big[\, weight\,\big]$ |
| | $\big[\ ,\ model\_options$ svy $diparm\_options\,\big]$ |

ml model in noninteractive mode

| ml <u>mod</u>el | *method progname eq* $\big[\, eq\, \dots\,\big]\ \big[\, if\,\big]\ \big[\, in\,\big]\ \big[\, weight\,\big]$, <u>maximize</u> |
| | $\big[\, model\_options$ svy $diparm\_options\ noninteractive\_options\,\big]$ |

| ml <u>mod</u>el | *method funcname*() *eq* $\big[\, eq\, \dots\,\big]\ \big[\, if\,\big]\ \big[\, in\,\big]\ \big[\, weight\,\big]$, <u>maximize</u> |
| | $\big[\, model\_options$ svy $diparm\_options\ noninteractive\_options\,\big]$ |

Noninteractive mode is invoked by specifying the maximize option. Use maximize when ml will be used as a subroutine of another ado-file or program and you want to carry forth the problem, from definition to posting of final results, in one command.

ml clear

ml <u>q</u>uery

ml check

| ml <u>search</u> | $\big[\ \big[\,/\,\big]\, eqname\big[\,:\,\big]\ \#_{lb}\ \#_{ub}\ \big]\ \big[\, \dots\,\big]\ \big[\ ,\ search\_options\,\big]$ |

| ml <u>pl</u>ot | $\big[\, eqname\,:\,\big]\, name\ \big[\,\#\ \big[\,\#\ \big[\,\#\,\big]\big]\big]$ |
| | $\big[\ ,\ \underline{saving}(\,filename\big[\ ,\ \texttt{replace}\,\big])\ \big]$ |

| ml <u>init</u> | $\big\{\ \big[\, eqname\,:\,\big]\, name\texttt{=}\#\ \mid\ /eqname\texttt{=}\#\ \big\}\ \big[\, \dots\,\big]$ |

285

```
ml init       # [ # ... ] , copy

ml init       matname [ , skip copy ]

ml report

ml trace      {on | off}

ml count      [ clear | on | off ]

ml maximize   [ , ml_maximize_options display_options eform_option ]

ml graph      [ # ] [ , saving( filename [ , replace ]) ]

ml display    [ , display_options eform_option ]

ml footnote

ml score      newvar [ if ] [ in ] [ , equation(eqname) missing ]

ml score      newvarlist [ if ] [ in ] [ , missing ]

ml score      [ type ] stub* [ if ] [ in ] [ , missing ]
```

where *method* is one of

lf	d0	lf0	gf0
	d1	lf1	
	d1debug	lf1debug	
	d2	lf2	
	d2debug	lf2debug	

method can be specified using one of the longer, more descriptive names if desired:

method	Longer name
lf	linearform
d0	derivative0
d1	derivative1
d1debug	derivative1debug
d2	derivative2
d2debug	derivative2debug
lf0	linearform0
lf1	linearform1
lf1debug	linearform1debug
lf2	linearform2
lf2debug	linearform2debug
gf0	generalform0

eq is the equation to be estimated, enclosed in parentheses, and optionally with a name to be given to the equation, preceded by a colon:

$$([\ eqname:\ ]\ [\ varlist_y\ =\ ]\ [\ varlist_x\ ]\ [\ ,\ eq\_options\ ])$$

Or *eq* is the name of a parameter, such as `sigma`, with a slash in front:

$$/\,eqname \qquad \text{which is equivalent to} \qquad (eqname:)$$

diparm_options is one or more `diparm(`*diparm_args*`)` options where *diparm_args* is either `__sep__` or anything accepted by the `_diparm` command; see `help _diparm`.

eq_options	Description
<u>nocons</u>tant	do not include intercept in the equation
<u>off</u>set(*varname_o*)	include *varname_o* in model with coefficient constrained to be 1
exposure(*varname_e*)	include ln(*varname_e*) in model with coefficient constrained to be 1

model_options	Description
group(*varname*)	use *varname* to identify groups
vce(*vcetype*)	*vcetype* may be <u>r</u>obust, <u>c</u>luster *clustvar*, oim, or opg
<u>constra</u>ints(*numlist*)	constraints by number to be applied
<u>constra</u>ints(*matname*)	matrix that contains the constraints to be applied
nocnsnotes	do not display notes when constraints are dropped
<u>titl</u>e(*string*)	place title on the estimation output
nopreserve	do not preserve the estimation subsample in memory
<u>collin</u>ear	keep collinear variables within equations
<u>miss</u>ing	keep observations containing variables with missing values
lf0($\#_k$ $\#_{ll}$)	number of parameters and log-likelihood value of "constant-only" model
<u>cont</u>inue	used to specify that model has been fit and set initial values $\mathbf{b}_0$ for model to be fit based on those results
<u>wald</u>test($\#$)	perform Wald test; see *Options for use with ml model in interactive or noninteractive mode* below
obs($\#$)	number of observations
crittype(*string*)	describe the criterion optimized by ml
<u>subpop</u>(*varname*)	compute estimates for the single subpopulation
nosvyadjust	carry out Wald test as $W/k \sim F(k,d)$
<u>techn</u>ique(*algorithm_spec*)	specify how the likelihood function is to be maximized

noninteractive_options	Description
<u>init</u>(*ml_init_args*)	set the initial values $\mathbf{b}_0$
<u>search</u>(on)	equivalent to ml search, repeat(0); default
<u>search</u>(norescale)	equivalent to ml search, repeat(0) norescale
<u>search</u>(quietly)	same as search(on) except that output is suppressed
<u>search</u>(off)	prevents calling ml search
<u>repeat</u>(#)	ml search's repeat() option; see below
<u>bounds</u>(*ml_search_bounds*)	specify bounds for ml search
<u>nowarn</u>ing	suppress "convergence not achieved" message of iterate(0)
novce	substitute the zero matrix for the variance matrix
negh	indicate that the evaluator return the negative Hessian matrix
<u>score</u>(*newvars*)	new variables containing the contribution to the score
maximize_options	control the maximization process; seldom used

search_options	Description
<u>re</u>peat(#)	number of random attempts to find better initial-value vector; default is repeat(10) in interactive mode and repeat(0) in noninteractive mode
<u>rest</u>art	use random actions to find starting values; not recommended
<u>noresc</u>ale	do not rescale to improve parameter vector; not recommended
maximize_options	control the maximization process; seldom used

ml_maximize_options	Description
<u>nowarn</u>ing	suppress "convergence not achieved" message of iterate(0)
novce	substitute the zero matrix for the variance matrix
negh	indicate that the evaluator return the negative Hessian matrix
<u>score</u>(*newvars* | *stub**)	new variables containing the contribution to the score
<u>noo</u>utput	suppress the display of the final results
<u>nocl</u>ear	do not clear ml problem definition after model has converged
maximize_options	control the maximization process; seldom used

display_options	Description
<u>no</u>header	suppress display of the header above the coefficient table
<u>no</u>footnote	suppress display of the footnote below the coefficient table
<u>l</u>evel(*#*)	set confidence level; default is level(95)
<u>f</u>irst	display coefficient table reporting results for first equation only
neq(*#*)	display coefficient table reporting first *#* equations
showeqns	display equation names in the coefficient table
<u>p</u>lus	display coefficient table ending in dashes–plus-sign–dashes
nocnsreport	suppress constraints display above the coefficient table
<u>no</u>omitted	suppress display of omitted variables
vsquish	suppress blank space separating factor-variable terms or time-series–operated variables from other variables
noemptycells	suppress empty cells for interactions of factor variables
<u>base</u>levels	report base levels of factor variables and interactions
<u>allbase</u>levels	display all base levels of factor variables and interactions
cformat(%*fmt*)	format the coefficients, standard errors, and confidence limits in the coefficient table
pformat(%*fmt*)	format the *p*-values in the coefficient table
sformat(%*fmt*)	format the test statistics in the coefficient table
<u>coefl</u>egend	display legend instead of statistics

eform_option	Description
<u>ef</u>orm(*string*)	display exponentiated coefficients; column title is "*string*"
<u>ef</u>orm	display exponentiated coefficients; column title is "exp(b)"
hr	report hazard ratios
shr	report subhazard ratios
<u>ir</u>r	report incidence-rate ratios
or	report odds ratios
<u>rr</u>r	report relative-risk ratios

fweights, pweights, aweights, and iweights are allowed; see [U] **11.1.6 weight**. With all but method lf, you must write your likelihood-evaluation program carefully if pweights are to be specified, and pweights may not be specified with method d0, d1, d1debug, d2, or d2debug.

See [U] **20 Estimation and postestimation commands** for more capabilities of estimation commands.

To redisplay results, type ml display.

Syntax of subroutines for use by evaluator programs

mleval   *newvar* = *vecname* $\big[$, eq(*#*) $\big]$

mleval   *scalarname* = *vecname* , scalar $\big[$ eq(*#*) $\big]$

mlsum   *scalarname*$_{\mathrm{lnf}}$ = *exp* $\big[$ *if* $\big]$ $\big[$, <u>no</u>weight $\big]$

mlvecsum   *scalarname*$_{\mathrm{lnf}}$ *rowvecname* = *exp* $\big[$ *if* $\big]$ $\big[$, eq(*#*) $\big]$

mlmatsum   *scalarname*$_{\mathrm{lnf}}$ *matrixname* = *exp* $\big[$ *if* $\big]$ $\big[$, eq(*#*$\big[$,*#*$\big]$) $\big]$

mlmatbysum   *scalarname*$_{\mathrm{lnf}}$ *matrixname* *varname*$_a$ *varname*$_b$ $\big[$ *varname*$_c$ $\big]$
              $\big[$ *if* $\big]$, by(*varname*) $\big[$ eq(*#*$\big[$,*#*$\big]$) $\big]$

Syntax of user-written evaluator

Summary of notation

The log-likelihood function is $\ln L(\theta_{1j}, \theta_{2j}, \ldots, \theta_{Ej})$, where $\theta_{ij} = \mathbf{x}_{ij}\mathbf{b}_i$, $j = 1, \ldots, N$ indexes observations, and $i = 1, \ldots, E$ indexes the linear equations defined by ml model. If the likelihood satisfies the linear-form restrictions, it can be decomposed as $\ln L = \sum_{j=1}^{N} \ln \ell(\theta_{1j}, \theta_{2j}, \ldots, \theta_{Ej})$.

Method-lf evaluators

```
program progname
        version 11
        args lnfj theta1 theta2 ...
        // if you need to create any intermediate results:
        tempvar tmp1 tmp2 ...
        quietly gen double 'tmp1' = ...
        ...
        quietly replace 'lnfj' = ...
end
```

where

'lnfj'	variable to be filled in with the values of $\ln \ell_j$
'theta1'	variable containing the evaluation of the first equation $\theta_{1j} = \mathbf{x}_{1j}\boldsymbol{\beta}_1$
'theta2'	variable containing the evaluation of the second equation $\theta_{2j} = \mathbf{x}_{2j}\boldsymbol{\beta}_2$
...	

Method-d0 evaluators

```
program progname
        version 11
        args todo b lnf

        tempvar theta1 theta2 ...
        mleval 'theta1' = 'b', eq(1)
        mleval 'theta2' = 'b', eq(2) // if there is a θ_2
        ...

        // if you need to create any intermediate results:
        tempvar tmp1 tmp2 ...
        gen double 'tmp1' = ...
        ...

        mlsum 'lnf' = ...
end
```

where

'todo'	always contains 0 (may be ignored)
'b'	full parameter row vector $\mathbf{b}=(\mathbf{b}_1,\mathbf{b}_2,...,\mathbf{b}_E)$
'lnf'	scalar to be filled in with overall $\ln L$

Method-d1 evaluators

```
program progname
        version 11
        args todo b lnf g

        tempvar theta1 theta2 ...
        mleval 'theta1' = 'b', eq(1)
        mleval 'theta2' = 'b', eq(2) // if there is a θ_2
        ...

        // if you need to create any intermediate results:
        tempvar tmp1 tmp2 ...
        gen double 'tmp1' = ...
        ...

        mlsum 'lnf' = ...
        if ('todo'==0 | 'lnf'>=.) exit

        tempname d1 d2 ...
        mlvecsum 'lnf' 'd1' = formula for ∂ ln ℓ_j/∂θ_1j, eq(1)
        mlvecsum 'lnf' 'd2' = formula for ∂ ln ℓ_j/∂θ_2j, eq(2)
        ...
        matrix 'g' = ('d1','d2', ... )
end
```

where

'todo'	contains 0 or 1
	0⇒'lnf' to be filled in;
	1⇒'lnf' and 'g' to be filled in
'b'	full parameter row vector $\mathbf{b}=(\mathbf{b}_1,\mathbf{b}_2,...,\mathbf{b}_E)$
'lnf'	scalar to be filled in with overall $\ln L$
'g'	row vector to be filled in with overall $\mathbf{g}=\partial \ln L/\partial \mathbf{b}$

Method-d2 evaluators

```
program progname
        version 11
        args todo b lnf g H

        tempvar theta1 theta2 ...
        mleval 'theta1' = 'b', eq(1)
        mleval 'theta2' = 'b', eq(2) // if there is a θ_2
        ...

        // if you need to create any intermediate results:
        tempvar tmp1 tmp2 ...
        gen double 'tmp1' = ...
        ...

        mlsum 'lnf' = ...
        if ('todo'==0 | 'lnf'>=.) exit

        tempname d1 d2 ...
        mlvecsum 'lnf' 'd1' = formula for ∂ln ℓ_j/∂θ_1j, eq(1)
        mlvecsum 'lnf' 'd2' = formula for ∂ln ℓ_j/∂θ_2j, eq(2)
        ...

        matrix 'g' = ('d1','d2', ... )
        if ('todo'==1 | 'lnf'>=.) exit

        tempname d11 d12 d22 ...
        mlmatsum 'lnf' 'd11' = formula for ∂^2ln ℓ_j/∂θ_1j^2, eq(1)
        mlmatsum 'lnf' 'd12' = formula for ∂^2ln ℓ_j/∂θ_1j∂θ_2j, eq(1,2)
        mlmatsum 'lnf' 'd22' = formula for ∂^2ln ℓ_j/∂θ_2j^2, eq(2)
        ...

        matrix 'H' = ('d11','d12', ... \ 'd12'','d22', ... )
    end
```

where

'todo'	contains 0, 1, or 2
	$0 \Rightarrow$ 'lnf' to be filled in;
	$1 \Rightarrow$ 'lnf' and 'g' to be filled in;
	$2 \Rightarrow$ 'lnf', 'g', and 'H' to be filled in
'b'	full parameter row vector $\mathbf{b}=(\mathbf{b}_1, \mathbf{b}_2, ..., \mathbf{b}_E)$
'lnf'	scalar to be filled in with overall $\ln L$
'g'	row vector to be filled in with overall $\mathbf{g}=\partial \ln L / \partial \mathbf{b}$
'H'	matrix to be filled in with overall Hessian $\mathbf{H}=\partial^2 \ln L / \partial \mathbf{b} \partial \mathbf{b}'$

Method-lf0 evaluators

```
program progname
        version 11
        args todo b lnfj

        tempvar theta1 theta2 ...
        mleval 'theta1' = 'b', eq(1)
        mleval 'theta2' = 'b', eq(2) // if there is a θ_2
        ...
        // if you need to create any intermediate results:
        tempvar tmp1 tmp2 ...
        gen double 'tmp1' = ...
        ...
        quietly replace 'lnfj' = ...
end
```

where

'todo'	always contains 0 (may be ignored)
'b'	full parameter row vector $\mathbf{b}=(\mathbf{b}_1,\mathbf{b}_2,...,\mathbf{b}_E)$
'lnfj'	variable to be filled in with observation-by-observation values of $\ln \ell_j$

Method-lf1 evaluators

```
program progname
        version 11
        args todo b lnfj g1 g2 ...

        tempvar theta1 theta2 ...
        mleval 'theta1' = 'b', eq(1)
        mleval 'theta2' = 'b', eq(2) // if there is a θ_2
        ...
        // if you need to create any intermediate results:
        tempvar tmp1 tmp2 ...
        gen double 'tmp1' = ...
        ...
        quietly replace 'lnfj' = ...
        if ('todo'==0) exit

        quietly replace 'g1' = formula for ∂ln ℓ_j/∂θ_1j
        quietly replace 'g2' = formula for ∂ln ℓ_j/∂θ_2j
        ...
end
```

where

'todo'	contains 0 or 1
	$0 \Rightarrow$ 'lnfj' to be filled in;
	$1 \Rightarrow$ 'lnfj', 'g1', 'g2', ... to be filled in
'b'	full parameter row vector $\mathbf{b}=(\mathbf{b}_1,\mathbf{b}_2,...,\mathbf{b}_E)$
'lnfj'	variable to be filled in with observation-by-observation values of $\ln \ell_j$
'g1'	variable to be filled in with $\partial \ln \ell_j/\partial \theta_{1j}$
'g2'	variable to be filled in with $\partial \ln \ell_j/\partial \theta_{2j}$
...	

Method-lf2 evaluators

```
program progname
        version 11
        args todo b lnfj g1 g2 ... H
        tempvar theta1 theta2 ...
        mleval 'theta1' = 'b', eq(1)
        mleval 'theta2' = 'b', eq(2) // if there is a θ_2
        ...
        // if you need to create any intermediate results:
        tempvar tmp1 tmp2 ...
        gen double 'tmp1' = ...
        ...
        quietly replace 'lnfj' = ...
        if ('todo'==0) exit
        quietly replace 'g1' = formula for ∂ ln ℓ_j/∂θ_1j
        quietly replace 'g2' = formula for ∂ ln ℓ_j/∂θ_2j
        ...
        if ('todo'==1) exit
        tempname d11 d12 d22 lnf ...
        mlmatsum 'lnf' 'd11' = formula for ∂^2 ln ℓ_j/∂θ_1j^2, eq(1)
        mlmatsum 'lnf' 'd12' = formula for ∂^2 ln ℓ_j/∂θ_1j∂θ_2j, eq(1,2)
        mlmatsum 'lnf' 'd22' = formula for ∂^2 ln ℓ_j/∂θ_2j^2, eq(2)
        ...
        matrix 'H' = ('d11','d12', ... \ 'd12'','d22', ... )
    end
```

where

'todo'	contains 0 or 1
	$0 \Rightarrow$ 'lnfj' to be filled in;
	$1 \Rightarrow$ 'lnfj', 'g1', 'g2', ... to be filled in
	$2 \Rightarrow$ 'lnfj', 'g1', 'g2', ... and 'H' to be filled in
'b'	full parameter row vector $\mathbf{b} = (\mathbf{b}_1, \mathbf{b}_2, ..., \mathbf{b}_E)$
'lnfj'	scalar to be filled in with observation-by-observation $\ln L$
'g1'	variable to be filled in with $\partial \ln \ell_j / \partial \theta_{1j}$
'g2'	variable to be filled in with $\partial \ln \ell_j / \partial \theta_{2j}$
...	
'H'	matrix to be filled in with overall Hessian $\mathbf{H} = \partial^2 \ln L / \partial \mathbf{b} \partial \mathbf{b}'$

Method-gf0 evaluators

```
program progname
        version 11
        args todo b lnfj

        tempvar theta1 theta2 ...
        mleval 'theta1' = 'b', eq(1)
        mleval 'theta2' = 'b', eq(2) // if there is a θ_2
        ...
        // if you need to create any intermediate results:
        tempvar tmp1 tmp2 ...
        gen double 'tmp1' = ...
        ...
        quietly replace 'lnfj' = ...
end
```

where

'todo'	always contains 0 (may be ignored)
'b'	full parameter row vector $\mathbf{b}=(\mathbf{b}_1,\mathbf{b}_2,...,\mathbf{b}_E)$
'lnfj'	variable to be filled in with the values of the log likelihood $\ln \ell_j$

Global macros for use by all evaluators

$ML_y1	name of first dependent variable
$ML_y2	name of second dependent variable, if any
...	
$ML_samp	variable containing 1 if observation to be used; 0 otherwise
$ML_w	variable containing weight associated with observation or 1 if no weights specified

Method lf can ignore $ML_samp, but restricting calculations to the $ML_samp==1 subsample will speed execution. Method-lf evaluators must ignore $ML_w; application of weights is handled by the method itself.

Methods d0, d1, d2, lf0, lf1, lf2, and gf0 can ignore $ML_samp as long as ml model's nopreserve option is not specified. These methods will run more quickly if nopreserve is specified. These evaluators can ignore $ML_w only if they use mlsum, mlvecsum, mlmatsum, and mlmatbysum to produce all final results.

Description

ml model defines the current problem.

ml clear clears the current problem definition. This command is rarely used because when you type ml model, any previous problem is automatically cleared.

ml query displays a description of the current problem.

ml check verifies that the log-likelihood evaluator you have written works. We strongly recommend using this command.

`ml search` searches for (better) initial values. We recommend using this command.

`ml plot` provides a graphical way of searching for (better) initial values.

`ml init` provides a way to specify initial values.

`ml report` reports the values of ln L, its gradient, and its Hessian at the initial values or current parameter estimates β_0.

`ml trace` traces the execution of the user-defined log-likelihood–evaluation program.

`ml count` counts the number of times the user-defined log-likelihood–evaluation program is called; the command is seldom used. `ml count clear` clears the counter. `ml count on` turns on the counter. `ml count` without arguments reports the current values of the counters. `ml count off` stops counting calls.

`ml maximize` maximizes the likelihood function and reports final results. Once `ml maximize` has successfully completed, the previously mentioned `ml` commands may no longer be used unless `noclear` is specified. `ml graph` and `ml display` may be used whether or not `noclear` is specified.

`ml graph` graphs the log-likelihood values against the iteration number.

`ml display` redisplays final results.

`ml footnote` displays a warning message when the model did not converge within the specified number of iterations.

`ml score` creates new variables containing the equation-level scores. The variables generated by `ml score` are equivalent to those generated by specifying the `score()` option of `ml maximize` (and `ml model ...,  ... maximize`).

progname is the name of a program you write to evaluate the log-likelihood function.

funcname() is the name of a Mata function you write to evaluate the log-likelihood function.

In this documentation, *progname* and *funcname()* are referred to as the *user-written evaluator*, the *likelihood evaluator*, or sometimes simply as the *evaluator*. The program you write is written in the style required by the method you choose. The methods are `lf`, `d0`, `d1`, `d2`, `lf0`, `lf1`, `lf2`, and `gf0`. Thus if you choose to use method `lf`, your program is called a method-`lf` evaluator.

Method-`lf` evaluators are required to evaluate the observation-by-observation log likelihood ln ℓ_j, $j = 1, \ldots, N$.

Method-`d0` evaluators are required to evaluate the overall log likelihood ln L. Method-`d1` evaluators are required to evaluate the overall log likelihood and its gradient vector $\mathbf{g} = \partial \ln L / \partial \mathbf{b}$. Method-`d2` evaluators are required to evaluate the overall log likelihood, its gradient, and its Hessian matrix $H = \partial^2 \ln L / \partial \mathbf{b} \partial \mathbf{b}'$.

Method-`lf0` evaluators are required to evaluate the observation-by-observation log likelihood ln ℓ_j, $j = 1, \ldots, N$. Method-`lf1` evaluators are required to evaluate the observation-by-observation log likelihood and its equation-level scores $g_{ji} = \partial \ln \ell / \partial \mathbf{x}_{ji} \mathbf{b}_i$. Method-lf2 evaluators are required to evaluate the observation-by-observation log likelihood, its equation-level scores, and its Hessian matrix $H = \partial^2 \ln \ell / \partial \mathbf{b} \partial \mathbf{b}'$.

Method-gf0 evaluators are required to evaluate the summable pieces of the log likelihood $\ln \ell_k$, $k = 1, \ldots, K$.

mleval is a subroutine used by evaluators of methods d0, d1, and d2 to evaluate the coefficient vector $\boldsymbol{\beta}$ that they are passed.

mlsum is a subroutine used by evaluators of methods d0, d1, and d2 to define the value $\ln L$ that is to be returned.

mlvecsum is a subroutine used by evaluators of methods d1 and d2 to define the gradient vector, $\mathbf{g}$, that is to be returned. It is suitable for use only when the likelihood function meets the linear-form restrictions.

mlmatsum is a subroutine used by evaluators of methods d2 and lf2 to define the Hessian matrix, $\mathbf{H}$, that is to be returned. It is suitable for use only when the likelihood function meets the linear-form restrictions.

mlmatbysum is a subroutine used by evaluators of method d2 to help define the Hessian matrix, $\mathbf{H}$, that is to be returned. It is suitable for use when the likelihood function contains terms made up of grouped sums, such as in panel-data models. For such models, use mlmatsum to compute the observation-level outer products and use mlmatbysum to compute the group-level outer products. mlmatbysum requires that the data be sorted by the variable identified in the by() option.

Options for use with ml model in interactive or noninteractive mode

group(*varname*) specifies the numeric variable that identifies groups. This option is typically used to identify panels for panel-data models.

vce(*vcetype*) specifies the type of standard error reported, which includes types that are robust to some kinds of misspecification, that allow for intragroup correlation, and that are derived from asymptotic theory; see [R] *vce_option*.

vce(robust), vce(cluster *clustvar*), pweight, and svy will work with evaluators of methods lf, lf0, lf1, lf2, and gf0; all you need to do is specify them.

These options will not work with evaluators of methods d0, d1, or d2, and specifying these options will produce an error message.

constraints(*numlist* | *matname*) specifies the linear constraints to be applied during estimation.

constraints(*numlist*) specifies the constraints by number. Constraints are defined
    using the `constraint` command; see [R] **constraint**.

constraint(*matname*) specifies a matrix that contains the constraints.

nocnsnotes prevents notes from being displayed when constraints are dropped. A
    constraint will be dropped if it is inconsistent, contradicts other constraints, or
    causes some other error when the constraint matrix is being built. Constraints are
    checked in the order they are specified.

title(*string*) specifies the title to be placed on the estimation output when results are
    complete.

nopreserve specifies that it is not necessary for `ml` to ensure that only the estima-
    tion subsample is in memory when the user-written likelihood evaluator is called.
    nopreserve is irrelevant when you use method `lf`.

    For the other methods, if nopreserve is not specified, `ml` saves the data in a file
    (preserves the original dataset) and drops the irrelevant observations before calling
    the user-written evaluator. This way, even if the evaluator does not restrict its
    attentions to the $ML_samp==1 subsample, results will still be correct. Later, `ml`
    automatically restores the original dataset.

    `ml` need not go through these machinations in the case of method `lf` because the
    user-written evaluator calculates observation-by-observation values, and `ml` itself
    sums the components.

    `ml` goes through these machinations if and only if the estimation sample is a sub-
    sample of the data in memory. If the estimation sample includes every observation
    in memory, `ml` does not preserve the original dataset. Thus programmers must not
    alter the original dataset unless they preserve the data themselves.

    We recommend that interactive users of `ml` not specify nopreserve; the speed gain
    is not worth the possibility of getting incorrect results.

    We recommend that programmers specify nopreserve, but only after verifying that
    their evaluator really does restrict its attentions solely to the $ML_samp==1 subsam-
    ple.

collinear specifies that `ml` not remove the collinear variables within equations. There
    is no reason you would want to leave collinear variables in place, but this option is of
    interest to programmers who, in their code, have already removed collinear variables
    and do not want `ml` to waste computer time checking again.

missing specifies that observations containing variables with missing values not be
    eliminated from the estimation sample. There are two reasons you might want to
    specify missing:

    Programmers may wish to specify missing because, in other parts of their code,
    they have already eliminated observations with missing values and do not want `ml`
    to waste computer time looking again.

You may wish to specify `missing` if your model explicitly deals with missing values. Stata's `heckman` command is a good example of this. In such cases, there will be observations where missing values are allowed and other observations where they are not—where their presence should cause the observation to be eliminated. If you specify `missing`, it is your responsibility to specify an `if` *exp* that eliminates the irrelevant observations.

`lf0(`$\#_k$ $\#_{ll}$`)` is typically used by programmers. It specifies the number of parameters and log-likelihood value of the "constant-only" model so that `ml` can report a likelihood-ratio test rather than a Wald test. These values may have been analytically determined or they may have been determined by a previous fitting of the constant-only model on the estimation sample.

Also see the `continue` option directly below.

If you specify `lf0()`, it must be safe for you to specify the `missing` option, too, else how did you calculate the log likelihood for the constant-only model on the same sample? You must have identified the estimation sample, and done so correctly, so there is no reason for `ml` to waste time rechecking your results. All of which is to say, do not specify `lf0()` unless you are certain your code identifies the estimation sample correctly.

`lf0()`, even if specified, is ignored if `vce(robust)`, `vce(cluster` *varname*`)`, `pweight`, or `svy` is specified because, in that case, a likelihood-ratio test would be inappropriate.

Do not confuse the `lf0()` option with evaluator method `lf0`. They are unrelated.

`continue` is typically specified by programmers and does two things:

First, it specifies that a model has just been fit by either `ml` or some other estimation command, such as `logit`, and that the likelihood value stored in `e(ll)` and the number of parameters stored in `e(b)` as of that instant are the relevant values of the constant-only model. The current value of the log likelihood is used to present a likelihood-ratio test unless `pweight`, `svy`, `vce(robust)`, `vce(cluster` *clustvar*`)`, or `constraints()` is specified. A likelihood-ratio test is inappropriate when `pweight`, `svy`, `vce(robust)`, or `vce(cluster` *clustvar*`)` is specified. We suggest using `lrtest` when `constraints()` is specified.

Second, `continue` sets the initial values, $\mathbf{b}_0$, for the model about to be fit according to the `e(b)` currently stored.

The comments made about specifying `missing` with `lf0()` apply equally well here.

`waldtest(`$\#$`)` is typically specified by programmers. By default, `ml` presents a Wald test, but that is overridden if the `lf0()` or `continue` option is specified. A Wald test is performed if `vce(robust)`, `vce(cluster` *clustvar*`)`, or `pweight` is specified.

`waldtest(0)` prevents even the Wald test from being reported.

waldtest(-1) is the default. It specifies that a Wald test be performed by constraining all coefficients except the intercept to 0 in the first equation. Remaining equations are to be unconstrained. A Wald test is performed if neither lf0() nor continue was specified, and a Wald test is forced if vce(robust), vce(cluster *clustvar*), or pweight was specified.

waldtest(k) for $k \leq -1$ specifies that a Wald test be performed by constraining all coefficients except intercepts to 0 in the first $|k|$ equations; remaining equations are to be unconstrained. A Wald test is performed if neither lf0() nor continue was specified, and a Wald test is forced if vce(robust), vce(cluster *clustvar*), or pweight was specified.

waldtest(k) for $k \geq 1$ works like the options above, except that it forces a Wald test to be reported even if the information to perform the likelihood-ratio test is available and even if none of vce(robust), vce(cluster *clustvar*), or pweight was specified. waldtest(k), $k \geq 1$, may not be specified with lf0().

obs(#) is used mostly by programmers. It specifies that the number of observations reported and ultimately stored in e(N) be #. Ordinarily, ml works that out for itself. Programmers may want to specify this option when, for the likelihood-evaluator to work for N observations, they first had to modify the dataset so that it contained a different number of observations.

crittype(*string*) is used mostly by programmers. It allows programmers to supply a string (up to 32 characters long) that describes the criterion that is being optimized by ml. The default is "log likelihood" for nonrobust and "log pseudolikelihood" for robust estimation.

svy indicates that ml is to pick up the svy settings set by svyset and use the robust variance estimator. This option requires the data to be svyset. svy may not be specified with *weights*, vce(robust), vce(opg), or vce(cluster *varname*).

subpop(*varname*) specifies that estimates be computed for the single subpopulation defined by the observations for which *varname* $\neq$ 0. Typically, *varname* = 1 defines the subpopulation, and *varname* = 0 indicates observations not belonging to the subpopulation. For observations whose subpopulation status is uncertain, *varname* should be set to missing ('.'). This option requires the svy option.

nosvyadjust specifies that the model Wald test be carried out as $W/k \sim F(k,d)$, where W is the Wald test statistic, k is the number of terms in the model excluding the constant term, d is the total number of sampled primary sampling units minus the total number of strata, and $F(k,d)$ is an F distribution with k numerator degrees of freedom and d denominator degrees of freedom. By default, an adjusted Wald test is conducted: $(d - k + 1)W/(kd) \sim F(k, d - k + 1)$. See Korn and Graubard (1990) for a discussion of the Wald test and the adjustments thereof. This option requires the svy option.

technique(*algorithm_spec*) specifies how the likelihood function is to be maximized. The following algorithms are currently implemented in ml.

technique(nr) specifies Stata's modified Newton–Raphson (NR) algorithm.

technique(bhhh) specifies the Berndt–Hall–Hall–Hausman (BHHH) algorithm.

technique(dfp) specifies the Davidon–Fletcher–Powell (DFP) algorithm.

technique(bfgs) specifies the Broyden–Fletcher–Goldfarb–Shanno (BFGS) algorithm.

The default is technique(nr).

You can switch between algorithms by specifying more than one in the technique() option. By default, ml will use an algorithm for five iterations before switching to the next algorithm. To specify a different number of iterations, include the number after the technique in the option. For example, specifying technique(bhhh 10 nr 1000) requests that ml perform 10 iterations using the BHHH algorithm followed by 1,000 iterations using the NR algorithm, and then switch back to BHHH for 10 iterations, and so on. The process continues until convergence or until the maximum number of iterations is reached.

Options for use with ml model in noninteractive mode

The following additional options are for use with ml model in noninteractive mode. Noninteractive mode is for programmers who use ml as a subroutine and want to issue a single command that will carry forth the estimation from start to finish.

maximize is required. It specifies noninteractive mode.

init(*ml_init_args*) sets the initial values $\mathbf{b}_0$. *ml_init_args* are whatever you would type after the ml init command.

search(on|norescale|quietly|off) specifies whether ml search is to be used to improve the initial values. search(on) is the default and is equivalent to running separately ml search, repeat(0). search(norescale) is equivalent to running separately ml search, repeat(0) norescale. search(quietly) is equivalent to search(on), except that it suppresses ml search's output. search(off) prevents the calling of ml search altogether.

repeat(#) is ml search's repeat() option. repeat(0) is the default.

bounds(*ml_search_bounds*) specifies the search bounds. The command ml model issues is ml search *ml_search_bounds*, repeat(#). Specifying search bounds is optional.

nowarning, novce, negh, and score() are ml maximize's equivalent options.

maximize_options: difficult, technique(*algorithm_spec*), iterate(#), [no]log, trace, gradient, showstep, hessian, showtolerance, tolerance(#), ltolerance(#), nrtolerance(#), nonrtolerance, and from(*init_specs*); see [R] **maximize**. These options are seldom used.

Options for use when specifying equations

noconstant specifies that the equation not include an intercept.

offset(*varname_o*) specifies that the equation be $\mathbf{x}\boldsymbol{\beta} + varname\_o$—that it include *varname_o* with the coefficient constrained to be 1.

exposure(*varname_e*) is an alternative to offset(*varname_o*); it specifies that the equation be $\mathbf{x}\boldsymbol{\beta} + \ln(varname\_e)$. The equation is to include $\ln(varname\_e)$ with the coefficient constrained to be 1.

Options for use with ml search

repeat(#) specifies the number of random attempts that are to be made to find a better initial-value vector. The default is repeat(10).

> repeat(0) specifies that no random attempts be made. More correctly, repeat(0) specifies that no random attempts be made if the first initial-value vector is a feasible starting point. If it is not, ml search will make random attempts, even if you specify repeat(0), because it has no alternative. The repeat() option refers to the number of random attempts to be made to improve the initial values. When the initial starting value vector is not feasible, ml search will make up to 1,000 random attempts to find starting values. It stops when it finds one set of values that works and then moves into its improve-initial-values logic.

> repeat(*k*), $k > 0$, specifies the number of random attempts to be made to improve the initial values.

restart specifies that random actions be taken to obtain starting values and that the resulting starting values not be a deterministic function of the current values. Generally, you should not specify this option because, with restart, ml search intentionally does not produce as good a set of starting values as it could. restart is included for use by the optimizer when it gets into serious trouble. The random actions ensure that the actions of the optimizer and ml search, working together, do not result in a long, endless loop.

> restart implies norescale, which is why we recommend that you do not specify restart. In testing, sometimes rescale worked so well that, even after randomization, the rescaler would bring the starting values right back to where they had been the first time and thus defeat the intended randomization.

norescale specifies that ml search not engage in its rescaling actions to improve the parameter vector. We do not recommend specifying this option because rescaling tends to work so well.

maximize_options: [no]log, trace; see [R] **maximize**. These options are seldom used.

Options for use with ml plot

saving(*filename*[, replace]) specifies that the graph be saved in *filename*.gph.
See [G] *saving_option*.

Options for use with ml init

copy specifies that the list of numbers or the initialization vector be copied into the
initial-value vector by position rather than by name.

skip specifies that any parameters found in the specified initialization vector that are
not also found in the model be ignored. The default action is to issue an error
message.

Options for use with ml maximize

nowarning is allowed only with iterate(0). nowarning suppresses the "convergence
not achieved" message. Programmers might specify iterate(0) nowarning when
they have a vector **b** already containing the final estimates and want ml to calculate
the variance matrix and post final estimation results. In that case, specify init(b)
search(off) iterate(0) nowarning nolog.

novce is allowed only with iterate(0). novce substitutes the zero matrix for the
variance matrix, which in effect posts estimation results as fixed constants.

negh specifies that the evaluator return the negative Hessian matrix. By default, ml
assumes d2 and lf2 evaluators return the Hessian matrix.

score(*newvars* | *stub**) creates new variables containing the contributions to the score
for each equation and ancillary parameter in the model; see [U] **20.17 Obtaining
scores**.

If score(*newvars*) is specified, the *newvars* must contain k new variables. For evaluators of methods lf, lf0, lf1, and lf2, k is the number of equations. For evaluators
of method gf0, k is the number of parameters. If score(*stub**) is specified, variables
named *stub*1, *stub*2, ..., *stubk* are created.

For evaluators of methods lf, lf0, lf1, and lf2, the first variable contains
$\partial \ln \ell_j / \partial(\mathbf{x}_{1j}\mathbf{b}_1)$, the second variable contains $\partial \ln \ell_j / \partial(\mathbf{x}_{2j}\mathbf{b}_2)$, and so on.

For evaluators of method gf0, the first variable contains $\partial \ln \ell_j / \partial \mathbf{b}_1$, the second
variable contains $\partial \ln \ell_j / \partial \mathbf{b}_2$, and so on.

nooutput suppresses display of the final results. This is different from prefixing ml
maximize with quietly in that the iteration log is still displayed (assuming that
nolog is not specified).

noclear specifies that after the model has converged, the ml problem definition not be
cleared. Perhaps you are having convergence problems and intend to run the model

to convergence. If so, use ml search to see if those values can be improved, and then start the estimation again.

maximize_options: difficult, iterate(#), [no]log, trace, gradient, showstep, hessian, showtolerance, tolerance(#), ltolerance(#), nrtolerance(#), and nonrtolerance; see [R] **maximize**. These options are seldom used.

display_options; see *Options for use with ml display* below.

eform_option; see *Options for use with ml display* below.

Options for use with ml graph

saving(*filename*[, replace]) specifies that the graph be saved in *filename*.gph. See [G] *saving_option*.

Options for use with ml display

noheader suppresses, above the coefficient table, the header display that displays the final log-likelihood value, the number of observations, and the model significance test.

nofootnote suppresses the footnote display below the coefficient table. The footnote displays a warning if the model fit did not converge within the specified number of iterations. Use ml footnote to display the warning if 1) you add to the coefficient table using the plus option or 2) you have your own footnotes and want the warning to be last.

level(#) is the standard confidence-level option. It specifies the confidence level, as a percentage, for confidence intervals of the coefficients. The default is level(95) or as set by set level; see [U] **20.7 Specifying the width of confidence intervals**.

first displays a coefficient table reporting results for the first equation only, and the report makes it appear that the first equation is the only equation. This option is used by programmers who estimate ancillary parameters in the second and subsequent equations and who wish to report the values of such parameters themselves.

neq(#) is an alternative to first. neq(#) displays a coefficient table reporting results for the first # equations. This option is used by programmers who estimate ancillary parameters in the #+1 and subsequent equations and who wish to report the values of such parameters themselves.

showeqns is a seldom-used option that displays the equation names in the coefficient table. ml display uses the numbers stored in e(k_eq) and e(k_aux) to determine how to display the coefficient table. e(k_eq) identifies the number of equations, and e(k_aux) identifies how many of these are for ancillary parameters. The first option is implied when showeqns is not specified and all but the first equation are for ancillary parameters.

plus displays the coefficient table, but rather than ending the table in a line of dashes, ends it in dashes–plus-sign–dashes. This is so that programmers can write additional display code to add more results to the table and make it appear as if the combined result is one table. Programmers typically specify plus with the first or neq() option. plus implies nofootnote.

nocnsreport suppresses the display of constraints above the coefficient table. This option is ignored if constraints were not used to fit the model.

noomitted specifies that variables that were omitted because of collinearity not be displayed. The default is to include in the table any variables omitted because of collinearity and to label them as "(omitted)".

vsquish specifies that the blank space separating factor-variable terms or time-series–operated variables from other variables in the model be suppressed.

noemptycells specifies that empty cells for interactions of factor variables not be displayed. The default is to include in the table interaction cells that do not occur in the estimation sample and to label them as "(empty)".

baselevels and allbaselevels control whether the base levels of factor variables and interactions are displayed. The default is to exclude from the table all base categories.

baselevels specifies that base levels be reported for factor variables and for interactions whose bases cannot be inferred from their component factor variables.

allbaselevels specifies that all base levels of factor variables and interactions be reported.

cformat(%fmt) specifies how to format coefficients, standard errors, and confidence limits in the coefficient table.

pformat(%fmt) specifies how to format p-values in the coefficient table.

sformat(%fmt) specifies how to format test statistics in the coefficient table.

coeflegend specifies that the legend of the coefficients and how to specify them in an expression be displayed rather than the coefficient table.

eform_option: eform(string), eform, hr, shr, irr, or, and rrr display the coefficient table in exponentiated form: for each coefficient, $\exp(b)$ rather than b is displayed, and standard errors and confidence intervals are transformed. Display of the intercept, if any, is suppressed. string is the table header that will be displayed above the transformed coefficients and must be 11 characters or shorter in length—for example, eform("Odds ratio"). The options eform, hr, shr, irr, or, and rrr provide a default string equivalent to "exp(b)", "Haz. Ratio", "SHR", "IRR", "Odds Ratio", and "RRR", respectively. These options may not be combined.

ml display looks at e(k_eform) to determine how many equations are affected by an eform_option; by default, only the first equation is affected.

Options for use with mleval

eq(#) specifies the equation number i for which $\theta_{ij} = \mathbf{x}_{ij}\boldsymbol{\beta}_i$ is to be evaluated. eq(1) is assumed if eq() is not specified.

scalar asserts that the ith equation is known to evaluate to a constant; the equation was specified as (), (*name*:), or /*name* on the ml model statement. If you specify this option, the new variable is created as a scalar. If the ith equation does not evaluate to a scalar, an error message is issued.

Options for use with mlsum

noweight specifies that weights ($ML_w) be ignored when summing the likelihood function.

Options for use with mlvecsum

eq(#) specifies the equation for which a gradient vector $\partial \ln L/\partial \boldsymbol{\beta}_i$ is to be constructed. The default is eq(1).

Options for use with mlmatsum

eq(#[,#]) specifies the equations for which the Hessian matrix is to be constructed. The default is eq(1), which means the same as eq(1,1), which in turn means $\partial^2 \ln L/\partial \boldsymbol{\beta}_1 \partial \boldsymbol{\beta}'_1$. Specifying eq($i$,$j$) results in $\partial^2 \ln L/\partial \boldsymbol{\beta}_i \partial \boldsymbol{\beta}'_j$.

Options for use with mlmatbysum

by(*varname*) is required and specifies the group variable.

eq(#[,#]) specifies the equations for which the Hessian matrix is to be constructed. The default is eq(1), which means the same as eq(1,1), which in turn means $\partial^2 \ln L/\partial \boldsymbol{\beta}_1 \partial \boldsymbol{\beta}'_1$. Specifying eq($i$,$j$) results in $\partial^2 \ln L/\partial \boldsymbol{\beta}_i \partial \boldsymbol{\beta}'_j$.

Options for use with ml score

equation(*eqname*) identifies from which equation the observation-level scores should come. This option may be used only when generating one variable.

missing specifies that observations containing variables with missing values not be eliminated from the estimation sample.

Saved results

For results saved by ml without the svy option, see [R] **maximize**.

For results saved by ml with the svy option, see [SVY] **svy**.

B Likelihood-evaluator checklists

B.1 Method lf

1. To use method `lf`, the likelihood function must meet the linear-form restrictions; otherwise, you must use method `d0`.

2. Write a program to evaluate $\ln \ell_j$, the contribution to the overall likelihood for an observation. The outline for the program is

   ```
   program myprog
           version 11
           args lnfj theta1 theta2 ...
           // if you need to create any intermediate results:
           tempvar tmp1 tmp2 ...
           quietly gen double 'tmp1' = ...
           ...
           quietly replace 'lnfj' = ...
   end
   ```

   Access the dependent variables (if any) by typing $ML_y1, $ML_y2, ... Use $ML_y1, $ML_y2, ... just as you would any existing variable in the data.

   Final results are to be saved in 'lnfj'. ('lnfj' is just a double-precision variable `ml` created for you. When `ml` calls your program, 'lnfj' contains missing values.)

   You must create any temporary variables for intermediate results as `doubles`: type `gen double` *name* = ...; do not omit the word `double`.

3. Issue the `ml model` statement as

   `. ml model lf` *myprog* *(equation for θ_1)* *(equation for θ_2)* ...

   Specify `if`, `in`, weights, and the `vce(opg)`, `vce(robust)`, `vce(cluster` *varname)*, ... options at this point.

4. Verify that your program works; type

   `. ml check`

5. Find starting values; type

   `. ml search`

   or use `ml init` or `ml plot`. We suggest you use `ml search`. In most cases, `ml search` can be omitted because `ml maximize` calls `ml search`, but in that case, `ml search` will not do as thorough a job.

6. Fit the model; type

```
. ml maximize
```

7. If later you wish to redisplay results, type

```
. ml display
```

B.2   Method d0

1. Method d0 may be used with any likelihood function.

2. Write a program to evaluate $\ln L$, the overall log-likelihood function. The outline for the evaluator is

```
program myprog
        version 11
        args todo b lnf

        tempvar theta1 theta2 ...
        mleval 'theta1' = 'b'
        mleval 'theta2' = 'b', eq(2) // if there is a θ2
        ...
        // if you need to create any intermediate results:
        tempvar tmp1 tmp2 ...
        gen double 'tmp1' = ...
        ...
        mlsum 'lnf' = ...
end
```

Argument 'todo' will always contain 0; it can be ignored.

You are passed the parameter row vector in 'b'. Use mleval to obtain the linear equation values $\theta_{ij} = \mathbf{x}_{ij}\boldsymbol{\beta}_i$ from it.

Access the dependent variables (if any) by typing $ML_y1, $ML_y2, ... Use $ML_y1, $ML_y2, ... just as you would any existing variable in the data.

Final results are to be saved in scalar 'lnf'. Use mlsum to produce it.

You must create any temporary variables for intermediate results as doubles: type gen double *name* = ...; do not omit the word double.

The estimation subsample is $ML_samp==1. You may safely ignore this if you do not specify ml model's nopreserve option. When your program is called, only relevant data will be in memory. If you do specify nopreserve, understand that mleval and mlsum automatically restrict themselves to the estimation subsample; it is not necessary to code if $ML_samp==1 on these commands. You merely need to restrict other commands to the $ML_samp==1 subsample.

The weights are stored in variable $ML_w, which contains 1 in every observation if no weights are specified. If you use mlsum to produce the log likelihood and mlvecsum to produce the gradient vector, you may ignore this because these commands handle that themselves.

3. Issue the `ml model` statement as

> . ml model d0 *myprog* (*equation for θ_1*) (*equation for θ_2*) ...

Specify any `if` *exp* or weights at this point. Although method-`d0` evaluators can be used with `fweights`, `aweights`, and `iweights`, `pweights` may not be specified. Use method `lf` or method `d1` if this is important. Similarly, you may not specify `ml model`'s `vce(opg)`, `vce(robust)`, or `vce(cluster` *varname*`)` option. Use method `lf` or method `d1` if this is important. The BHHH algorithm may not be used with method `d0`.

4. Verify that your program works; type

> . ml check

5. Find starting values; type

> . ml search

or use `ml init` or `ml plot`. We suggest that you use `ml search`. In most cases, `ml search` can be omitted because `ml maximize` calls `ml search`, but in that case `ml search` will not do as thorough a job.

6. Fit the model; type

> . ml maximize

7. If later you wish to redisplay results, type

> . ml display

B.3   Method d1

1. Method `d1` may be used with any likelihood function. This will be a considerable effort, however, if the function does not meet the linear-form restrictions.

2. Write a program to evaluate $\ln L$, the overall log-likelihood function and its derivatives. The outline for the evaluator is

(Continued on next page)

```
program myprog
        version 11
        args todo b lnf g

        tempvar theta1 theta2 ...
        mleval 'theta1' = 'b'
        mleval 'theta2' = 'b', eq(2) // if there is a θ₂
        ...

        // if you need to create any intermediate results:
        tempvar tmp1 tmp2 ...
        gen double 'tmp1' = ...
        ...

        mlsum 'lnf' = ...
        if ('todo'==0 | 'lnf'==.) exit

        tempname d1 d2 ...
        mlvecsum 'lnf' 'd1' = formula for ∂ln ℓⱼ/∂θ₁ⱼ, eq(1)
        mlvecsum 'lnf' 'd2' = formula for ∂ln ℓⱼ/∂θ₂ⱼ, eq(2)
        ...
        matrix 'g' = ('d1', 'd2', ... )
end
```

Argument `todo` will contain 0 or 1. You are to fill in the log likelihood if `todo`==0 or fill in the log likelihood and gradient otherwise.

You are passed the parameter row vector in `b`. Use mleval to obtain the linear equation values $\theta_{ij} = \mathbf{x}_{ij}\boldsymbol{\beta}_i$ from it.

Access the dependent variables (if any) by referring to \$ML_y1, \$ML_y2, ... Use \$ML_y1, \$ML_y2, ... just as you would any existing variable in the data.

The overall log-likelihood value is to be saved in scalar `lnf`. Use mlsum to produce it.

The gradient vector is to be saved in vector `g`. Use mlvecsum to produce it. You issue one mlvecsum per theta and then use matrix *name* = ... to put the results together.

You must create any temporary variables for intermediate results as doubles: type gen double *name* = ...; do not omit the word double.

The estimation subsample is \$ML_samp==1. You may safely ignore this if you do not specify ml model's nopreserve option. When your program is called, only relevant data will be in memory. If you do specify nopreserve, understand that mleval, mlsum, and mlvecsum automatically restrict themselves to the estimation subsample; it is not necessary to code if \$ML_samp==1 on these commands. You merely need to restrict other commands to the \$ML_samp==1 subsample.

The weights are stored in variable \$ML_w, which contains 1 if no weights are specified. If you use mlsum to produce the log likelihood and mlvecsum to produce the gradient vector, you may ignore this because these commands handle the weights themselves.

3. Issue the `ml model` statement as

     . ml model d1debug *myprog* (*equation for* θ_1) (*equation for* θ_2) ...

Note: Begin with method **d1debug** and not method **d1**. This allows testing whether the program produces correct first derivatives.

4. Verify that your program works; type

     . ml check

5. Find starting values; type

     . ml search

or use `ml init` or `ml plot`. We suggest that you use `ml search`. In most cases, `ml search` can be omitted because `ml maximize` calls `ml search`, but in that case `ml search` will not do as thorough a job.

6. Fit the model; type

     . ml maximize

Review the reported `mreldif()`s in the iteration log. Even if you have programmed the derivatives correctly, it is typical to observe the `mreldif()`s falling and then rising and even observe the `mreldif()` being largest in the final iteration, when the gradient vector is near zero. This result should not concern you. The reported `mreldif()`s should be 1e–3 or smaller in the middle iterations. If you suspect problems, reissue the `ml model` statement and then run `ml maximize` with the `gradient` option.

7. Now that you have verified that your program is calculating derivatives correctly, switch to using method **d1** by typing

     . ml model d1 *myprog* (*equation for* θ_1) (*equation for* θ_2) ...

Repeat steps 4, 5, and 6. You should obtain approximately the same results. The iteration log will probably differ; method **d1** may require fewer or more iterations than method **d1debug**. This does not indicate problems. Method **d1debug** used the numerically calculated derivatives, and maximization is a chaotic process.

8. If later you wish to redisplay results, type

     . ml display

B.4   Method d2

1. Method **d2** may be used with any likelihood function. This will be a considerable effort, however, if the function does not meet the linear-form restrictions.

2. Write a program to evaluate $\ln L$, the overall log-likelihood function, and its first and second derivatives. The outline for the evaluator is

```
program myprog
        version 11
        args todo b lnf g H

        tempvar theta1 theta2 ...
        mleval 'theta1' = 'b'
        mleval 'theta2' = 'b', eq(2) // if there is a θ₂
        ...
        // if you need to create any intermediate results:
        tempvar tmp1 tmp2 ...
        gen double 'tmp1' = ...
        ...

        mlsum 'lnf' = ...
        if ('todo'==0 | 'lnf'==.) exit

        tempname d1 d2 ...
        mlvecsum 'lnf' 'd1' = formula for ∂ ln ℓⱼ/∂θ₁ⱼ , eq(1)
        mlvecsum 'lnf' 'd2' = formula for ∂ ln ℓⱼ/∂θ₂ⱼ , eq(2)
        ...
        matrix 'g' = ('d1', 'd2', ... )
        if ('todo'==1 | 'lnf'==.) exit

        tempname d11 d12 d22 ...
        mlmatsum 'lnf' 'd11' = formula for ∂² ln ℓⱼ/∂θ²₁ⱼ , eq(1)
        mlmatsum 'lnf' 'd12' = formula for ∂² ln ℓⱼ/∂θ₁ⱼ∂θ₂ⱼ , eq(1,2)
        mlmatsum 'lnf' 'd22' = formula for ∂² ln ℓⱼ/∂θ²₂ⱼ , eq(2)
        ...
        matrix 'H' = ('d11','d12',... \ 'd12'','d22',... )
end
```

Argument 'todo' will contain 0, 1, or 2. You are to fill in the log likelihood if 'todo'==0; the log likelihood and gradient if 'todo'==1; or the log likelihood, gradient, and negative Hessian otherwise.

You are passed the parameter row vector in 'b'. Use mleval to obtain the linear-equation values $\theta_{ij} = \mathbf{x}_{ij}\boldsymbol{\beta}_i$ from it.

Access the dependent variables (if any) by referring to $ML_y1, $ML_y2, ... Use $ML_y1, $ML_y2, ... just as you would any existing variable in the data.

The overall log-likelihood value is to be saved in scalar 'lnf'. Use mlsum to produce it.

The gradient vector is to be saved in vector 'g'. Use mlvecsum to produce it. You issue one mlvecsum per theta and then use matrix *name* = ... to put the results together.

The Hessian is to be saved in matrix 'H'. Use mlmatsum to produce it. You issue one mlmatsum per equation and equation pair and then use matrix *name* = ... to put the results together.

You must create any temporary variables for intermediate results as doubles: type gen double *name* = ...; do not omit the word double.

The estimation subsample is $ML_samp==1. You may safely ignore this if you do not specify ml model's nopreserve option. When your program is called, only relevant data will be in memory. If you do specify nopreserve, understand that `mleval`, `mlsum`, `mlvecsum`, and `mlmatsum` automatically restrict themselves to the estimation subsample; it is not necessary to code `if $ML_samp==1` on these commands. You merely need to restrict other commands to the `$ML_samp==1` subsample.

The weights are stored in variable `$ML_w`, which contains 1 if no weights are specified. If you use `mlsum` to produce the log likelihood and `mlvecsum` to produce the gradient vector, you may ignore this because these commands handle the weights themselves.

3. Issue the `ml model` statement as

    . ml model d2debug *myprog* (*equation for* θ_1) (*equation for* θ_2) ...

Note: Begin with method **d2debug** and not method **d2**. This allows testing whether the program produces correct first and second derivatives.

4. Verify that your program works; type

    . ml check

5. Find starting values; type

    . ml search

or use `ml init` or `ml plot`. We suggest that you use `ml search`. In most cases, `ml search` can be omitted because `ml maximize` calls `ml search`, but in that case `ml search` will not do as thorough a job.

6. Fit the model; type

    . ml maximize

Review the reported `mreldif()`s in the iteration log. Even if you have programmed the derivatives correctly, it is typical to observe the `mreldif()`s for the gradient vector falling and then rising and even observe the `mreldif()` being largest in the final iteration, when the gradient vector is near zero. This should not concern you. The reported `mreldif()`s should be 1e–3 or smaller in the middle iterations.

For the negative Hessian, early iterations often report large `mreldif()`s but eventually fall. The reported `mreldif()` should be 1e–5 or smaller in the last iteration.

If you suspect problems, reissue the `ml model` statement and then run `ml maximize` with the **gradient** option or the **hessian** option.

7. Now that you have verified that your program is calculating derivatives correctly, switch to using method **d2** by typing

    . ml model d2 *myprog* (*equation for* θ_1) (*equation for* θ_2) ...

Repeat steps 4, 5, and 6. You should obtain approximately the same results (relative differences are 1e–8 for coefficients and smaller than 1e–5 for standard errors). The iteration log will probably differ; method **d2** may require fewer or more iterations than method **d2debug**. This does not indicate problems. Method d2debug used the numerically calculated derivatives, and maximization is a chaotic process.

8. If later you wish to redisplay results, type

```
. ml display
```

B.5   Method lf0

1. To use method lf0, the likelihood function must meet the linear-form restrictions.

2. Write a program to evaluate $\ln \ell_j$, the contribution to the overall likelihood function for an observation. The outline for the evaluator is

```
program myprog
        version 11
        args todo b lnfj

        tempvar theta1 theta2 ...
        mleval 'theta1' = 'b'
        mleval 'theta2' = 'b', eq(2) // if there is a θ₂
        ...

        // if you need to create any intermediate results:
        tempvar tmp1 tmp2 ...
        gen double 'tmp1' = ...
        ...

        quietly replace 'lnfj' = ...
end
```

Argument 'todo' will always contain 0; it can be ignored.

You are passed the parameter row vector in 'b'. Use mleval to obtain the linear equation values $\theta_{ij} = \mathbf{x}_{ij}\boldsymbol{\beta}_i$ from it.

Access the dependent variables (if any) by typing $ML_y1, $ML_y2, ... Use $ML_y1, $ML_y2, ... just as you would any existing variable in the data.

Final results are to be saved in 'lnfj'. ('lnfj' is just a temporary double-precision variable ml created for you. When ml calls your program, 'lnfj' contains missing values.)

You must create any temporary variables for intermediate results as doubles: type gen double *name* = ...; do not omit the word double.

The estimation subsample is $ML_samp==1. You may safely ignore this if you do not specify ml model's nopreserve option. When your program is called, only relevant data will be in memory. If you do specify nopreserve, understand that mleval automatically restricts itself to the estimation subsample; it is not necessary to code if $ML_samp==1 on these commands. You merely need to restrict other commands to the $ML_samp==1 subsample.

The weights are stored in variable $ML_w, which contains 1 in every observation if no weights are specified.

3. Issue the `ml model` statement as

> . ml model lf0 *myprog* (*equation for θ_1*) (*equation for θ_2*) ...

Specify any `if` *exp* or weights at this point. Although method-`d0` evaluators can be used with `fweights`, `aweights`, and `iweights`, `pweights` may not be specified.

Specify `if`, `in`, weights, and the `vce(opg)`, `vce(robust)`, `vce(cluster` *var-name*), ... options at this point.

4. Verify that your program works; type

> . ml check

5. Find starting values; type

> . ml search

or use `ml init` or `ml plot`. We suggest that you use `ml search`. In most cases, `ml search` can be omitted because `ml maximize` calls `ml search`, but in that case `ml search` will not do as thorough a job.

6. Fit the model; type

> . ml maximize

7. If later you wish to redisplay results, type

> . ml display

B.6   Method lf1

1. To use method `lf1`, the likelihood function must meet the linear-form restrictions.

2. Write a program to evaluate $\ln \ell_j$, the contribution to the overall likelihood function for an observation, and to calculate the equation-level score variables. The outline for the evaluator is

```
program myprog
        version 11
        args todo b lnfj g1 g2 ...
        tempvar theta1 theta2 ...
        mleval 'theta1' = 'b'
        mleval 'theta2' = 'b', eq(2) // if there is a θ_2
        ...
        // if you need to create any intermediate results:
        tempvar tmp1 tmp2 ...
        gen double 'tmp1' = ...
        ...
        quietly replace 'lnfj' = ...
        if ('todo'==0) exit
        quietly replace 'g1' = formula for ∂ln ℓ_j/∂θ_{1j}, eq(1)
        quietly replace 'g2' = formula for ∂ln ℓ_j/∂θ_{2j}, eq(2)
        ...
end
```

Argument 'todo' will contain 0 or 1. You are to fill in the log likelihood if 'todo'==0 or fill in the log likelihood and scores otherwise.

You are passed the parameter row vector in 'b'. Use mleval to obtain the linear equation values $\theta_{ij} = \mathbf{x}_{ij}\boldsymbol{\beta}_i$ from it.

Access the dependent variables (if any) by referring to \$ML_y1, \$ML_y2, ... Use \$ML_y1, \$ML_y2, ... just as you would any existing variable in the data.

The observation-level log-likelihood values are to be saved in 'lnfj'.

The scores are to be saved in g1, g2, ...

You must create any temporary variables for intermediate results as doubles: type gen double *name* = ...; do not omit the word double.

The estimation subsample is \$ML_samp==1. You may safely ignore this if you do not specify ml model's nopreserve option. When your program is called, only relevant data will be in memory. If you do specify nopreserve, understand that mleval automatically restricts itself to the estimation subsample; it is not necessary to code if \$ML_samp==1 on these commands. You merely need to restrict other commands to the \$ML_samp==1 subsample.

The weights are stored in variable \$ML_w, which contains 1 if no weights are specified.

3. Issue the ml model statement as

> . ml model lf1debug *myprog* (equation for θ_1) (equation for θ_2) ...

Note: Begin with method lf1debug and not method lf1. This allows testing whether the program produces correct scores.

4. Verify that your program works; type

> . ml check

5. Find starting values; type

> . ml search

or use ml init or ml plot. We suggest that you use ml search. In most cases, ml search can be omitted because ml maximize calls ml search, but in that case ml search will not do as thorough a job.

6. Fit the model; type

> . ml maximize

Review the reported mreldif()s in the iteration log. Even if you have programmed the derivatives correctly, it is typical to observe the mreldif()s falling and then rising and even observe the mreldif() being largest in the final iteration, when the gradient vector is near zero. This result should not concern you. The reported mreldif()s should be 1e–3 or smaller in the middle iterations. If you suspect problems, reissue the ml model statement and then run ml maximize with the gradient option.

7. Now that you have verified that your program is calculating derivatives correctly, switch to using method lf1 by typing

    . ml model lf1 *myprog* (*equation for* θ_1) (*equation for* θ_2) ...

Repeat steps 4, 5, and 6. You should obtain approximately the same results. The iteration log will probably differ; method lf1 may require fewer or more iterations than method lf1debug. This does not indicate problems. Method lf1debug used the numerically calculated derivatives, and maximization is a chaotic process.

8. If later you wish to redisplay results, type

    . ml display

B.7   Method lf2

1. To use method lf2, the likelihood function must meet the linear-form restrictions.

2. Write a program to evaluate $\ln \ell_j$, the contribution to the overall likelihood function for an observation, to calculate the equation-level score variables, and to calculate the Hessian of the overall likelihood function. The outline for the evaluator is

```
program myprog
        version 11
        args todo b lnfj g1 g2 ... H
        tempvar theta1 theta2 ...
        mleval 'theta1' = 'b'
        mleval 'theta2' = 'b', eq(2) // if there is a θ2
        ...
        // if you need to create any intermediate results:
        tempvar tmp1 tmp2 ...
        gen double 'tmp1' = ...
        ...
        quietly replace 'lnfj' = ...
        if ('todo'==0) exit
        quietly replace 'g1' = formula for ∂ln ℓj/∂θ1j, eq(1)
        quietly replace 'g2' = formula for ∂ln ℓj/∂θ2j, eq(2)
        ...
        if ('todo'==1) exit
        tempname d11 d12 d22 ...
        mlmatsum 'lnf' 'd11' = formula for ∂²ln ℓj/∂θ²1j, eq(1)
        mlmatsum 'lnf' 'd12' = formula for ∂²ln ℓj/∂θ1j∂θ2j, eq(1,2)
        mlmatsum 'lnf' 'd22' = formula for ∂²ln ℓj/∂θ²2j, eq(2)
        ...
        matrix 'H' = ('d11','d12',... \ 'd12'','d22',... )
    end
```

Argument 'todo' will contain 0, 1, or 2. You are to fill in the log likelihood if 'todo'==0; the log likelihood and scores if 'todo'==1; or the log likelihood, scores, and Hessian if 'todo'==2.

You are passed the parameter row vector in 'b'. Use mleval to obtain the linear-equation values $\theta_{ij} = \mathbf{x}_{ij}\boldsymbol{\beta}_i$ from it.

Access the dependent variables (if any) by referring to $ML_y1, $ML_y2, ... Use $ML_y1, $ML_y2, ... just as you would any existing variable in the data.

The observation-level log-likelihood values are to be saved in 'lnfj'.

The scores are to be saved in g1, g2, ...

The Hessian is to be saved in H. Use mlmatsum to produce it. You issue one mlmatsum per equation and equation pair and then use matrix *name* = ... to put the results together.

You must create any temporary variables for intermediate results as doubles: type gen double *name* = ...; do not omit the word double.

The estimation subsample is $ML_samp==1. You may safely ignore this if you do not specify ml model's nopreserve option. When your program is called, only relevant data will be in memory. If you do specify nopreserve, understand that mleval and mlmatsum automatically restrict themselves to the estimation subsample; it is not necessary to code if $ML_samp==1 on these commands. You merely need to restrict other commands to the $ML_samp==1 subsample.

The weights are stored in variable $ML_w, which contains 1 if no weights are specified.

3. Issue the ml model statement as

> . ml model lf2debug *myprog* (*equation for* θ_1) (*equation for* θ_2) ...

Note: Begin with method lf2debug and not method lf2. This allows testing whether the program produces correct first and second derivatives.

4. Verify that your program works; type

> . ml check

5. Find starting values; type

> . ml search

or use ml init or ml plot. We suggest that you use ml search. In most cases, ml search can be omitted because ml maximize calls ml search, but in that case ml search will not do as thorough a job.

6. Fit the model; type

> . ml maximize

Review the reported mreldif()s in the iteration log. Even if you have programmed the derivatives correctly, it is typical to observe the mreldif()s for the gradient vector falling and then rising and even observe the mreldif() being largest in the final iteration, when the gradient vector is near zero. This should not concern you. The reported mreldif()s should be 1e–3 or smaller in the middle iterations.

For the negative Hessian, early iterations often report large `mreldif()`s but eventually fall. The reported `mreldif()` should be 1e–5 or smaller in the last iteration.

If you suspect problems, reissue the `ml model` statement and then run `ml maximize`
with the `gradient` option or the `hessian` option.

7. Now that you have verified that your program is calculating derivatives correctly,
   switch to using method `lf2` by typing

   > . ml model lf2 *myprog* (*equation for* θ_1) (*equation for* θ_2) ...

   Repeat steps 4, 5, and 6. You should obtain approximately the same results (relative differences are 1e–8 for coefficients and smaller than 1e–5 for standard errors).
   The iteration log will probably differ; method `lf2` may require fewer or more iterations than method `lf2debug`. This does not indicate problems. Method `lf2debug`
   used the numerically calculated derivatives, and maximization is a chaotic process.

8. If later you wish to redisplay results, type

   > . ml display

C Listing of estimation commands

C.1 The logit model

begin mylogit.ado

```
program mylogit, properties(or svyb svyj svyr swml)
        version 11
        if replay() {
                if ('"'e(cmd)'"' != "mylogit") error 301
                Replay '0'
        }
        else    Estimate '0'
end
program Estimate, eclass sortpreserve
        syntax varlist(fv) [if] [in]            ///
                [fweight pweight iweight] [,    ///
                noLOg noCONStant                /// -ml model- options
                noLRTEST                        ///
                vce(passthru)                   ///
                OFFset(varname numeric)         ///
                EXPosure(varname numeric)       ///
                Level(cilevel)                  /// -Replay- options
                OR                              ///
                *                               /// -mlopts- options
        ]

        // mark the estimation sample
        marksample touse

        // check syntax
        mlopts mlopts, 'options'
        local cns 's(constraints)'
        gettoken lhs rhs : varlist
        if "'weight'" != "" {
                local wgt "['weight''exp']"
                // for initial value calculations
                if "'weight'" == "pweight" {
                        local awgt "[aw'exp']"
                }
                else    local awgt "'wgt'"
        }
        if "'log'" != "" {
                local qui quietly
        }
        if "'offset'" != "" {
                local offopt "offset('offset')"
        }
        if "'exposure'" != "" {
                local expopt "exposure('exposure')"
        }
```

```
        // markout missing values from the estimation sample
        markout 'touse' 'offset' 'exposure'
        _vce_parse 'touse', opt(Robust oim opg) argopt(CLuster): 'wgt', 'vce'

        if "'constant'" == "" {
                // initial value
                sum 'lhs' 'awgt' if 'touse', mean
                local b0 = logit(r(mean))
                if "'weight'" == "iweight" {
                        local n = r(sum_w)
                }
                else    local n = r(N)
                local initopt init(_cons='b0') search(quietly)

                if "'lrtest'" == "" {
                        local n1 = r(sum)
                        local lf0 = 'n1'*ln(invlogit('b0'))        ///
                                +('n'-'n1')*ln(invlogit(-'b0'))
                        local initopt 'initopt' lf0(1 'lf0')
                }
        }

        // fit the full model
        'qui' di as txt _n "Fitting full model:"
        ml model lf2 mylogit_lf2                        ///
                (xb: 'lhs' = 'rhs',                     ///
                        'constant' 'offopt' 'expopt'    ///
                )                                       ///
                'wgt' if 'touse',                       ///
                'log' 'mlopts' 'vce'                    ///
                'modopts' 'initopt'                     ///
                missing maximize

        // save a title for -Replay- and the name of this command
        ereturn local title "My logit estimates"
        ereturn local cmd mylogit

        Replay , level('level') 'or'
end

program Replay
        syntax [, Level(cilevel) OR ]
        ml display , level('level') 'or'
end
```

─── end `mylogit.ado` ─────────

─────────────────────────────────────── begin `mylogit_lf2.ado` ───────────

```
program mylogit_lf2
        version 11
        args todo b lnfj g1 H
        tempvar xb lj
        mleval `xb' = `b'
        quietly {
                gen double `lj' = invlogit( `xb')   if $ML_y1 == 1
                replace    `lj' = invlogit(-`xb')   if $ML_y1 == 0
                replace `lnfj' = ln(`lj')
                if (`todo'==0) exit

                replace `g1' =  invlogit(-`xb')   if $ML_y1 == 1
                replace `g1' = -invlogit( `xb')   if $ML_y1 == 0
                if (`todo'==1) exit

                mlmatsum `lnfj' `H' = -1*abs(`g1')*`lj', eq(1,1)
        }
end
```

─────────────────────────────────────── end `mylogit_lf2.ado` ───────────

C.2   The probit model

─────────────────────────────────────── begin `myprobit.ado` ───────────

```
program myprobit, properties(svyb svyj svyr swml)
        version 11
        if replay() {
                if (`"`e(cmd)'"' != "myprobit") error 301
                Replay `0'
        }
        else    Estimate `0'
end

program Estimate, eclass sortpreserve
        syntax varlist [if] [in]            ///
                [fweight pweight iweight] [, ///
                noLOg noCONStant            /// -ml model- options
                noLRTEST                    ///
                vce(passthru)               ///
                OFFset(varname numeric)     ///
                EXPosure(varname numeric)   ///
                Level(cilevel)              /// -Replay- options
                *                           /// -mlopts- options
        ]
        // check syntax
        mlopts mlopts, `options'
        local cns `s(constraints)'
        gettoken lhs rhs : varlist
        if "`weight'" != "" {
                local wgt "[`weight'`exp']"
                // for initial value calculations
                if "`weight'" == "pweight" {
                        local awgt "[aw`exp']"
                }
                else    local awgt "`wgt'"
        }
```

```
        if "`log'" != "" {
                local qui quietly
        }
        if "`offset'" != "" {
                local offopt "offset(`offset')"
        }
        if "`exposure'" != "" {
                local expopt "exposure(`exposure')"
        }
        // mark the estimation sample
        marksample touse
        markout `touse' `offset' `exposure'
        _vce_parse `touse', opt(Robust oim opg) argopt(CLuster): `wgt', `vce'
        if "`constant'" == "" {
                // initial value
                sum `lhs' `awgt' if `touse', mean
                local b0 = invnormal(r(mean))
                if "`weight'" == "iweight" {
                        local n = r(sum_w)
                }
                else    local n = r(N)
                local initopt init(_cons=`b0') search(quietly)
                if "`lrtest'" == "" {
                        local n1 = r(sum)
                        local lf0 = `n1'*ln(normal(`b0'))         ///
                                +(`n'-`n1')*ln(normal(-`b0'))
                        local initopt `initopt' lf0(1 `lf0')
                }
        }
        // fit the full model
        `qui' di as txt _n "Fitting full model:"
        ml model lf2 myprobit_lf2                      ///
                (xb: `lhs' = `rhs',                    ///
                        `constant' `offopt' `expopt'   ///
                )                                      ///
                `wgt' if `touse',                      ///
                `log' `mlopts' `vce'                   ///
                `modopts' `initopt' missing maximize
        // save a title for -Replay- and the name of this command
        ereturn local title "My probit estimates"
        ereturn local cmd myprobit
        Replay , level(`level')
end
program Replay
        syntax [, Level(cilevel) ]
        ml display , level(`level')
end
```

––– end myprobit.ado –––––––––––

```
                                              begin myprobit_lf2.ado
program myprobit_lf2
        version 11
        args todo b lnfj g1 H
        tempvar xb lj
        mleval 'xb' = 'b'
        quietly {
                // Create temporary variable used in both likelihood
                // and scores
                gen double 'lj'  = normal( 'xb')  if $ML_y1 == 1
                replace    'lj'  = normal(-'xb')  if $ML_y1 == 0
                replace   'lnfj' = log('lj')
                if ('todo'==0) exit

                replace 'g1' =  normalden('xb')/'lj'  if $ML_y1 == 1
                replace 'g1' = -normalden('xb')/'lj'  if $ML_y1 == 0
                if ('todo'==1) exit

                mlmatsum 'lnfj' 'H' = -'g1'*('g1'+'xb'), eq(1,1)
        }
end
                                              end myprobit_lf2.ado
```

C.3   The normal model

```
                                              begin mynormal.ado
program mynormal, properties(svyb svyj svyr swml)
        version 11
        if replay() {
                if ('"'e(cmd)'"' != "mynormal") error 301
                Replay '0'
        }
        else    Estimate '0'
end
program Estimate, eclass sortpreserve
        syntax varlist [if] [in]                ///
                [fweight pweight iweight] [,     ///
                noLOg noCONStant                 /// -ml model- options
                HETero(varlist)                  ///
                noLRTEST                         ///
                vce(passthru)                    ///
                OFFset(varname numeric)          ///
                EXPosure(varname numeric)        ///
                Level(cilevel)                   /// -Replay- options
                EForm                            ///
                *                                /// -mlopts- options
                ]
        // check syntax
        mlopts mlopts, 'options'
        local cns 's(constraints)'
        gettoken lhs rhs : varlist
        if "'weight'" != "" {
                local wgt "['weight''exp']"
                // for initial value calculations
                if "'weight'" == "pweight" {
                        local awgt "[aw'exp']"
                }
```

```
                else    local awgt "'wgt'"
        }
        if "'log'" != "" {
                local qui quietly
        }
        if "'offset'" != "" {
                local offopt "offset('offset')"
        }
        if "'exposure'" != "" {
                local expopt "exposure('exposure')"
        }
        // mark the estimation sample
        marksample touse
        markout 'touse' 'offset' 'exposure'
        _vce_parse 'touse', opt(Robust oim opg) argopt(CLuster): 'wgt', 'vce'
        if "'constant'" == "" {
                // initial values
                quietly sum 'lhs' 'awgt' if 'touse'
                local mean = r(mean)
                local lnsd = ln(r(sd))+ln((r(N)-1)/r(N))/2
                local initopt init(/mu='mean' /lnsigma='lnsd') search(off)

                'qui' di as txt _n "Fitting constant-only model:"
                ml model lf2 mynormal_lf2                       ///
                        (mu: 'lhs' =, 'offopt' 'expopt' )       ///
                        (lnsigma: 'hetero')                     ///
                        'wgt' if 'touse',                       ///
                        'log' 'mlopts' 'initopt'                ///
                        nocnsnotes missing maximize
                if "'lrtest'" == "" {
                        local contin continue search(off)
                }
                else {
                        tempname b0
                        mat 'b0' = e(b)
                        local contin init('b0') search(off)
                }
        }

        // fit the full model
        'qui' di as txt _n "Fitting full model:"
        ml model lf2 mynormal_lf2                       ///
                (mu: 'lhs' = 'rhs',                     ///
                        'constant' 'offopt' 'expopt'    ///
                )                                       ///
                (lnsigma: 'hetero')                     ///
                'wgt' if 'touse',                       ///
                'log' 'mlopts' 'vce'                    ///
                'contin' missing maximize

        if "'hetero'" == "" {
                ereturn scalar k_aux = 1
        }
        else    ereturn scalar k_aux = 0
        // save a title for -Replay- and the name of this command
        ereturn local title "My normal estimates"
        ereturn local predict mynormal_p
        ereturn local cmd mynormal

        Replay , level('level') 'eform'
end
```

```
program Replay
        syntax [, Level(cilevel) EForm ]
        if `e(k_aux)' {
                local sigma diparm(lnsigma, exp label("sigma"))
        }
        ml display , level(`level') `sigma' `eform'
end
```

———————————————————————————————— end `mynormal.ado` ————————

———————————————————————————————— begin `mynormal_lf2.ado` ————————
```
program mynormal_lf2
        version 11
        args todo b lnfj g1 g2 H
        tempvar mu lnsigma sigma
        mleval `mu' = `b', eq(1)
        mleval `lnsigma' = `b', eq(2)
        quietly {
                gen double `sigma' = exp(`lnsigma')
                replace `lnfj' = ln( normalden($ML_y1,`mu',`sigma') )
                if (`todo'==0) exit

                tempvar z
                tempname dmu dlnsigma
                gen double `z' = ($ML_y1-`mu')/`sigma'
                replace `g1' = `z'/`sigma'
                replace `g2' = `z'*`z'-1
                if (`todo'==1) exit

                tempname d11 d12 d22
                mlmatsum `lnfj' `d11' = -1/`sigma'^2      , eq(1)
                mlmatsum `lnfj' `d12' = -2*`z'/`sigma'    , eq(1,2)
                mlmatsum `lnfj' `d22' = -2*`z'*`z'        , eq(2)
                matrix `H' = (`d11', `d12' \ `d12', `d22')
        }
end
```

———————————————————————————————— end `mynormal_lf2.ado` ————————

C.4   The Weibull model

———————————————————————————————— begin `myweibull.ado` ————————
```
program myweibull, properties(hr svyb svyj svyr swml)
        version 11
        if replay() {
                if (`""`e(cmd)'"'' != "myweibull") error 301
                Replay `0'
        }
        else    Estimate `0'
end

program Estimate, eclass sortpreserve
        syntax varlist(fv) [if] [in]              ///
                [fweight pweight iweight] [,      ///
                FAILure(varname numeric)         ///
                noLOg noCONStant                 /// -ml model- options
                noLRTEST                         ///
                vce(passthru)                    ///
                OFFset(varname numeric)          ///
                EXPosure(varname numeric)        ///
```

```
            Level(cilevel)                      /// -Replay- options
            HR                                  ///
            *                                   /// -mlopts- options
]
// check syntax
mlopts mlopts, 'options'
local cns 's(constraints)'
gettoken time rhs : varlist
if "'failure'" == "" {
        tempvar failure
        gen byte 'failure' = 1
}
if "'weight'" != "" {
        local wgt "['weight''exp']"
        // for initial value calculations
        if "'weight'" == "pweight" {
                local awgt "[aw'exp']"
        }
        else    local awgt "'wgt'"
}
if "'log'" != "" {
        local qui quietly
}
if "'offset'" != "" {
        local offopt "offset('offset')"
}
if "'exposure'" != "" {
        local expopt "exposure('exposure')"
}
// mark the estimation sample
marksample touse
markout 'touse' 'failure' 'offset' 'exposure'
_vce_parse 'touse', opt(Robust oim opg) argopt(CLuster): 'wgt', 'vce'
if "'constant'" == "" {
        // initial value
        sum 'time' 'awgt' if 'touse', mean
        local b0 = r(mean)
        sum 'failure' 'awgt' if 'touse', mean
        local b0 = ln('b0'/r(mean))
        local initopt init(_cons='b0') search(quietly)

        'qui' di as txt _n "Fitting constant-only model:"
        ml model lf2 myweibull_lf2                      ///
                (lneta: 'time' 'failure' = ,            ///
                        'offopt' 'expopt'               ///
                )                                       ///
                /lngamma                                ///
                'wgt' if 'touse',                       ///
                'log' 'mlopts' 'initopt'                ///
                nocnsnotes missing maximize
        if "'lrtest'" == "" {
                local contin continue search(off)
        }
        else {
                tempname b0
                mat 'b0' = e(b)
                local contin init('b0') search(off)
        }
}
```

```
        // fit the full model
        `qui' di as txt _n "Fitting full model:"
        ml model lf2 myweibull_lf2
                (lneta: `time' `failure'= `rhs',        ///
                        `constant' `offopt' `expopt'     ///
                )                                        ///
                /lngamma                                 ///
                `wgt' if `touse',                        ///
                `log' `mlopts' `vce'                     ///
                `contin' missing maximize                ///
                diparm(lngamma, exp label("gamma"))      ///
                diparm(lngamma,                          ///
                        function(exp(-@))                ///
                        derivative(exp(-@))              ///
                        label("1/gamma")                 ///
                )
        ereturn scalar k_aux = 1
        ereturn local failure `failure'
        ereturn local depvar `time' `failure'
        // save a title for -Replay- and the name of this command
        ereturn local title "My Weibull estimates"
        ereturn local cmd myweibull

        Replay , level(`level') `hr'
end

program Replay
        syntax [, Level(cilevel) HR ]
        ml display , level(`level') `hr'
end
```

────────────────────────────── end myweibull.ado ──────────

────────────────────────────── begin myweibull_lf2.ado ──────────

```
program myweibull_lf2
        version 11
        args todo b lnf g1 g2 H
        tempvar leta lgam p M R
        mleval `leta' = `b', eq(1)
        mleval `lgam' = `b', eq(2)
        local t "$ML_y1"
        local d "$ML_y2"
        quietly {
                gen double `p' = exp(`lgam')
                gen double `M' = (`t'*exp(-`leta'))^`p'
                gen double `R' = ln(`t')-`leta'
                replace `lnf' = -`M' + `d'*(`lgam'-`leta' + (`p'-1)*`R')
                if (`todo'==0) exit

                replace `g1' = `p'*(`M'-`d')
                replace `g2' = `d' - `R'*`p'*(`M'-`d')
                if (`todo'==1) exit

                tempname d11 d12 d22
                mlmatsum `lnf' `d11' = -`p'^2 * `M'                      , eq(1)
                mlmatsum `lnf' `d12' = `p'*(`M'-`d'+`R'*`p'*`M')     , eq(1,2)
                mlmatsum `lnf' `d22' = -`p'*`R'*(`R'*`p'*`M'+`M'-`d') , eq(2)
                matrix `H' = (`d11',`d12' \ `d12'',`d22')
        }
end
```

────────────────────────────── end myweibull_lf2.ado ──────────

C.5   The Cox proportional hazards model

── begin `mycox.ado` ────────────

```
program mycox, properties(hr swml)
        version 11
        if replay() {
                if (`"`e(cmd)'"' != "mycox") error 301
                Replay `0'
        }
        else    Estimate `0'
end

program Estimate, eclass sortpreserve
        syntax varlist(fv min=2) [if] [in]         ///
                [fweight] [,                        ///
                FAILure(varname numeric)            ///
                noLOg                               /// -ml model- options
                noLRTEST                            ///
                vce(passthru)                       ///
                Level(cilevel)                      /// -Replay- options
                HR                                  ///
                *                                   /// -mlopts- options
                ]
        // check syntax
        local diopts level(`level') `hr'
        mlopts mlopts , `options'
        local cns `s(constraints)'
        gettoken time rhs : varlist
        tempvar negt
        quietly gen double `negt' = -`time'
        sort `negt' `failure'
        if "`failure'" == "" {
                tempvar failure
                gen byte `failure' = 1
        }
        if "`weight'" != "" {
                local wgt "[`weight'`exp']"
        }
        if "`cns'" != "" {
                local lrtest nolrtest
        }

        // mark the estimation sample
        marksample touse
        markout `touse' `failure'
        _vce_parse, optlist(OIM) :, `vce'

        // fit constant-only model
        if "`lrtest'" == "" {
                quietly ml model d0 mycox2_d0                     ///
                        (xb: `negt' `failure' = `rhs',            ///
                                noconstant                        ///
                        )                                         ///
                        `wgt' if `touse',                         ///
                        iterate(0)                                ///
                        nocnsnotes missing maximize
                local initopt lf0(0 `e(ll)')
        }

        // fit the full model
        `qui' di as txt _n "Fitting full model:"
        ml model d0 mycox2_d0                            ///
```

```
                        (xb: `negt' `failure' = `rhs',           ///
                             noconstant                          ///
                        )                                        ///
                        `wgt' if `touse',                        ///
                        `log' `mlopts' `vce' `initopt'           ///
                        missing maximize
        // dependent variables
        ereturn local depvar `time' `failure'
        // save a title for -Replay- and the name of this command
        ereturn local title "My cox estimates"
        ereturn local cmd mycox

        Replay , `diopts'
end
program Replay
        syntax [, Level(cilevel) HR ]
        ml display , level(`level') `hr'
end
```
── end `mycox.ado` ──────────

────────────────────────────────────── begin `mycox2_d0.ado` ──────────
```
program mycox2_d0
        version 11
        args todo b lnf
        local negt "$ML_y1"                // $ML_y1 is -t
        local d "$ML_y2"                   // $ML_y2 is d
        tempvar xb B A sumd last L
        mleval `xb' = `b'
        // data assumed already sorted by `negt' and `d'
        local byby "by `negt' `d'"
        local wxb "$ML_w*`xb'"
        quietly {
                gen double `B' = sum($ML_w*exp(`xb'))
                `byby': gen double `A' = cond(_n==_N, sum(`wxb'), .) if `d'==1
                `byby': gen `sumd' = cond(_n==_N, sum($ML_w*`d'), .)
                `byby': gen byte `last' = (_n==_N & `d' == 1)
                gen double `L' = `A' - `sumd'*ln(`B') if `last'
                mlsum `lnf' = `L' if `last', noweight
        }
end
```
─────────────────────────────────────── end `mycox2_d0.ado` ──────────

C.6   The random-effects regression model

—————————————————————— begin `myrereg.ado` ——————

```
program myrereg
        version 11
        if replay() {
                if (`"`e(cmd)'"' != "myrereg") error 301
                Replay `0'
        }
        else    Estimate `0'
end
program Estimate, eclass sortpreserve
        syntax varlist(fv) [if] [in] ,          ///
                panel(varname)                  ///
                [                               ///
                noLOg noCONStant                /// -ml model- options
                noLRTEST                        ///
                vce(passthru)                   ///
                Level(cilevel)                  /// -Replay- options
                *                               /// -mlopts- options
                ]
        // check syntax
        local diopts level(`level')
        mlopts mlopts , `options'
        local cns `s(constraints)'
        gettoken lhs rhs : varlist
        if "`cns'" != "" {
                local lrtest nolrtest
        }
        if "`log'" != "" {
                local qui quietly
        }

        // mark the estimation sample
        marksample touse
        markout `touse' `panel', strok
        _vce_parse, optlist(OIM) :, `vce'

        // capture block to ensure removal of global macro
capture noisily {
        // identify the panel variable for the evaluator
        global MY_panel `panel'
        sort `panel'

        if "`constant'" == "" {
                // initial values: variance components from one-way ANOVA
                sum `lhs' if `touse', mean
                local mean = r(mean)
                quietly oneway `lhs' `panel' if `touse'
                local np  = r(df_m) + 1
                local N   = r(N)
                local bms = r(mss)/r(df_m)       // between mean squares
                local wms = r(rss)/r(df_r)       // within mean squares
                local lns_u = log( (`bms'-`wms')*`np'/`N' )/2
                if missing(`lns_u') {
                        local lns_u = 0
                }
                local lns_e = log(`wms')/2
                local   initopt search(off)             ///
                        init(/xb=`mean' /lns_u=`lns_u' /lns_e=`lns_e')
```

```
                        'qui' di as txt _n "Fitting constant-only model:"
                        ml model d2 myrereg2_d2                        ///
                                (xb: 'lhs' = , 'offopt' 'expopt' )     ///
                                /lns_u                                 ///
                                /lns_e                                 ///
                                if 'touse',                           ///
                                'log' 'mlopts' 'initopt'              ///
                                nocnsnotes missing maximize
                        if "'lrtest'" == "" {
                                local contin continue search(off)
                        }
                        else {
                                tempname b0
                                mat 'b0' = e(b)
                                local contin init('b0') search(off)
                        }
                }
                // fit the full model
                'qui' di as txt _n "Fitting full model:"
                ml model d2 myrereg2_d2                        ///
                        (xb: 'lhs' = 'rhs',                   ///
                                'constant' 'offopt' 'expopt'   ///
                        )                                     ///
                        /lns_u                                ///
                        /lns_e                                ///
                        if 'touse',                          ///
                        'log' 'mlopts' 'contin'              ///
                        missing maximize
                // clear MY global
                global MY_panel
        } // capture noisily
                // exit in case of error
                if c(rc) exit 'c(rc)'

                ereturn scalar k_aux = 2
                // save the panel variable
                ereturn local panel 'panel'
                // save a title for -Replay- and the name of this command
                ereturn local title "My rereg estimates"
                ereturn local cmd myrereg

                Replay , 'diopts'
        end

        program Replay
                syntax [, Level(cilevel) ]
                ml display , level('level')                  ///
                        diparm(lns_u, exp label("sigma_u"))   ///
                        diparm(lns_e, exp label("sigma_e"))
        end
```

─────────────────────────────────────── end `myrereg.ado` ───────────

```
program myrereg2_d2
        version 11
        args todo b lnf g H
        tempvar xb z T S_z2 Sz_2 a last
        tempname s_u s_e
        mleval 'xb'  = 'b', eq(1)
        mleval 's_u' = 'b', eq(2) scalar
        mleval 's_e' = 'b', eq(3) scalar
        scalar 's_u' = exp('s_u')
        scalar 's_e' = exp('s_e')
        // MY_panel contains the panel ID
        local by $MY_panel
        local y $ML_y1
        quietly {
                gen double 'z' = 'y' - 'xb'
                by 'by': gen 'last' = _n==_N
                by 'by': gen 'T' = _N
                by 'by': gen double 'S_z2' = sum('z'^2)
                by 'by': replace    'S_z2' = 'S_z2'[_N]
                by 'by': gen double 'Sz_2' = sum('z')^2
                by 'by': replace    'Sz_2' = 'Sz_2'[_N]
                gen double 'a' = 's_u'^2 / ('T'*'s_u'^2 + 's_e'^2)
                mlsum 'lnf'= -.5 *                              ///
                        (                                       ///
                            ('S_z2'-'a'*'Sz_2')/'s_e'^2 +       ///
                            ln('T'*'s_u'^2/'s_e'^2 + 1) +       ///
                            'T'*ln(2*c(pi)*'s_e'^2)             ///
                        )                                       ///
                        if 'last' == 1
                if ('todo'==0 | 'lnf'>=.) exit

                // compute the gradient
                tempvar S_z
                tempname dxb du de
                by 'by': gen double 'S_z' = sum('z')
                by 'by': replace 'S_z' = 'S_z'[_N]
                mlvecsum 'lnf' 'dxb' = ('z'-'a'*'S_z')/'s_e'^2  , eq(1)
                mlvecsum 'lnf' 'du' = 'a'^2*'Sz_2'/'s_u'^2      ///
                        -'T'*'a'                                ///
                        if 'last'==1                            , eq(2)
                mlvecsum 'lnf' 'de' = 'S_z2'/'s_e'^2 -          ///
                        'a'*'Sz_2'/'s_e'^2 -                     ///
                        'a'^2*'Sz_2'/'s_u'^2 -                   ///
                        'T'+1-'a'*'s_e'^2/'s_u'^2                ///
                        if 'last'==1                            , eq(3)
                mat 'g' = ('dxb','du','de')
                if ('todo'==1 | 'lnf'>=.) exit

                // compute the Hessian
                tempname d2xb1 d2xb2 d2xb d2u d2e dxbdu dxbde dude one
                mlmatsum 'lnf' 'd2u' = 2*'a'^2*'Sz_2'/'s_u'^2 - ///
                        4*'s_e'^2*'a'^3*'Sz_2'/'s_u'^4 +        ///
                        2*'T'*'a'^2*'s_e'^2/'s_u'^2             ///
                        if 'last'==1                            , eq(2)
                mlmatsum 'lnf' 'd2e' =                          ///
                        2*('S_z2'-'a'*'Sz_2')/'s_e'^2 -         ///
                        2*'a'^2*'Sz_2'/'s_u'^2 -                ///
                        4*'a'^3*'Sz_2'*'s_e'^2/'s_u'^4 +        ///
                        2*'a'*'s_e'^2/'s_u'^2 -                 ///
```

```
                          2*'a'^2*'s_e'^4/'s_u'^4              ///
                          if 'last'==1                        , eq(3)
              mlmatsum 'lnf' 'dude' =                         ///
                          4*'a'^3*'Sz_2'*'s_e'^2/'s_u'^4 -    ///
                          2*'T'*'a'^2*'s_e'^2/'s_u'^2         ///
                          if 'last'==1                        , eq(2,3)
              mlmatsum 'lnf' 'dxbdu' = 2*'a'^2*'S_z'/'s_u'^2  , eq(1,2)
              mlmatsum 'lnf' 'dxbde' =                        ///
                          2*('z'-'a'*'S_z')/'s_e'^2 -         ///
                          2*'a'^2*'S_z'/'s_u'^2               , eq(1,3)
              // 'a' is constant within panel; and
              // -mlmatbysum- treats missing as 0 for 'a'
              by 'by': replace 'a' = . if !'last'
              mlmatsum 'lnf' 'd2xb2' = 1                      , eq(1)
              gen double 'one' = 1
              mlmatbysum 'lnf' 'd2xb1' 'a' 'one', by($MY_panel) eq(1)
              mat 'd2xb' = ('d2xb2'-'d2xb1')/'s_e'^2
              mat 'H' = -(                                    ///
                  'd2xb',   'dxbdu',  'dxbde'       \         ///
                  'dxbdu'', 'd2u',    'dude'        \         ///
                  'dxbde'', 'dude',   'd2e'                   ///
              )
          }
   end
```

—— end `myrereg2_d2.ado` ——————————

C.7   The seemingly unrelated regression model

—— begin `mysureg.ado` ——————————————

```
program mysureg, properties(svyb svyj svyr)
        version 11
        if replay() {
                if ('""'e(cmd)'"'' != "mysureg") error 301
                Replay '0'
        }
        else    Estimate '0'
end

program Estimate, eclass sortpreserve
        syntax anything(id="equations" equalok) ///
                [if] [in]                        ///
                [fweight pweight iweight] [,      ///
                noLOg                            /// -ml model- options
                noLRTEST                         ///
                vce(passthru)                    ///
                Level(cilevel)                   /// -Replay- options
                corr                             ///
                *                                /// -mlopts- options
        ]
        // parse equations
        ParseEqns 'anything'
        local p     's(p)'
        local k_eq  's(k_eq)'
        local yvars 's(yvars)'
        local xvars 's(xvars)'
        local eqns  's(eqns)'
        local eqns0 's(eqns0)'
        local ceqns 's(ceqns)'
```

```
local myeqns `anything'
local nocons 0
forval i = 1/`p' {
        local xvars`i' `s(xvars`i')'
        local xvars`i'_cns `s(xvars`i'_cns)'
        if "`xvars`i'_cns'" != "" {
                local nocons 1
        }
}

// check syntax
mlopts mlopts, `options'
local cns `s(constraints)'
if "`weight'" != "" {
        local wgt "[`weight'`exp']"
        // for initial value calculations
        local iwgt "[iw`exp']"
}
if "`log'" != "" {
        local qui quietly
}

// mark the estimation sample
marksample touse
markout `touse' `yvars' `xvars'
markout `touse' `offset' `exposure'
_vce_parse `touse', opt(Robust oim opg) argopt(CLuster): `wgt', `vce'

// initial values
tempname b0
InitialValues `yvars' `iwgt' if `touse' , init(`b0')
local initopt init(`b0', skip) search(quietly)

`qui' di as txt _n "Fitting constant-only model:"
ml model d2 mysuregc_d2 `eqns0'              ///
        `iwgt' if `touse',                   ///
        `log' `mlopts' `initopt'             ///
        nocnsnotes missing maximize
if `nocons' {
        mat `b0' = e(b)
        local initopt init(`b0', skip) search(off)
}
else    local initopt continue search(off)
if "`lrtest'" == "" & `nocons' == 0 {
        local lf0 lf0(`k_eq' `e(ll)')
}

// iterate to solution using concentrated likelihood
`qui' di as txt _n "Fitting concentrated model:"
ml model d2 mysuregc_d2 `ceqns'                      ///
        `iwgt' if `touse',                           ///
        `log' `mlopts'                               ///
        `initopt'                                    ///
        missing maximize
UnConcentrate `yvars' `iwgt' if `touse', b(`b0')
local initopt init(`b0', copy) search(quietly)

// fit the full unconcentrated likelihood
`qui' di as txt _n "Fitting full model:"
ml model lf0 mysureg_lf0 `eqns'                      ///
        `wgt' if `touse',                            ///
        `log' `mlopts' `vce'                         ///
        `initopt' `lf0' missing maximize waldtest(-`k_eq')
```

```
            // build saved results
            tempname Sigma sd
            matrix 'Sigma' = I('p')
            matrix 'sd'  = I('p')
            matrix colnames 'Sigma' = 'yvars'
            matrix rownames 'Sigma' = 'yvars'
            forval i = 'p'(-1)1 {
                    mat 'Sigma'['i','i'] = [sigma'i'_'i']_b[_cons]
                    forval j = '='i'+1'/'p' {
                            mat 'Sigma'['i','j'] = [sigma'i'_'j']_b[_cons]
                            mat 'Sigma'['j','i'] = 'Sigma'['i','j']
                    }
                    ereturn local xvars'i' 'xvars'i''
                    ereturn local xvars'i'_cns 'xvars'i'_cns'
            }
            ereturn local xvars 'xvars'
            ereturn local mleqns 'ceqns'
            ereturn local myeqns 'myeqns'
            if "'weight'" != "" {
                    ereturn local wexp "= 'exp'"
            }
            ereturn matrix Sigma = 'Sigma'
            ereturn matrix Vars = 'sd'
            ereturn scalar k_aux = 'k_eq'-'p'
            // save a title for -Replay- and the name of this command
            ereturn local title "My sureg estimates"
            ereturn local cmd mysureg
            Replay , 'diopts'
    end
    program Replay
            syntax [, Level(cilevel) notable corr ]
            if "'table'" == "" {
                    ml display , level('level')
            }
            if "'corr'" != "" {
                    di
                    di as txt "Correlation matrix of residuals:"
                    tempname corr
                    matrix 'corr' = corr(e(Sigma))
                    matrix list 'corr', noheader format(%6.4g)
            }
    end
    // Syntax:
    //     ParseEqns (y1 xvars1) (y2 xvars2) ...
    //
    // Build the -ml- equivalent equations, including the equations for the
    // variance matrix.
    //
    // Saved results:
    //     s(p)           number of dependent variables
    //     s(k_eq)        number of -ml- equations : unconcentrated model
    //     s(yvars)       the dependent variables
    //     s(xvars)       the independent variables from all the equations
    //     s(xvars'i')    the independent variables from equation 'i'
    //     s(xvars'i'_cns) "noconstant" or ""
    //     s(eqns)        -ml- equations : unconcentrated model
    //     s(ceqns)       -ml- equations : concentrated model
    //     s(eqns0)       -ml- equations : constant-only concentrated model
```

```
program ParseEqns, sclass
        gettoken eq eqns : 0 , match(paren) parse("()") bind
        local p 0
        while "`eq'" != "" {
                local ++p
                local 0 `eq'
                syntax varlist [, noCONStant]
                gettoken yi xvarsi : varlist
                local xvars`p' `xvarsi'
                local xvars`p'_cns `constant'
                local meqns `meqns' (`yi' : `yi' = `xvarsi', `constant')
                local meqns0 `meqns0' (`yi' : `yi' = )
                local yvars `yvars' `yi'
                local xvars `xvars' `xvarsi'
                gettoken eq eqns : eqns , match(paren) parse("()") bind
        }
        forval i = 1/`p' {
                forval j = `i'/`p' {
                        local seqns `seqns' /sigma`i'_`j'
                }
        }
        sreturn clear
        sreturn local p `p'
        sreturn local k_eq = `p'*(`p'+3)/2
        sreturn local yvars        `yvars'
        sreturn local xvars        `: list uniq xvars'
        sreturn local eqns         `meqns' `seqns'
        sreturn local ceqns        `meqns'
        sreturn local myeqns       `myeqns'
        sreturn local eqns0        `meqns0'
        forval i = 1/`p' {
                sreturn local xvars`i' `xvars`i''
                sreturn local xvars`i'_cns `xvars`i'_cns'
        }
end

// Take e(b) from the concentrated model and return the equivalent
// unconcentrated coefficents vector.
program UnConcentrate
        syntax varlist if [iw] , b(name)

        local wgt [`weight'`exp']

        local p : word count `varlist'
        tempname b0 s0 sv
        mat `b0' = e(b)
        forval i = 1/`p' {
                tempvar r`i'
                local y : word `i' of `varlist'
                matrix score `r`i'' = `b0' `if', eq(`y')
                quietly replace `r`i'' = `y' - `r`i''
                local resids `resids' `r`i''
        }
        quietly matrix accum `s0' = `resids' `wgt' `if', noconstant
        matrix `s0' = `s0'/r(N)
        forval i = 1/`p' {
                matrix `sv' = nullmat(`sv') , `s0'[`i',`i'...]
        }
        matrix `b' = e(b) , `sv'
end

// Compute the initial values for the concentrated constant-only model.
```

```
program InitialValues
        syntax varlist if [iw] , init(name)
        tempname m
        tempvar one
        quietly gen double 'one' = 1
        quietly matrix vecaccum 'm' = 'one' 'varlist'    ///
                ['weight''exp'] 'if', noconstant
        matrix 'init' = 'm'/r(N)
        matrix colnames 'init' = _cons
        matrix coleq    'init' = 'varlist'
end
```
─── end mysureg.ado ─────────────

─── begin mysureg_d1.ado ─────────
```
program mysureg_d1
        version 11
        args todo b lnf g H
        local p : word count $ML_y
        local k = 'p'*('p'+1) / 2 + 'p'

        tempname S iS sij isij isi ip
        matrix 'S' = J('p','p',0)
quietly {
        // get residuals and build variance matrix
        local e 1
        forval i = 1/'p' {
                tempvar r'i'
                mleval 'r'i'' = 'b', eq('i')
                replace 'r'i'' = ${ML_y'i'} - 'r'i''
                local resids 'resids' 'r'i'
                mleval 'sij' = 'b', eq('='p'+'e'') scalar
                matrix 'S'['i','i'] = 'sij'
                local ++e
                forval j = '='i'+1'/'p' {
                        mleval 'sij' = 'b', eq('='p'+'e'') scalar
                        matrix 'S'['i','j'] = 'sij'
                        matrix 'S'['j','i'] = 'sij'
                        local ++e
                }
        }
        matrix 'iS' = invsym('S')
        // get score variables
        forval i = 1/'k' {
                tempvar g'i'
                quietly gen double 'g'i'' = .
        }
        gen double 'ip' = 0
        forval i = 1/'p' {
                matrix 'isi' = 'iS'['i',1...]
                matrix colnames 'isi' = 'resids'
                matrix score 'g'i'' = 'isi', replace
                replace 'ip' = 'ip' + 'r'i''*'g'i''
        }
        mlsum 'lnf' = -0.5*('p'*ln(2*c(pi))+ln(det('S'))+'ip')
} // quietly
        if ('todo' == 0 | missing('lnf')) exit

        // compute the scores and gradient
        tempname gi gs
```

```
                capture matrix drop `g'
        quietly {
                local e = `p'+1
                forval i = 1/`p' {
                        mlvecsum `lnf' `gi' = `g`i'', eq(`i')
                        matrix `g' = nullmat(`g'), `gi'
                        replace `g`e'' = 0.5*(`g`i''*`g`i''-`iS'[`i',`i'])
                        mlvecsum `lnf' `gi' = `g`e'' , eq(`e')
                        matrix `gs' = nullmat(`gs'), `gi'
                        local ++e
                        forval j = `=`i'+1'/`p' {
                                replace `g`e'' = `g`i''*`g`j'' - `iS'[`i',`j']
                                mlvecsum `lnf' `gi' = `g`e'' , eq(`e')
                                matrix `gs' = nullmat(`gs'), `gi'
                                local ++e
                        }
                }
                matrix `g' = `g' , `gs'
        } // quietly
        end
```

——————————————————————————— end `mysureg_d1.ado` ———————

——————————————————————————— begin `mysuregc_d2.ado` ———————

```
        program mysuregc_d2
                // concentrated likelihood for the SUR model
                version 11
                args todo b lnf g H
                local p : word count $ML_y
                tempname W sumw

                // get residuals and build variance matrix
                forval i = 1/`p' {
                        tempvar r`i'
                        mleval `r`i'' = `b', eq(`i')
                        quietly replace `r`i'' = ${ML_y`i'} - `r`i''
                        local resids `resids' `r`i''
                }
                quietly matrix accum `W' = `resids' [iw=$ML_w] if $ML_samp, noconstant
                scalar `sumw' = r(N)
                matrix `W' = `W'/`sumw'
                scalar `lnf' = -0.5*`sumw'*(`p'*(ln(2*c(pi))+1)+ln(det(`W')))

                if (`todo' == 0 | missing(`lnf')) exit

                // compute gradient
                tempname gi iW iWi
                tempvar  scorei
                quietly generate double `scorei' = .
                capture matrix drop `g'
                matrix `iW' = invsym(`W')
                forval i = 1/`p' {
                        matrix `iWi' = `iW'[`i',1...]
                        matrix colnames `iWi' = `resids'
                        matrix score `scorei' = `iWi', replace
                        mlvecsum `lnf' `gi' = `scorei', eq(`i')
                        matrix `g' = nullmat(`g'), `gi'
                }

                if (`todo' == 1 | missing(`lnf')) exit

                // compute the Hessian, as if we were near the solution
```

```
            tempname hij
            local k = colsof('b')
            matrix 'H' = J('k','k',0)
            local r 1
            forval i = 1/'p' {
                    local c 'r'
                    mlmatsum 'lnf' 'hij' = -1*'iW'['i','i'], eq('i')
                    matrix 'H'['r','c'] = 'hij'
                    local c = 'c' + colsof('hij')
                    forval j = '='i'+1'/'p' {
                            mlmatsum 'lnf' 'hij' = -1*'iW'['i','j'], eq('i','j')
                            matrix 'H'['r','c'] = 'hij'
                            matrix 'H'['c','r'] = 'hij''
                            local c = 'c' + colsof('hij')
                    }
                    local r = 'r' + rowsof('hij')
            }
    end
```

── end `mysuregc_d2.ado` ──────

── begin `mysureg_lf0.ado` ──────

```
program mysureg_lf0
        version 11
        args todo b lnfj
        local p : word count $ML_y
        local k = 'p'*('p'+1) / 2 + 'p'

        tempname S iS sij isij isi
        tempvar ip
        matrix 'S' = J('p','p',0)
quietly {
        // get residuals and build variance matrix
        local e 1
        forval i = 1/'p' {
                tempvar r'i'
                mleval 'r'i'' = 'b', eq('i')
                replace 'r'i'' = ${ML_y'i'} - 'r'i''
                local resids 'resids' 'r'i''
                mleval 'sij' = 'b', eq('='p'+'e'') scalar
                matrix 'S'['i','i'] = 'sij'
                local ++e
                forval j = '='i'+1'/'p' {
                        mleval 'sij' = 'b', eq('='p'+'e'') scalar
                        matrix 'S'['i','j'] = 'sij'
                        matrix 'S'['j','i'] = 'sij'
                        local ++e
                }
        }
        matrix 'iS' = invsym('S')
        // get score variables
        tempvar scorei
        gen double 'scorei' = .
        gen double 'ip' = 0
        forval i = 1/'p' {
                matrix 'isi' = 'iS'['i',1...]
                matrix colnames 'isi' = 'resids'
                matrix score 'scorei' = 'isi', replace
                replace 'ip' = 'ip' + 'r'i''*'scorei'
        }
```

```
        replace 'lnfj' = -0.5*('p'*ln(2*c(pi))+ln(det('S'))+'ip')
} // quietly
end
```

———————————————————————————————— end `mysureg_lf0.ado` ————————

References

Berndt, E. K., B. H. Hall, R. E. Hall, and J. A. Hausman. 1974. Estimation and inference in nonlinear structural models. *Annals of Economic and Social Measurement* 3/4: 653–665.

Binder, D. A. 1983. On the variances of asymptotically normal estimators from complex surveys. *International Statistical Review* 51: 279–292.

Breslow, N. E. 1974. Covariance analysis of censored survival data. *Biometrics* 30: 89–99.

Broyden, C. G. 1967. Quasi-Newton methods and their application to function minimization. *Mathematics of Computation* 21: 368–381.

Cleves, M., W. Gould, R. G. Gutierrez, and Y. V. Marchenko. 2010. *An Introduction to Survival Analysis Using Stata*. 3rd ed. College Station, TX: Stata Press.

Davidon, W. C. 1959. Variable metric method for minimization. Technical Report ANL-5990, Argonne National Laboratory, U.S. Department of Energy, Argonne, IL.

Davidson, R., and J. G. MacKinnon. 1993. *Estimation and Inference in Econometrics*. New York: Oxford University Press.

Fletcher, R. 1970. A new approach to variable metric algorithms. *Computer Journal* 13: 317–322.

———. 1987. *Practical Methods of Optimization*. 2nd ed. New York: Wiley.

Fletcher, R., and M. J. D. Powell. 1963. A rapidly convergent descent method for minimization. *Computer Journal* 6: 163–168.

Fuller, W. A. 1975. Regression analysis for sample survey. *Sankhyā, Series C* 37: 117–132.

Gail, M. H., W. Y. Tan, and S. Piantadosi. 1988. Tests for no treatment effect in randomized clinical trials. *Biometrika* 75: 57–64.

Goldfarb, D. 1970. A family of variable-metric methods derived by variational means. *Mathematics of Computation* 24: 23–26.

Gould, W. 2001. Statistical software certification. *Stata Journal* 1: 29–50.

Greene, W. H. 2008. *Econometric Analysis*. 6th ed. Upper Saddle River, NJ: Prentice Hall.

Grunfeld, Y., and Z. Griliches. 1960. Is aggregation necessarily bad? *Review of Economics and Statistics* 42: 1–13.

Huber, P. J. 1967. The behavior of maximum likelihood estimates under nonstandard conditions. In Vol. 1 of *Proceedings of the Fifth Berkeley Symposium on Mathematical Statistics and Probability*, 221–233. Berkeley: University of California Press.

Kent, J. T. 1982. Robust properties of likelihood ratio tests. *Biometrika* 69: 19–27.

Kish, L., and M. R. Frankel. 1974. Inference from complex samples. *Journal of the Royal Statistical Society, Series B* 36: 1–37.

Korn, E. L., and B. I. Graubard. 1990. Simultaneous testing of regression coefficients with complex survey data: Use of Bonferroni t statistics. *American Statistician* 44: 270–276.

Lin, D. Y., and L. J. Wei. 1989. The robust inference for the Cox proportional hazards model. *Journal of the American Statistical Association* 84: 1074–1078.

Lütkepohl, H. 1996. *Handbook of Matrices*. New York: Wiley.

Marquardt, D. W. 1963. An algorithm for least-squares estimation of nonlinear parameters. *Journal of the Society for Industrial and Applied Mathematics* 11: 431–441.

Nash, J. C. 1990. *Compact Numerical Methods for Computers: Linear Algebra and Function Minimization*. 2nd ed. New York: Adam Hilger.

Peto, R. 1972. Contribution to the discussion of paper by D. R. Cox. *Journal of the Royal Statistical Society, Series B* 34: 205–207.

Press, W. H., S. A. Teukolsky, W. T. Vetterling, and B. P. Flannery. 2007. *Numerical Recipes in C: The Art of Scientific Computing*. 3rd ed. Cambridge: Cambridge University Press.

Rogers, W. H. 1993. sg17: Regression standard errors in clustered samples. *Stata Technical Bulletin* 13: 19–23. Reprinted in *Stata Technical Bulletin Reprints*, vol. 3, pp. 88–94. College Station, TX: Stata Press.

Royall, R. M. 1986. Model robust confidence intervals using maximum likelihood estimators. *International Statistical Review* 54: 221–226.

Shanno, D. F. 1970. Conditioning of quasi-Newton methods for function minimization. *Mathematics of Computation* 24: 647–656.

Stuart, A., and J. K. Ord. 1991. *Kendall's Advanced Theory of Statistics. Volume 2: Classical Inference*. 5th ed. London: Arnold.

Welsh, A. H. 1996. *Aspects of Statistical Inference*. New York: Wiley.

White, H. 1980. A heteroskedasticity-consistent covariance matrix estimator and a direct test for heteroskedasticity. *Econometrica* 48: 817–838.

————. 1982. Maximum likelihood estimation of misspecified models. *Econometrica* 50: 1–25.

Williams, R. L. 2000. A note on robust variance estimation for cluster-correlated data. *Biometrics* 56: 645–646.

Wooldridge, J. M. 2002. *Econometric Analysis of Cross Section and Panel Data*. Cambridge, MA: MIT Press.

Author index

Subject index